Firms, Governments and Climate Policy

CENTRE FOR
EUROPEAN
POLICY
STUDIES

ESRI STUDIES SERIES ON THE ENVIRONMENT

In April 2000 the Japanese government launched a series of comprehensive, interdisciplinary and international research projects called 'the Millennium Projects' and as part of this initiative the Economic and Social Research Institute (ESRI) of the Cabinet Office of Japan initiated a two year project entitled 'A Study on Sustainable Economic and Social Structures in the 21st Century' which focuses on ageing and environmental problems in the Japanese and international context.

The 'ESRI Studies Series on the Environment' will provide a forum for the publication of a limited number of books which are the result of this international collaboration on four main issues: research on solid waste management; the analysis of waste recycling and the conservation of resources and energy; research on the compatibility of environmental protection and macroeconomic policy and the analysis of problems related to climate change. The series will be invaluable to students and scholars of environment and ecology as well as consultants and practitioners involved in environmental policymaking.

Firms, Governments and Climate Policy

Incentive-based Policies for Long-term Climate Change

Edited by
Carlo Carraro

Professor of Econometrics and Environmental Economics, University of Venice, Italy, Research Director, Fondazione Eni Enrico Mattei (FEEM), Milan, Italy and Associate Fellow, Centre for European Policy Studies (CEPS), Brussels, Belgium

Christian Egenhofer

Senior Fellow, Centre for European Policy Studies (CEPS), Brussels, Belgium and Senior Research Fellow, Centre for Energy, Petroleum and Mineral Law and Policy, University of Dundee, Scotland, UK

ESRI STUDIES SERIES ON THE ENVIRONMENT

Edward Elgar
Cheltenham, UK • Northampton, MA, USA

Published by
Edward Elgar Publishing Limited
Glensanda House
Montpellier Parade
Cheltenham
Glos GL50 1UA
UK

Edward Elgar Publishing, Inc.
136 West Street
Suite 202
Northampton
Massachusetts 01060
USA

A catalogue record for this book
is available from the British Library

Library of Congress Cataloging in Publication Data
Firms, governments and climate policy : incentive-based policies for long-term climate change / edited by Carlo Carraro, Christian Egenhofer.
p. cm. — (ESRI studies series on the environment)
Includes index.
1. Environmental policy. 2. Climatic changes—Government policy. 3. Age distribution (Demography)—Environmental aspects. 4. Sustainable development. 5. Energy consumption—Environmental aspects. I. Carraro, Carlo. II. Egenhofer, Christian, 1960– III. Series.

GE170 .F5 2003
363.738'74'05610973—dc21

2002024586

ISBN 1 84376 055 X

Typeset by Manton Typesetters, Louth, Lincolnshire, UK.
Printed and bound in Great Britain by MPG Books Ltd, Bodmin, Cornwall.

Contents

List of boxes

List of figures

List of tables

Contributors

Patrick ten Brink, Senior Fellow and Head of Brussels Office, Institute for European Environmental Policy (IEEP), Belgium

Carlo Carraro, Professor of Econometrics and Environmental Economics, University of Venice, Research Director, Fondazione Eni Enrico Mattei (FEEM), Milan, Italy and Associate Fellow at the Centre for European Policy Studies (CEPS), Brussels, Belgium

Christian Egenhofer, Senior Fellow at the Centre for European Policy Studies (CEPS), Brussels, Belgium and Senior Research Fellow, Centre for Energy, Petroleum and Mineral Law and Policy, University of Dundee, Scotland, UK.

Marzio Galeotti, Professor of Economics, University of Bergamo, Italy and Research Program Co-ordinator, Climate Change Modelling and Policy Program, FEEM, Italy

Koichi Hamada, President, Economic and Social Research Institute, Japanese Cabinet Office, Japan and Professor of Economics, Yale University, US

Josef Janssen, Head of Emissions Trading and Climate Policy Unit, Institute for Economy and the Environment, University of St Gallen, Switzerland, and Managing Partner, etsg Emissions Trading Solutions, St Gallen Ltd

Marina Morère, Independent Economist, France

Jane Wallace-Jones, Veneto Regional Agency for the Environment (ARPAV)

List of abbreviations

ABT	averaging, banking and trading
AGE	applied general equilibrium (model)
AIJ	activities implemented jointly
BAT	best available technology
BAU	business-as-usual
CAC	command-and-control
CCL	climate change levy
CDM	clean development mechanism
CEERP	Canadian Early Emission Reduction Program
CFC	chlorofluorocarbon
COP	Conference of Parties
DEA	Danish Energy Authority
DETR	Department of Environment, Transport and the Regions (UK)
EA	environmental agreement
EBRD	European Bank for Reconstruction and Development
ECCP	European Climate Change Programme
EEA	European Environment Agency
EIA	environmental impact assessment
EIT	economy in transition
EMAS	Eco-Management and Audit Scheme
EPCOR	Edmonton Power Corporation
EpE	Entreprises pour l'Environnement
ESRI	Economic and Social Research Institute (Japan)
ET	emissions trading
GHG	greenhouse gas
IAM	integrated assessment model
ICT	information and communication technology
IEA	International Energy Agency
IET	international emissions trading
IPCC	International Panel for Climate Change
JI	joint implementation
LbD	learning-by-doing
LbU	learning-by-using
LTA	long-term agreement

MEA	multilateral environmental agreement
NEA	negotiated environmental agreement
NGO	non-governmental organization
NPV	net present value
PAM	Policies and Measures
PPA	power purchase agreement
PPP	polluter pays principle
RACT	Reasonably Achievable Control Technology
SIP	State Implementation Plan
UNFCCC	UN Framework Convention on Climate Change
US EPA	US Environmental Protection Agency
VOC	volatile organic compound

Preface

The world at the beginning of the twenty-first century must place the highest priority on constructing a sustainable socioeconomic system that can cope with the rapid ageing of populations in developed countries and with the limited environmental resources available in those countries. At first glance, the ageing problem and the environment problem may seem to be quite separate issues. However, they share a common feature: they both deal with intergenerational problems. The essence of the ageing problem is how to find effective ways for a smaller, younger generation to support a larger, ageing generation. The crux of the environmental problem is to find a feasible way to leave environmental resources to future generations. Moreover, in terms of consumption, slower population growth may slow consumption and help environmental problems. On the other hand, a rapidly ageing society may use more energy-intensive technology to compensate for the inevitable labour shortage, and deteriorate the natural environment by doing so.

Today, these concerns are highly applicable in Japan. The pressure created by the rapid ageing of the Japanese population is becoming acute; Japan must construct a sustainable society that does not create intergenerational inequity or deteriorate the public welfare. At the same time, Japan cannot deplete its environmental resources and energy, which would leave future generations with an unbearably heavy burden.

The government of Japan has recognized the vital importance of both problems. To explore and implement solutions for this difficult task, in April 2000 former Prime Minister Keizo Obuchi launched several comprehensive, interdisciplinary and international research projects that he called the 'Millennium Projects'. These projects are now being carried out through collaboration among Japan's governmental ministries. As a part of these projects, the Economic and Social Research Institute (ESRI), Cabinet Office, Government of Japan, initiated a two-year project entitled 'A Study on Sustainable Economic and Social Structures in the 21st Century' in April 2000. While the Millennium Projects cover a wide range of topics and a wide range of disciplines such as natural science and technological innovation, the project conducted by ESRI places major emphasis on social science. While taking into account technological innovation and feasibility, it focuses on ageing and environmental problems. It aims to design a desirable socioeconomic structure under the pressure of an ageing population and environmental constraints by identifying the necessary policy tools to attain stable and sustainable growth.

This project is being implemented with close collaboration among Japanese as well as foreign scholars and research institutions. Besides Japanese scholars and institutions, foreign participants have joined from, among others, the USA, the UK, Norway, Germany, Austria, Italy, Australia, Korea and Thailand. They amount to ten countries and 30 working groups.

The ageing problems studied in this project can be divided into three main sections. The first concerns macroeconomic problems caused by an ageing population and diminishing young workers. The second concerns problems in the labour market, and the third focuses on problems with the social security system, and emphasizes the reform of the pension system.

The problem of sustainable economic growth was explored in the section on macroeconomic issues. Few researchers are confident of the possibility of future high-speed economic growth, but many are optimistic about the possibility of sustainable economic growth that is supported by an increase in total factor productivity through the information and communications technology (ITC) revolution. Indeed, this view has often emphasized the possibility of correcting the labour force shortage by increasing the employment of females and immigrants. If the economy does develop according to this optimistic scenario, it will achieve the simultaneous resolution of problems of social security and that of the ageing working population. If the growth falls short of our expectations, however, the reform of the social security system will become more difficult.

Until now, developed countries have employed simplified economic welfare indicators such as economic growth and GDP for policy management. According to the convergence of economic growth, developed countries will be expected to slow down their pace of growth. Thus we need an economic and social structure that is capable, even under relatively low economic growth, of alleviating social and economic difficulties such as inequality and the conflict of interests between stakeholders. From a different perspective, this is a classical problem of the division of roles between government and the private sector. The government should provide a minimum safety net for the risks that cannot be covered by individuals without reducing the degree of economic vitality based on private incentives. It is necessary to re-examine simplified quantitative standards such as GDP and growth rates as measures of society's overall well-being, and to explore multidimensional criteria that incorporate concepts like happiness, good lifestyles, peace and healthy environments, none of which can be accurately measured in monetary terms.

Research on environmental problems aimed at the creation of a closed-loop type of economic society addresses four main issues: (1) research on solid waste management; (2) analysis of waste recycling that incorporates the technological possibility, sustainability and conservation of resources and energy; (3) research on the compatibility of environment protection and macroeconomic policy, especially policies aimed at increasing employment; and (4) an analysis of problems related to climate changes. It should be noted that much empirical investigation has been conducted in the field of waste management.

Japan's environmental problems began with pollution problems. In the early Meiji era (1868–1912), the development of copper mines such as Ashio, Beshi and others was accompanied by the production of noxious gases that polluted the environs, but the problem was resolved by the conclusion of various agreements concerning environmental protection. During the high-growth period following World War II, different pollution problems emerged: mad hatter's disease, water contamination caused by factory operations, and air contamination caused by car exhaust emissions are three examples. Local administrations tried to curtail these bad effects by imposing limits on emissions of harmful substances, concluding pollution control agreements that featured voluntary efforts from companies. This anti-pollution policy did not constitute a constraint on industrial activity and growth. Actually, it had just the opposite effect: the policy and various pollution control restrictions stimulated creativity and technological progress, leading to the further development of industry. A remarkable example of such an effect was the Muskie Act, established to prevent air pollution in the US state of California. It offered effective incentives for technological progress and entailed precedence of the Japanese automobile industry over other countries. Pollution problems were thus controlled.

Today, however, the improvement of living standards and development of the world economy is leading to global environmental degradation. Moreover, economic growth has caused new environmental problems such as increases in waste.

Japan – taking into account its geographic location in the monsoon climatic zone and its sanitation – opted for the landfill method of waste disposal. However, its landfill site capacity was not enough to absorb the sharp increase in waste the country was producing. The incineration method was thus applied as the intermediate disposal procedure to reduce the quantity of landfill waste. After the quantity of waste had been reduced by burning, the landfill method was reapplied. But at the end of twentieth century it became obvious that landfills were facing an impasse, and that this policy was little more than a stop-gap measure. From the 1990s onward, the government of Japan explored different measures and encouraged efforts to cope with the waste problem by implementing closed-loop social measures, that is, using fewer resources, recycling used materials and reusing the waste as energy and other inputs. Some people have argued that the creation of closed-loop social measures is a specific problem for Japan, faced with the shortage of landfill sites. Nevertheless, the waste problem will sooner or later become acute for high-growth countries. Even in countries that have vast territories, the adoption of these closed-loop measures is meaningful because of the scarcity of resources. Moreover, closed-loop measures that effectively dispose of waste are a way to reduce the emission of greenhouse gases. By employing resources-efficient and energy-efficient technologies, at least according to our optimistic scenarios, we may eventually attain sustainability.

By forming this kind of closed-loop socioeconomic system, it will become possible to save energy and to reduce emission of greenhouse gases through the reuse of natural resources. Moreover, it will become possible to prevent the greenhouse effect by more effective use of fossil energy and other alternative energy sources. By employing this 'bottom-up' approach, we will create jobs in environmental protection, and in the use of alternative sources of energy such as wind. As we shift to the closed-loop type economic society, the beneficial effects on the labour market will become evident.

Some argue that environmental problems constitute constraints on macroeconomic growth. However, employment of resource-effective technologies and the reuse of waste to the fullest extent possible will lower the possibility of resource shortages. At the same time, it will contribute to the minimization of environmental impacts. The necessity to build a closed-loop type economic society was triggered by Japan's urgent need to examine the problem of waste management. In this sense, building a closed-loop society is a way to attain Japanese-style sustainable development.

To repeat, the problems examined here should not be considered separately. Research on the problems of population ageing and environmental degradation has a common thread. That is, these studies are concerned with how to use the market mechanism to solve them. Many propose the use of the market mechanism and economic tools to solve problems in the medical care system, in pensions, and in the labour market that are caused by an ageing population, climate change, and waste disposal. Unfortunately, many of these problems arise in areas that are noted for their frequent market failures. Here, the market mechanism in its pure form often fails to produce efficient outcomes. Therefore, making the market mechanism work effectively is difficult and preconditioned by the exact identification of externalities. At the same time, we must face classical trade-off problems between efficiency and equity, as well as the problem of the best relationship between the public and private sectors.

In this project, we explore optimal solutions to problems in terms of social science. After taking into account the political and social constraints we face, and after alignment and coordination with the results of other studies, we will sketch out an ideal design and examine the possible direction of further research.

This project came to an end in March 2002. It solved many theoretical and empirical issues, but has created new debates. Twice a year, all the members of the project, along with selected participants, met to discuss the results of the research. Regretfully, it has not been possible to reproduce these fruitful discussions in the present volume.

Overall, the papers presented in this project were extremely challenging, and covered a wide range of topics. In the near future I strongly hope we will have a chance to discuss the research once more from a common standpoint. I would like to acknowledge the ceaseless efforts of the members of ESRI, especially those of the Department of Administrative Affairs, headed by Mr Mitsuo Hosen.

The results of this research will appear in print from Edward Elgar Publishing Ltd as an ESRI study series, available to policy makers, academics and business people with a keen interest in these subjects. The series on environmental problems covers climate change, waste disposal problems, sources of energy and technology, and environmental and macroeconomic policy. Unfortunately, because of space limitations we regret that we are able to publish only selected papers from the total research effort.

The timing seems right to publish the first volume of the ESRI Studies Series on the Environment, *Firms, Governments and Climate Policy*, while research in this field is encouraged by the Marrakesh Agreement.

The research papers to be published were selected by Carlo Carraro of Venice University, Frank Convery of University College Dublin, Christian Egenhofer of CEPS, and Koichi Hamada and Hiromi Kato of ESRI. We are especially grateful to Professor Carraro, who initiated our association with Edward Elgar. Last but not least, we would like to thank Dymphna Evans from Edward Elgar.

Koichi Hamada
ESRI

Acknowledgements

This book is the result of work that was undertaken in the context of a two-year research project by the Economic and Social Research Institute (ESRI), Cabinet Office, Government of Japan, entitled 'A Study on Sustainable Economic and Social Structures in the 21st Century'. We would like to thank all staff of the Economic and Social Research Institute for their support throughout our work and especially during the subsequent meetings we had in Osaka and Tokyo. We are particularly grateful to Koichi Hamada, Hiromi Kato and Misuo Hosen of ESRI for the opportunities to discuss and receive many comments on our ongoing research. We would also like to thank the London offices of both Mitsubishi Research Institute and Nomura Research Institute for ensuring the continuous smooth running of the project. The research project could not have been completed in such an effective and coordinated way without the cooperation and commitment of all researchers who participated in the preparation of the chapters of this book. We are therefore grateful to Patrick ten Brink, Marina Morère, Jane Wallace-Jones, Josef Janssen and Marzio Galeotti for their scientific input and dedicated work. Finally, the excellent organization of CEPS, and in particular the support received by Thomas Legge, Sally Scott and Isabelle Tenaerts, must be gratefully acknowledged.

Carlo Carraro
Christian Egenhofer

Introduction

Carlo Carraro and Christian Egenhofer

The current international climate debate is dominated by conflicts about the efficiency and equity effects of the international climate regime with an emphasis on the economic and legal aspects on the one hand, and on the environmental on the other. No balanced compromise over the complex trade-offs that characterize the climate change problem is likely to appear in the short run, irrespective of the results of the seventh climate summit in Marrakesh in November 2001. Marrakesh has kept the international climate change process alive and arguably is a major step towards entry into force of the Kyoto Protocol. But the Kyoto Protocol is only one (small) step on the way to an effective climate policy, which will call for far greater greenhouse gas reductions and the full engagement of the USA and ultimately developing countries.

With these difficulties as background, this book approaches climate change policy from the following perspective: rather than focusing on short-term issues dealing with the implementation of the Kyoto Protocol (ceilings, sinks, monitoring, enforcement, liability and so on), this study looks at the fundamentals of the climate problem. It takes a long-term perspective on climate change and concentrates on identifying incentives both for governments and industry to adopt climate change policies. The book aims to determine whether certain incentives will induce the adoption of a climate policy, and if so, which incentives are provided by different policy instruments. Hence the focus of the study is on the tools that provide incentives for firms and for governments to act in order to reduce or control their greenhouse gas (GHG) emissions. More concretely, the book asks: what are these incentives? How can short-term incentives be combined with long-term incentives? Are there enough incentives also for long-term policies, namely those designed to foster R&D, innovation and its diffusion? The analysis does not stop there, however, but moves on to 'policy mixes'. It finds that incentives to act can be optimized by blending different instruments so that they mutually reinforce incentives. Hence, this study addresses the question: what policy mix is most likely to increase the motivation of firms and governments to act? To answer this question, we will adopt a European perspective on policy making.

Our work focuses on firms and governments. One may argue that this neglects other actors in the climate change arena, for example, NGOs, trade unions and households. However, the book stresses those incentive mechanisms that are likely to provide the most significant results in terms of emissions reductions even in the first commitment period, thus enabling countries to attain the Kyoto targets set in 1997. The role of NGOs and households is certainly important in this process. For example, a change in household consumption patterns could be very effective. Nevertheless, the 2012 objectives are most likely to be achieved if firms and governments are presented with the appropriate incentives to reduce GHG emissions. Moreover, even in terms of burden sharing, this seems to be the most equitable option.

SCOPE OF THE BOOK

This book, composed of five integrated chapters, could logically be divided into three parts. The first part provides an overview of different climate policy instruments from both a theoretical and a historical perspective. Here the goal is to describe the whole set of instruments that governments and firms can adopt to reduce GHG emissions, and to argue in favour of a well-designed policy mix, which would provide the necessary incentives to undertake action. The second part of the book focuses on the behaviour of firms. Implicitly assuming that an agreement can be found by governments, the study analyses which instruments are most likely to induce industries and companies to act to reduce GHG emissions. Action may take the form of negotiated agreements, business opportunities in the trading market, investments implemented within joint implementation (JI) or the clean development mechanism (CDMs). But actions may also take the form of investments in R&D or in new climate-friendly technologies and products. Finally, the third part of the book analyses whether there are actually incentives for governments to achieve agreement on climate change control, and which policy instruments and institutional design are most likely to provide these incentives.

Each chapter concludes with a summary of the main findings. Appendices present a number of case studies supporting the analysis.

The structure of the book is similar to a three-stage (logical) game: in the first stage, governments decide whether or not to implement the Kyoto Protocol (or more generally, a climate treaty); in the second, governments implement the consequent policy mix; and in the third, firms respond to the governments' decisions with their own optimal strategy (independently or through negotiations and different forms of cooperation). As usual, the sub-game perfect equilibrium is found by solving the game backwards. Hence the book starts from firms' behaviour and ends with the strategic decision facing governments of whether or not to ratify the climate treaty.

Three criteria guide our analysis and comparison of different policy instruments:

- the incentives induced by different instruments and policy mixes;
- the actual historical trends (which seem to lead towards the so-called Kyoto mechanisms); and
- the practical compatibility of the Kyoto mechanisms with traditional environmental instruments, notably regulation, voluntary agreements and taxation, which have served as the backbone of environmental policy to date.

Against the background of the historical trends of, first, increasing diffusion of incentive-based instruments and, then, a combination of different instruments, the first chapter, written by Christian Egenhofer, addresses the issue of the compatibility of instruments to address climate change and policy mixes. It concentrates on the linkages between flexible mechanisms, notably emissions trading on the one hand and voluntary agreements, regulation and taxation on the other. It also concentrates on the potential incompatibility between domestic and international emissions trading, and explores ways of addressing these. It takes a practical approach. Rather than reviewing the results of theoretical models of the relative merits of different instruments, the first chapter reviews the literature on past experiences with the Kyoto mechanisms, defines the design options that could make these fit together, and evaluates a number of proposed or already running programmes as to how they would eventually fit with existing environmental policies. This chapter provides an analysis of design options that not only ensure compatibility but 'optimize' existing policy mixes.

The second chapter, by Patrick ten Brink, Marina Morère and Jane Wallace-Jones, analyses the incentives that firms have in adopting climate-related voluntary agreements. These voluntary agreements (or negotiated environmental agreements, NEAs) can be seen as a first step towards a more cost-effective and environmentally efficient policy. Either they are useful for negotiating the targets for the subsequent implementation of trading schemes, or they directly tackle the goal of emissions reduction by exploiting business opportunities (green consumerism). Alternatively, they aim to smooth the impacts of direct regulation or of other more costly (from the viewpoint of firms) environmental policies. What are the risks and rewards of adopting NEAs, and in particular what are the incentives for companies to participate in environmental agreements, and what are the incentives for action once companies become official participants? How can the environmental and economic impacts of NEAs be assessed? And what can one say, given current data availability? What are the costs of implementing the NEAs (for example, transaction costs)? Can NEAs be effective as stand-alone policy instruments, or is their effectiveness linked to the portfolio of policy instruments that governments use to address climate change? The second chapter tackles these questions and shows how voluntary agreements should be designed to provide the right incentives for firms to adopt them without losing their environmental effectiveness.

A similar approach is adopted in Chapter 3, in which Josef Janssen explores opportunities and barriers associated with the Kyoto mechanisms for industry. It focuses on two different categories of industry: the first type encompasses those companies that will act as final buyer or supplier of GHG emissions permits; the second type includes financial institutions. Chapter 3 provides a deeper understanding of both the general potential of the Kyoto mechanisms in terms of cost savings and business opportunities, and the barriers potentially hampering their application by industry. It also provides a comprehensive analysis and description of business opportunities, starting from a general assessment of the size of the market and ending with an analysis of the role of financial institutions.

The fourth chapter of this book, by Marzio Galeotti and Carlo Carraro, takes a longer-term perspective and focuses on the incentives for firms to undertake climate-related R&D and investments. This opens the way to an analysis of the dynamic efficiency of different policy instruments. Hence this chapter looks at the effect of environmental policies on the development and spread of new technologies, which, in the long run, may be the most important determinant of success or failure in climate change control. The chapter starts with an overview of the theoretical literature on the dynamic effects of environmental policy instruments on technological change, followed by the study of the role of policies in empirical set-ups. In particular, the chapter focuses on climate change models which explicitly incorporate a specification of technical change and diffusion. There, the working of flexibility mechanisms – international emissions trading in particular – alongside traditional policies and measures, is analysed. The results show how Kyoto mechanisms can play a double role. On the one hand, they minimize abatement costs in a static framework. On the other, they provide adequate incentives for investments in R&D and innovation.

The last chapter of this book, by Carlo Carraro and Marzio Galeotti, moves up along the game tree and analyses governments' incentives to ratify or to reject the Kyoto Protocol. Again, these incentives are highly modified by the actual climate policy that will be implemented. Hence the basic question is whether there are strategies that could enhance the prospects of the Kyoto

Protocol being implemented by a large group of highly polluting countries, and whether these are strategies that would induce countries to adopt emissions reduction policies even after the first commitment period. To answer these questions, it is crucial to identify which policies can be designed to lower the costs and increase the incentives to ratify a given international agreement on climate. This adds another dimension to the policy discussions of the previous four chapters of the book. Indeed, whereas the previous chapters focus on the environmental and cost-effectiveness of alternative policy mixes, Chapter 5 focuses on the effectiveness of different policy mixes in determining and increasing the number of countries that ratify and implement the Kyoto Protocol.

The importance of the structure of the prevailing international regime and its bi-directional links with the costs and benefits of different policy mixes leads to a second set of issues. If a larger number of participating countries reduces the implementation costs, thus opening the way to further extensions of the Protocol, what are the negotiating strategies that could help increase the number of countries willing to sign and ratify the Kyoto Protocol? Are there negotiating and institutional rules that favour countries' participation? Are there policy strategies that could provide the right incentives for countries to participate in the agreement? More specifically, should negotiations focus only on climate change policies or should they be linked to other policy issues? How can transfers help expand a coalition? The overall findings are discussed in the next section.

MAIN FINDINGS

Chapter 1 begins with an analysis of the long-trend shift from traditional command-and-control instruments towards incentive-based instruments in the European Union. This shift has been influenced by a diversity of motivations including economic (for example cost-effectiveness), sociopolitical (for example political acceptability), environmental (for example implementation and enforcement) or legal (for example competencies). This shift in government policies is also reflected in a shift of position by the regulated firms. While firms' behaviour under command-and-control regimes consisted of cost-avoidance strategies, the shift towards incentive-based instruments is mirrored by companies exploiting the environment in search of markets for new products to gain a competitive edge. The long-trend shift is advancing unevenly from one country to another, depending *inter alia* on the political conditions, the legal system, national preferences, the nature of the environmental problem and different environmental priorities. For example, while tradable permits have played a minor role in the EU, voluntary initiatives and negotiated agreements have flourished. In the USA, there has been more interest in tradable permits.

At the same time that policy is shifting towards the use of incentive-based instruments, we are also witnessing a decided shift towards combining instruments. Originally motivated by governments' desire to apply the most appropriate instrument to a given environmental problem, policy mixes are increasingly a result of practical solutions to a government's need to accommodate the 'real world', that is, to build on existing policies and instruments, whether taxation, regulation or NEAs. An example of this phenomenon are the newly emerging climate policies within the framework of the Kyoto Protocol, which has introduced three genuine international market-based instruments: international emissions trading, joint implementation and the clean development mechanism (IET, JI and CDM) for governments to use and combine with domestic policies into new policy mixes. Although there is a risk that policy mixes increase transaction costs,[1] they are increasingly designed to 'optimize' incentives for action. A good example is the combination of NEAs and emissions trading. While NEAs are seen as low-cost and flexible instruments, allowing for emissions trading in addition (as is proposed in the UK) will further increase both efficiency and effectiveness. There are a number of examples (see also Chapter 2), where the threat of regulation or taxation is meant to 'persuade' industry to negotiate an NEA (for example in Denmark, Germany, Belgium, the UK). The experience gained from negotiating and setting targets (for example benchmarking) from an NEA is expected to increase the incentives of firms to accept a cap-and-trade scheme. NEAs generally improve the knowledge of both the regulator and the regulated firm on real abatement potential and its costs.

Chapter 1 discusses in particular the role of the relative merits of different emissions trading programmes with a view to combining them with other instruments, notably NEAs. The economic literature expresses a clear preference for allowance (or cap-and-trade) emissions trading as opposed to baseline-and-credit programmes (or credit-based trading).[2] Allowance trading programmes have the double advantage over credit trading schemes in that they ensure environmental effectiveness and tend to have fewer transaction costs. These were the two main reasons why the US EPA, which implemented most actual emissions trading schemes, has moved over time from credit trading to cap-and-trade schemes, which by now can be said to be the 'traditional' approach. This contrasts with the situation in Europe, where numerous voluntary or negotiated agreements between government and industry exist that have set relative or specific targets. The main advantage of credit trading is that it can be combined at the same time with policies based on absolute and relative performance such as in NEAs and a mix thereof, which is the reality in Europe. Only rate-based credit trading programmes are compatible with the 'European model of voluntary approaches'. France and the UK but also Canada have started to experiment with policy mixes that blend cap-and-trade emissions trading with baseline-and-credit schemes (both of which are based on absolute and relative targets). The UK scheme for emissions trading is particularly interesting since it attempts to combine the two forms. Chapter 1 concludes that the advantages or disadvantages of allowance and credit trading in a 'real-world' setting are not absolute and can be addressed by institutional design, although the existing economic literature has paid little attention to this.

Baseline-and-credit programmes have the further advantage that they might be politically more acceptable[3] due to the fact that credits – unlike permits in allowance schemes that are allocated, most often for free – must be earned by demonstrating better-than-mandated behaviour. In this way, credit schemes can overcome popular criticism that allowance trading gives participants a 'licence to pollute'. Some maintain that credit trading avoids the contentious issue of initially allocating allowances and may therefore be more equitable than cap-and-trade systems with grandfathering. There is, however, very little discussion of the equity issues of emissions trading schemes. One should expect that equity considerations are likely to dominate the debate on emissions trading in the coming years.

Chapter 2 takes an in-depth look at the role of voluntary or, as they are called in Europe, negotiated environmental agreements (NEAs), both as a single instrument and as part of a policy mix. The overall conclusion is that NEAs are increasingly seen as offering a pragmatic and potentially valuable instrument to help address climate change. In several countries NEAs have been chosen by governments to be an important instrument to meet the Kyoto targets (for example France, the Netherlands, Italy and Germany – at least under the last government). However, the chapter also shows that significant further efforts will be required to ensure that the NEA tool can fulfil this promise and responsibility.

The theoretical analysis and the case studies reveal a number of emerging *patterns* regarding NEAs. First, an increasing number of countries look to NEAs. For some countries, they are the main instruments (for example Germany, the Netherlands and Italy), while in others, NEAs are often a significant part of a portfolio of instruments. Second, governments see NEAs as a means to preserve competitiveness of domestic firms. NEAs are perceived as a least-cost and flexible alternative to other instruments such as taxation or regulation.[4] Third, NEAs are increasingly integrated with emissions trading or the Kyoto mechanisms (for example in the UK and the Netherlands). Fourth, there is an increasing demand for more mature NEAs that are credible and offer a real potential to be successful. Early and somewhat weak NEAs have sometimes discredited the instrument, risking taxation or regulation instead, and generally could not guarantee certainty to business.

Early NEAs have entailed some degree of *regulatory capture*, but pressure by the public and NGOs led to more realistic and demanding targets. There is no conclusive evidence on the impact of NEAs. They are a relatively new instrument still under construction, and tend to be directed to the long term. Improving monitoring and measuring will be important in this respect.

The *linkage* to emissions trading and eventually the other Kyoto mechanisms will play an important role, although the nature and extent of this role will only become fully clear in the future (see also Chapter 1). To date, few NEAs are explicitly linked to emissions trading schemes and the Kyoto mechanisms. Some NEAs are linked to permitting (the Netherlands) and some to tax exemption (Denmark and the UK). Subsidies and regulation (for example standards or requirement for audits) are also usually linked to NEAs. It is clear that NEAs cannot realistically – without linkage to other instruments, and at least incentives and sanctions – offer any significant benefits in the area of emissions reduction and meeting GHG emission reduction targets.

It is far from clear at this stage whether NEAs are just a transition measure to address climate change or whether they will become the standard instrument, integrated into the portfolio of instruments designed to address climate change. While NEAs have the potential to become a long-term instrument, much depends on their achievements, on the culture and country-specific contexts, and on the institutional structure that develops around the climate change issue. Their future also depends on the actual motivation (and success) of stakeholders in the continued process of revision and improvement of NEAs. Finally, the uptake depends on the level of international commitment to and collaboration in combating climate change. It is clear at this stage, however, that the climate change NEAs can serve a real learning and regulatory capacity development role, but without clear and quantitative early successes that are identified and communicated through more credible monitoring schemes, the tool is unlikely to live up to its declared potential. To the contrary, in fact; it might instead live up to its threat of being a tool for exemptions rather than a tool for action.

Chapter 2 argues that it is impossible to judge the potential of an NEA without looking explicitly at specific sectors and the country-specific context. It concludes by posing a set of questions that need to be answered when the choice is made.

Chapter 3 takes a different angle in its analysis of business opportunities and barriers with respect to the use of flexible mechanisms as defined in the Kyoto Protocol. The role of firms is likely to become increasingly important in the emerging international regime governing climate change. From the outset, the international regime has focused on market-based solutions driven by economic incentives rather than regulation. The practical implications of the analysis developed in Chapter 3 are necessarily preliminary. Whether the Kyoto Protocol will be ratified or implemented at this stage – even after the conclusion of the seventh Conference of Parties (COP7) in Marrakesh – remains unclear. Similarly, unless the US role is clarified, any empirical analysis must remain preliminary. However, irrespective of the details of an international agreement including US participation, it is clear that both national and international emissions trading will play a key role in any future climate policy regime at the national, regional or international level. Although the lack of certainty about the international regime strongly affects the quantitative results, such as the size of the markets, it does not fundamentally affect the analysis of opportunities and barriers in principle.

This chapter reiterates the argument put forward in the previous two chapters (but from a company perspective) that governments might wish to implement a mix of different climate policies, thus not relying exclusively on allowance trading. It confirms that inventory-based emissions trading, or allowance trading, will be complemented by project-based emissions trading (baseline-and-credit trading) in those industries and sectors that are not regulated by allowance trading, but other policy instruments such as energy efficiency standards or negotiated environmental agreements (see Chapter 2 on the latter) will also play a role. Joint implementation and the clean development mechanism may then be interpreted as international baseline-and-credit trading.

The first and most basic opportunity is on the demand side of GHG markets and mainly involves significantly lower costs of compliance for industries that are regulated by climate policies. Companies with comparatively low abatement costs may benefit from emissions trading and the other Kyoto flexible mechanisms by marketing any surplus emissions reductions, thus making additional profits. Indeed, the value of the *international* market for emissions permits is estimated to amount to several tens of billions of dollars annually.[5] In addition to international markets, emerging national markets will provide good opportunities to sell profitably surplus emissions permits.

The main barriers that prevent companies from taking advantage of GHG emissions trading and the other Kyoto mechanisms are identified as market risks and regulatory barriers at the international and national level. Important market risks include uncertainties about the price of permits and liquidity risks. Regulatory barriers induced by international rules include possible import restrictions, levies on transactions, and eligibility requirements for countries. At the national level, the low compatibility of domestic climate policy regulations with the Kyoto mechanisms and the lack of incentives for market participants to engage in the Kyoto mechanisms may represent major barriers. The latter issue is also addressed in Chapter 1.

Price risks may be managed through derivatives. In contrast, regulatory barriers need to be removed through appropriate design of climate policies and regulations. In particular, governments need to provide incentives for private-sector companies to use the Kyoto flexible mechanisms for compliance. To this end, companies that apply the Kyoto mechanisms should be relieved from obligations incurred by domestic climate policy regulations by a corresponding amount. Transparent and stable regulations governing the Kyoto mechanisms are essential for their use by the private sector.

Financial institutions may be expected to facilitate market development and to enhance the efficiency of GHG markets. At present, the role for commercial insurance is rather limited since many risks in emerging GHG markets are regulatory, political or market risks, which are difficult to insure. In contrast, carbon investment funds have the potential to hedge some risks associated with investments in abatement activities that aim to result in marketable emission permits. Such funds may be set up and managed by investment banks and other financial institutions. Indeed, some carbon funds are already emerging in reality.

Chapters 1, 2 and 3 focus on short-term actions taken by a single industry or firm. In Chapter 4, the analysis covers incentive mechanisms that may also lead to effective emissions reductions in the long run, by stressing the role of climate-related technical change and by analysing which incentive mechanisms can be designed to foster its adoption and diffusion. In particular, Chapter 4 identifies which policy mixes are most suitable to induce firms to adopt long-term innovation policies aimed at reducing GHG emissions.

The perspective chosen in Chapter 4 is both a microeconomic and a macroeconomic one, because it also takes into account the systemic effects of growth and innovation, including their links with environmental degradation. Hence the impact of different climate policies on growth has been considered jointly with their impact on innovation and emissions.

Chapter 4 stresses that an understanding of the process of technological change is essential to any economic analysis of environmental issues. This is true for two main reasons. First, the environmental impact of social and economic activity is greatly affected by the rate and direction of technological change. This linkage occurs because new technologies may either create or mitigate pollution, and because many environmental problems and policy responses are evaluated over periods of time in which the cumulative impact of technological change is likely to be large. This observation is a natural bridge to the second broad linkage between technology and environment – the effect of environmental policy interventions on the process of technological change. The empirical evidence presented in Chapter 4 is generally consistent with theoretical findings that market-based instruments for environmental protection are likely to have significantly greater, positive impacts over time than command-and-control approaches on the invention, innovation and diffusion of desirable, environmentally friendly technologies.

The potential long-run consequences of today's policy choices (for example policy-induced incentive mechanisms) create a high priority for broadening and deepening our understanding of the effects of environmental policy on innovation and the diffusion of new technology. Unfortunately, these issues cannot be resolved at a purely theoretical level, or on the basis of aggregate empirical analyses. For both benefit–cost and cost-effectiveness analysis, modern economics relies on a set of powerful analytical and computational tools that support quantitative modelling of economy- and society-wide policies over the long run. In practice, such analysis is typically carried out through the construction and application of large-scale computer models that combine scientific and economic theories and data into unified quantitative frameworks. These 'integrated assessment' models have emerged as decision makers' primary tool for quantitative climate policy analysis.

Only a handful of such recent models, which are reviewed in Chapter 4, have tackled the difficult issue of explicitly modelling the process of technical change. Most of them have then been applied to study the effects of the implementation of the Kyoto Protocol, particularly as far as the consequences of the emission targets and of the Kyoto mechanisms are concerned. Chapter 4 shows that in those models in which technical change is better designed, the Kyoto mechanisms have a relevant impact on the incentives to carry out R&D and to develop energy-efficient technologies. In addition, emissions trading provides incentives for a strategic use of R&D, that is, R&D is also undertaken for the profit opportunities provided by trading emissions internationally. It is in addition shown that constraints on the use of the Kyoto mechanisms are of no help in increasing R&D and technical change.

The policy implications that can be drawn from the analysis presented in Chapter 4 hinge on the following basic conclusion: technological innovation and diffusion are the main engine of growth. Growth is correlated with increasing environmental efficiency. Hence innovation accomplishes a twofold task: it stimulates growth and competitiveness and reduces pollution levels.

However, several market failures characterize existing innovation systems. Spillover effects, inefficiency in the banking sector and financial markets (which have the task of channelling the funds to finance innovation), risk aversion, the nature of knowledge as a public good, and lack of coordination are all elements that reduce private incentives to undertake R&D and innovation and call for public intervention. These market failures are even stronger in the case of international and global environmental problems, for example climate change. Climate change is a global, long-term, uncertain phenomenon. These features crucially reduce firms' incentives to undertake R&D and innovation. Therefore policy measures are necessary. These policies – usually policy mixes based on a bundle of instruments, as shown in Chapter 4 – can be undertaken both nationally and internationally.

At the national level, the long-term dimension of climate change calls for action on R&D and innovation that enhances *long-run competitiveness* in environmentally sensitive sectors, increases *financial resources* provided by the banking sector (for example by partly involving banks in the liability process), guarantees *protection* for energy-efficient innovation (for example by introducing a climate-related licence policy), and supports *cooperation and research joint ventures* designed to undertake environmental R&D.

At the international level, another dimension of the problem must be taken into account. As further emphasized in Chapter 5, the global dimension of climate change and the increasing mobility of capital and goods call for action in both developed and developing countries. According to the analysis developed in Chapter 4, it is therefore crucial to increase:

- the diffusion of knowledge by disseminating information about new environmentally friendly technologies;
- basic learning and innovation by stimulating interactions between academic research and applied environmental innovation; and
- the diffusion of technology by improving the quality of the role of multinationals in the global economy and by implementing technological cooperation agreements.

In this framework, environmental innovation policies are structurally linked and interact with competition, industrial and trade policies. These should support R&D cooperation, increase the competitiveness of the banking system, protect innovation and introduce lender liability schemes, and support trade rules that favour technology transfers.

As stated above, the objective of Chapter 5 was to answer a set of questions which are crucial for understanding the future prospects of climate negotiations. In particular, the goal was not to understand whether the Kyoto Protocol is going to be ratified and implemented, but whether an *environmentally effective* climate agreement, possibly based on the Kyoto framework, is going to be adopted in the near future. Therefore the basic questions addressed in Chapter 5 are as follows: given the peculiar features of the climate change problem, what is the future of the United Nations Framework Convention on Climate Change (UNFCCC) and of the Kyoto Protocol? Is there any strategy that could enhance the chances of implementation of the Kyoto Protocol, and its extension beyond 2012, by the USA and other major polluters? Or should countries move to a different type of agreement, as recently suggested in the USA?

In answering these questions, Chapter 5 analyses which policy strategies and negotiation rules can help increase the number of countries engaging in a long-term international agreement to combat climate change by signing and ratifying the Kyoto Protocol and its possible extensions after 2012. This analysis identifies a set of policy guidelines that could help the future evolution of climate agreements.

Before summarizing these policy guidelines, it is important to stress a methodological point put forward in Chapter 5. Given the strict interdependence between the costs and benefits of different policy options (that is, the timing of optimal responses to climate change, the choice between mitigation and adaptation responses, the role of technological innovation and diffusion, the choice between domestic action and the adoption of 'flexible mechanisms', the importance of secondary benefits, and so on) and the characteristics of the international agreement on climate change that is adopted (for example the number of signatories, the size of their quantitative commitment to control GHG emissions), it is impossible to assess the costs and benefits of the Kyoto Protocol, or of other potential agreements on climate change, independently of the number of signatories to the agreement and of their abatement targets and/or policy commitment. But at the same time the number of signatories is endogenous and depends on the abatement targets and mitigation policies adopted in various countries. Hence it is important to analyse the environmental and cost-effectiveness of climate policy jointly with a country's decision to sign and ratify the agreement. Hence the analysis of which strategies can be proposed to reduce the costs of mitigation policies becomes quite different from the ones usually proposed in recent papers on climate policy (see Table 5A.1 in Appendix 2 of Chapter 5). The reason is that a country's goal is not only to identify a new climate-friendly technology or an adequate redistribution of costs across sectors. An important goal is also to affect other countries' behaviour in order to increase the number of countries sharing the burden, and the equity with which this burden is shared.

From this perspective, two main policy dimensions are identified in Chapter 5. The first moves from domestic and rigid (for example norms and measures) policies to global and increasingly flexible policies. The second moves from unilateral to increasingly cooperative actions. Chapter 5 emphasizes that it is not necessary to achieve the optimum along both policy dimensions. For example, even a suboptimal distribution of targets across countries can be consistent with optimal emissions reduction schemes if coupled with a global and flexible policy (for example a global system of emissions trading).

Chapter 5 also shows that it is quite difficult to characterize as 'likely' a scenario in which either a global coalition forms or a global and perfectly competitive emissions trading system is implemented. Hence we are left with a world in which only a subset of countries may decide to commit themselves to emissions control and only a subset of countries, not necessarily the same countries, may decide to participate in a freely competitive trading system. In this latter scenario, there are three policy options:

1. Design policies that induce countries that are not committed to emissions control to accept this commitment. This can be done through appropriate policies that, on the one hand, reduce the cost of the commitment (for example competitive emissions trading again) and on the other hand increase the benefit of the commitment by means of transfers or by linking the climate agreement to other economic agreements (for example on technological cooperation).
2. Recognize that most countries' main incentive is to form several coalitions that may adopt different policy measures. In this case, instead of a single global agreement, countries should be free to negotiate regional climate agreements. In this setting, it may be reasonable to pursue a climate agreement within the EU or between the EU, Eastern Europe and former Soviet republics, because other regional climate agreements are likely to emerge subsequently. Once several regional agreements enter into force, for example in Europe, East Asia and the Americas, then the three groups of countries can negotiate to achieve a global agreement.
3. Focus on treaty design and investigate whether there are negotiation and accession rules that could increase the incentives to sign and ratify a climate agreement.

Chapter 5 also addresses the equity issue. In particular, this chapter has analysed the conjecture that a more equitable distribution of the burden of reducing emissions could provide the right incentives for more countries – particularly long-run big polluters (for example the USA, China, India) – to accept an emissions reduction scheme defined within an international climate agreement. Chapter 5 only partly supports this conjecture. Even though more equitable burden-sharing rules provide better incentives to sign and ratify a climate agreement than the burden-sharing rule implicit in the Kyoto Protocol, a stable agreement cannot generally be achieved; that is, equity seems to enhance the profitability of an agreement but it does not offset the incentives to free-ride.

Therefore, Chapter 5 also analyses whether there exists a transfer mechanism that could help broaden an initially stable, but partial, coalition based on an agreement on an equitable burden-sharing scheme. Our results suggest that transfers can indeed help broaden a given coalition. However, the grand coalition could not be achieved with merely the three equity rules considered in Chapter 5 (equal average abatement costs, equal per capita abatement costs and equal abatement costs per unit of GDP). The only strategy that could achieve a global agreement without free-riding incentives is a policy mix in which global emissions trading is coupled with a transfer mechanism designed to offset *ex-post* incentives to free-ride. This policy mix can achieve a stable global agreement whatever the initially stable coalition and whatever the *ex-ante* burden-sharing rule.

As a consequence, Chapter 5 seems to suggest that an excessive focus on *ex-ante* equity rules is not fruitful. It is more effective to minimize overall abatement costs via emissions trading and then to use the resulting surplus to provide incentives for free-riding countries to join the initial coalition.

By combining the insights derived by analysing the efficiency of climate policies with those related to their equity, Chapter 5 leads to some basic policy guidelines:

- Policies, mainly policies that support research on climate change, should reduce uncertainty concerning climate phenomena, their origin and their impact (in particular the distribution of this impact).
- Policies should redistribute benefits from climate change control to make a climate agreement profitable to all participating countries. Relevant redistributive effects can be achieved both through the Kyoto mechanisms and through direct transfers or economic cooperation.
- Policies should address the free-riding incentives by focusing on equity, but also by implementing a wide range of transfers and linkage strategies.
- In the absence of supranational institutions to regulate emissions reductions and in the presence of strong negative incentives to participate in the climate agreement, negotiations should allow for regional and/or domestic cost-effective attempts to control GHG emissions (for example regional or domestic negotiated agreements or trading schemes).
- Policies should support the development of cheaper climate-friendly technologies and their diffusion through negotiated agreements and incentive mechanisms based on detailed policy mixes.
- Climate policies should not be undertaken independently of other economic policies. The best results on climate change control can be achieved by designing trade policies, industrial and R&D policies, insurance and competition policies, patent and property rights policies, technological cooperation and international monetary policies, with a view to climate issues and with a clear understanding of the climate-related benefits that these policies, if appropriately designed, can achieve.

NOTES

1. This is for example the case of the so-called 'gateway' in the UK emissions trading scheme (see Appendix 1 in Chapter 1), which avoids carbon leakage but increases transaction costs.
2. For definitions and a discussion of the differences, see Chapters 1 and 3.
3. There is some evidence for this in France.
4. This view is increasingly challenged by the emergence of emissions trading.
5. This figure is derived from models where the USA is a signatory to the Kyoto Protocol.

1. The compatibility of the Kyoto mechanisms with traditional environmental instruments

Christian Egenhofer

1. INTRODUCTION

Environmental policy has for long been relying on direct regulation, also referred to as command-and-control instruments. In the last decade or so this approach has been giving way to a new balance of environmental instruments. Not only have incentive-based instruments started to play a bigger role; at the same time, as a result of an enlarged toolbox of instruments, governments have started to combine instruments and develop policy mixes. Climate change policy is a good illustration of both trends. The Kyoto flexible mechanisms have introduced incentive-based instruments to be applied at a global level. Climate change policy will demand a mixture of domestic and global action, hence reinforcing the trend towards the combination of instruments. This raises issues of compatibility of instruments as well as 'optimizing' policy mixes to enhance incentives to abate pollution.

The starting-point of our analysis is that in the European Union (EU) there is a long-trend shift from regulation to incentive-based instruments as well as to a combination of instruments. This trend is likely to be accelerated through the Kyoto Protocol, as is argued below. Little academic work has been done, however, on the *practical* compatibility of emissions trading (ET) with traditional environmental instruments, both command-and-control (that is, regulation) and incentive-based instruments such as voluntary agreements and taxation. This is in stark contrast with what is happening 'on the ground'. As Appendix 1 shows, existing approaches to address climate change have developed sophisticated policy mixes, blending, for example, various models of emissions trading, voluntary instruments, taxation and regulation. This chapter analyses issues of potential incompatibility and explores ways of addressing these. It will take a practical approach. Rather than reviewing the results of economic modelling on the relative merits of the different instruments, it will concentrate on the following four steps: (1) description of the long-trend shift from regulation to incentive-based instruments and the combination of instruments; (2) review of the literature on past experiences with 'flexible mechanisms'; (3) definition of the design options that could make the flexible mechanisms fit; and (4) checking of a number of proposed or current programmes to see how they would eventually fit with existing environmental policy. The objective is to develop criteria that flexible mechanisms must meet in order to be implemented in the 'real world'. As a consequence, the subject of this chapter is the political economy of the Kyoto mechanism as exemplified in Europe.

2. THE GRADUAL RISE OF INCENTIVE-BASED INSTRUMENTS IN THE EU AND ITS MEMBER STATES

The EU's political system is highly decentralized, based on concurrent (shared) competencies between the EU and member states (see Pelkmans, 2001; Dinan, 1999). As a consequence, EU environmental policy is a combination of EU and member state action within the boundaries of the EU Treaty (Pelkmans and Egenhofer, 1993). By and large, most regulatory functions, if related to the four freedoms (goods, services, capital, persons), are exercised by the EU level of government, while implementation tends to be done at member state level. Local pollution with little or no effects on the four freedoms, however, is often entirely dealt with by member states. Decisions on the appropriate level of competency are dealt with on a case-by-case basis. This is enshrined in the Subsidiarity principle,[1] which was introduced at the time of the 1992 EU Treaty revision. This principle guides the assignment of competencies between the EU level of government and member states. In essence it proposes EU actions only for those cases where solutions cannot be reached at national level, typically for reasons of cross-border externalities and economies of scale (see Sun and Pelkmans, 1995; Jordan and Jeppesen, 2000; Shaw et al., 2000). This situation has added some specific twists to the political economy of the double shift introduced above.

2.1 Performance of Command-and-control Approaches

The environment chapter of the EU Treaty, which created an explicit EU competence in this field, is as recent as 1987. Nevertheless, before that the EU adopted more than 200 pieces of legislation dealing with environment, mainly as a side-effect of legislation relating to the EU's internal market (for example harmonization of product or emissions standards). With environmental legislation being embedded in harmonization of internal market regulation, the 'instrument of choice' was inevitably direct regulation. Moreover, this was very much in line with general trends of emerging environmental policy (and regulation) in other countries. Throughout the 1970s and much of the 1980s environmental policy in OECD countries focused mainly on management of environmental resources and the creation of institutional capacity directing the environmental performance of firms. Command-and-control instruments revolve around standards that can take different forms, such as ambient (for example air quality), technology (for example best-available technology, known as BAT) or emissions standards (for example emissions ceiling).[2] This allows firms to produce a certain level of pollution. Generally, non-compliance resulted in penalties, mostly fines in some cases (for example Germany), and liability for correction of environmental damage. In addition, command-and-control approaches can also aim at energy efficiency standards (for example fuel efficiency, insulation requirements of a specific product). More recently, governments have started to experiment with renewable energy quotas.

Regulation was governments' preferred approach to environmental management for reasons of effectiveness. The feeling was that regulation offers better protection since the environmental objectives are clearly specified in terms of physical limits that cannot be exceeded and the technology to reach such objectives is often prescribed. This preference also reflects the important role that engineers played in environmental decision making (Potier, 1996). Another reason for politicians to prefer regulation was their ability to hide the full costs and their distribution, avoiding difficult debates on equity. The regulators' (that is, governments') preferences have been mirrored by the regulated firms' preferences, which have for long seen environmental regulation as a cost factor and have chosen to minimize the cost of complying with command-and-control regulations. Firms relied primarily on problem avoidance and risk management rather than exploiting the opportunities in the form of markets for new products and processes or to gain a competitive advantage from environmental decisions (Reinhardt, 2000). As long as costs for environmental protection remained relatively limited and/or exposure to international competition was low, the need for instruments providing incentives to exploit the opportunities environmental legislation offered was small.

2.2 Higher Efficiency through Incentive-based Instruments

During the 1980s and the 1990s, not at least due to increased international competition and a general trend towards 'markets', economists, politicians and business, especially in the OECD, became interested in regulatory reform of environmental policies. This trend towards the increased use of incentive-based or fiscal and economic instruments (OECD, 1996, 1997a) and voluntary agreements (Storey, 1996) and, more generally, the broadening of environmental instruments, has also been mirrored in the EU. The Fifth EU Environmental Action Programme sketching the EU's environmental priorities for most of the previous decade consequently revolved *inter alia* around attempts to broaden the range of instruments. The EU regulatory reform agenda was further influenced by successive revisions of the EU Treaty, leading in 1992 to include sustainable development as an EU objective and mandating the integration of the environment into other policies.[3]

The OECD has identified five categories of incentive-based instruments: (1) charges and taxes; (2) deposit refund systems; (3) tradable permits; (4) financial enforcement incentives; and (5) subsidies (see OECD, 1994). Others include subsidy reform, voluntary agreements, liability regimes and information and labelling programmes (see Clinch, 2000 and Stavins, 2001).

- *Charges and taxes* put a price on pollution. While taxes are paid into the general budget, charges are payments for which the payer receives a benefit in return. In practice, however, the distinction is blurred. Charges and taxes can aim either at emissions or at products.
- *Deposit refund systems* levy a surcharge on the price of potentially polluting products which is returned if the pollution is avoided by, for example, returning the product.
- *Cap-and-trade emissions trading* schemes put a limit on emissions for a certain product or process. Polluters hold property rights for all allowances under the ceiling. Individual emissions rights (that is, allowances) are allocated either by auctioning or distribution free of charge (grandfathering). Grandfathering is often based on benchmarking to avoid punishing those who have abated in the past.
- *Baseline-and-credit or simply credit-based schemes* are another variant of emissions trading. In this case polluters have to earn credits by over-complying with a target. Credits are fully tradable.
- *Financial enforcement incentives* take the form of non-compliance fees (penalties), which are more or less proportional to the savings reaped by non-compliance. A special form is performance bonds, which are payments to authorities in expectation of compliance, refunded once compliance is achieved.
- *Subsidies* cover forms of financial assistance to provide incentives to the polluter to alter its behaviour. Subsidies can take the form of grants, soft loans, tax allowances or R&D subsidies.
- *Subsidy reform* aims at reducing the distorting effects of government policy that keep pollution control costs artificially low through direct subsidies or tax incentives.[4]

- *Environmental liability* relies on a careful definition of property rights over the environment. This allows consumers and producers hurt by pollution to ask for compensation by the polluter by either public or private law. This instrument is only applicable where both the source of the damage can be easily identified and the damage can be easily assessed.
- *Voluntary agreements* include many different approaches, including codes of conducts and responsible care programmes, voluntary measures such as voluntary restraint agreements or unilateral commitments, implementation of (accredited) environmental management systems (for example EMAS – Eco-Management and Audit Scheme), voluntary auditing, voluntary environmental reporting as well as more formalized negotiated agreements. For a more detailed discussion, see Chapter 2.
- *Renewable energy quotas* constitute obligations to produce (if targeted at electricity producers), supply (if targeted at electricity suppliers) or consume (if targeted at electricity consumers) a set amount of electricity produced from renewable sources. Increasingly this instrument is linked to tradable permit schemes.
- *Green government purchasing* attempts to create markets for more environmentally friendly (that is, green) products. Similar to renewable energy quotas, it aims at overcoming market barriers for environmental beneficial products.
- *Information programmes* play an indispensable role in the successful application of incentive-based instruments, which will only reach their full potential if accompanied by programmes aiming at informing the end-users and citizens in general about environmental standards and performance. Examples of such instruments include energy efficiency labelling requirements, eco-labels as well freedom of information acts, environmental management systems and environmental impact assessments.

The above instruments all have in common – contrary to direct regulations, which prescribe a specific mandatory action to economic agents – that they use, directly or indirectly, market signals to influence behaviour. These signals take the form of a modification of relative prices or a financial transfer or other incentives to get environmental performance rewarded by the market.

Policy makers became sympathetic to the application of incentive-based instruments due to a mixture of push and pull factors.[5] The push factor was that regulation had not been as effective as perceived in theory. It proved expensive and lengthy to bring offenders to court, thereby reducing the possibilities of enforcing effective compliance. The level and sometimes the

Table 1.1 The gains from least-cost air pollution control

Pollutants covered	Area of the USA	Command-and-control policy (CAC) benchmark	Ratio of CAC cost to least-cost policy
Particulates	St Louis	State Implementation Plan (SIP) regulations	6.00
Sulphur dioxide	Four Corners: Utah, Colorado, Arizona, New Mexico	State Implementation Plan (SIP) regulations	4.25
Sulphates standards	Los Angeles	California emissions standards	1.07
Nitrogen dioxide regulations	Baltimore	Proposed Reasonably Achievable Control Technology (RACT) (technology requirements)	5.96
Nitrogen dioxide regulations	Chicago	Proposed RACT (technology requirements)	14.40
Particulates	Baltimore	State Implementation Plan (SIP) regulations	4.18
Sulphur dioxide Particulates	Lower Delaware Valley	Uniform percentage reductions	1.78 22.00
Airport noise	USA	Mandatory Refit	1.72
Hydrocarbons	US DuPont plants	Uniform percentage reductions	4.15
CFC emissions ex. aerosols	USA	Proposed emissions standards	1.96

Source: Clinch (2000).

probability of a fine were not sufficient to ensure compliance. Specific to the EU was the widening implementation gap of environmental regulations among member states (CEC, 1992). Command-and-control regulations based on technology-based standards typically specify the method and the actual equipment that firms must use to comply. Procedures for permits and licensing have become increasingly cumbersome, pushing up administrative costs.

Pull factors included the prospect for less bureaucracy and better environmental performance due to more effective implementation (Pelkmans and Egenhofer, 1993; Delbeke, 1996) and the revenue-raising function – in the cases of taxation and tradable permits, if allowances are auctioned (EEA, 1996; Potier, 1996; OECD, 2001). Incentive-based instruments were also seen as better than uniform technology standards to deal with the sensitivities that environmental policy and regulation commands among the electorate. National governments felt a more urgent desire to follow their own environmental preferences (Lévêque, 1996). The strongest arguments in favour of incentive-based instruments are that they theoretically minimize costs to society – incentive-based instruments equalize the incremental amounts that firms spend to reduce pollution – and that they have dynamic efficiency, that is, incentives to innovate and improve environmental performance. In contrast, command-and-control regulations hold all firms to the same target. As a consequence, command-and-control measures equalize the pollution *levels* among firms. While standards – if adhered to and properly implemented – effectively limit pollution, they tend to increase costs by forcing some firms to apply expensive technologies to comply with a given standard. In addition to typically exerting high costs, technical standards tend to freeze the development of technologies in the absence of financial incentives to exceed a control target (see Clinch, 2000; Stavins, 2001).

Empirical evidence suggests that society's economic benefits from the application of incentive-based instruments are considerable. One survey of ten empirical studies of air pollution control in the USA found that the ratio of actual aggregate costs of command-and-control regulations to the aggregate costs of least-cost benchmarks ranged from 1.07 for sulphate emissions in Los Angeles to 22.0 for hydrocarbon emissions at all domestic DuPont plants (See Table 1.1).

2.3 Diffusion in the EU

Despite the broadening of the range of policy instruments with a greater use of taxes, voluntary agreements and information and transparency, EU environmental policy still relies on regulation both at member state and EU level. According to the European Environment Agency (EEA, 1999), currently 315 environmental directives are in force dealing primarily with industry, agricul-

Table 1.2 Summary of use of instruments in sectors within the EU

	Agriculture	Industry	Energy	Transport	Households
Regulations (315 directives)	30	40	5	14	9
Environmental taxes (134)	3	9	18	54	16
Voluntary agreements (305)	3	88	5	4	–
EIAs* per annum (7000)	16	26	8	30	20 (waste)
Environmental Management Systems	–	88	4	8	–
(3910 sites by October 2001)	NA	NA	NA	NA	NA

Note: * EIA = Environmental impact assessment.

Source: Adapted from Clinch (2000).

ture and transport, but gradually enlarging to energy and households.[6] For an overview, see Table 1.2. Implementation of incentive-based instruments was slowed down on grounds of lack of political acceptability due to equity concerns and trade implications, that is, fear of loss of competitiveness. An example of opposition for reasons of acceptability and distributive implications could recently be witnessed in Europe, where truck drivers were demonstrating for – and largely obtaining – reductions on fuel taxes partly levied on environmental grounds.

Tradable permit schemes have traditionally played a minor role in the EU, with the notable exception of a volatile organic compound (VOC) control scheme in Germany (Norregaard and Reppelin-Hill, 2000) and the use of trades to achieve reductions in the production of ozone-depleting substances to enable chemical companies to concentrate production and shift to substitutes (Klaassen and Nentjes, 1997a). Europe instead relied to a large extent on negotiated environmental agreements (NEAs), the most often applied form of voluntary approaches or initiatives for environmental protection in the EU (see CEC, 1996; Storey, 1996; EEA, 1997; Carraro and Lévêque, 1999), even being dubbed the 'European model of voluntary approaches' (Börkey and Lévêque, 1998). NEAs are commitments undertaken by firms and sectors as a result of negotiations with public authorities. Chapter 2 provides a comprehensive analysis of NEAs and other voluntary initiatives.

At the EU level of governance, reliance on command-and-control regulations is even more pronounced, mainly for constitutional reasons (see Table 1.3). In the case of taxation, the EU Treaty stipulates that measures for taxation need unanimous agreement by all member states, meaning *de facto* that each member state has a veto on taxation. Not surprisingly, taxation for environmental purposes is exclusively implemented at member state level (see EEA, 1996; Ekins, 1996; EEA, 2000). At the end of the 1990s the use of taxes increased, although the share of environmental taxes expressed as the total revenue from total taxes and social contributions is rising only slowly. While the share was 5.84 per cent in 1980, it rose to 6.19 per cent in 1990 and stayed at 6.71 per cent in 1997 (EEA, 2000). See also Tables 1.2 and 1.3 and the European Commission's website.[7] Most of the revenues relate to transport and energy.

Voluntary agreements have also had very little impact at the EU level of governance, again mainly for institutional reasons. Since voluntary agreements are negotiated between the European Commission and the industry concerned, the European Parliament has been opposed to agreements since effectively their adoption is withheld from parliamentary scrutiny and approval. The conclusion of voluntary agreements is also complicated because of a lack of a suitable interlocutor at industry level at the EU. Not all EU-wide industry associations can claim to represent all if not most of the

Table 1.3 Progress at EU and member state level in introducing environmental policy instruments

Instrument	EU level initiatives	Member state initiatives
Regulations	Approx. 315 directives	Approx. 90% of EU directives translated into national legislation
Environmental taxation	Proposal for harmonization of energy taxation and discussion of pesticide tax	Growth in environmental taxation in Nordic countries leading the way. €6 billion raised in pollution taxes in EU in 1996, a 100% increase since 1990; green tax reform was introduced in Germany
Voluntary agreements	Agreements on energy efficiency in washing machines and TVs and CO_2 emissions with auto industry	More than 300 voluntary agreements agreed from 1990 to 1996, mostly for industry, with about 100 in Germany and 100 in the Netherlands
Emissions trading	CO_2 trading being discussed in relation to implementation of the Kyoto Protocol; Commission proposal in October 2001	CO_2 trading being discussed in many member states and implemented in Denmark and UK; NO_x trading scheme perceived in the Netherlands
Subsidy reform	Reform of CAP, Common Fisheries Policy, Structural Funds, Cohesion Fund, European Investment Bank	Reform of domestic energy and industrial subsidies under way
Environmental impact assessment	Directive on EIA in 1985 (revised in 1997)	Approx. 7000 EIAs per annum conducted across the EU
Environmental management systems	Eco-Management and Audit Scheme (EMAS) from 1993	Approx. 1500 sites registered with EMAS by 1998
Research and development	Funding in EU Framework Programmes around €2 billion for the environment	Support for clean technology in many member states

Source: adapted from Clinch (2000).

sector, and it is increasingly doubtful whether European-wide industry associations can commit effectively to voluntary agreements due to the heterogeneity of their membership and their internal divisions. So far, only one binding agreement has been made at EU level, namely the agreement between the EU and the car industry to reduce CO_2 emissions of vehicles. Originally agreed between the European Commission and the European Car Manufacturers' Association (ACEA), it was extended to the Japanese and Korean manufacturers and importers. In addition, there have been a number of non-binding voluntary environmental agreements. These took the form of commitments by industry – in the aerosol, foam plastics, refrigerator and detergent and cleaning product sectors (EEA, 1997). This has, however, not prevented the EU member states from concluding a large number of voluntary agreements (Börkey and Lévêque, 1998 and 2000; ten Brink and Morère, 2000). Particularly Germany, Finland, Denmark and the Netherlands have relied on voluntary initiatives (Figure 1.1). For more empirical evidence see Chapter 2.

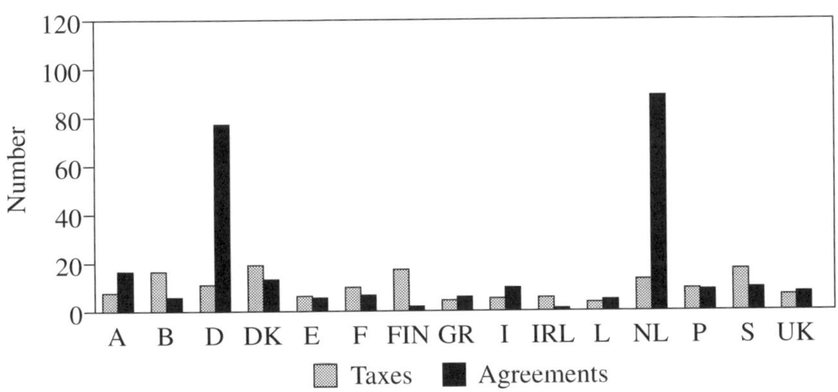

Source: Clinch (2000), based on OECD (1997a) and European Environment Agency (1997).

Figure 1.1 Environmental taxes and agreements in EU countries, 1996

In conclusion, regulation remains the dominant instrument applied to environmental policy. Taxation and voluntary agreements, although unevenly across member states, are becoming increasingly used. Regulation is predominantly applied at the European level of government, while both taxation and voluntary agreements have – mainly for constitutional reasons – been mostly restricted to member states' use.[8]

Tradable permit schemes have traditionally played a subordinate role in the EU, with the two minor exceptions mentioned above. The public at large and especially Green Parties in Europe, traditionally influential in the environmental instrument debate, have been opposed to tradable permits on the grounds that they allow those who can afford to pollute to do so ('licence to pollute'). This position has largely vanished. A more substantive explanation might be that the EU has preferred to rely on tested instruments such as command-and-control measures and taxation. The preference of taxation over tradable permit schemes might also be explained by the generally better revenue-raising potential of the former. That tradable permit schemes potentially create distortions to competition in the EU's internal market if implemented at either member state or EU level, as is evidenced by current national GHG emissions trading schemes (Egenhofer and Mullins, 2000), has further contributed to the EU's shunning of this instrument. The next section will analyse tradable permit schemes in more depth.

3. THE RISE OF TRADABLE PERMITS IN THE CONTEXT OF CLIMATE CHANGE

As we have shown in the previous section, incentive-based instruments are now considered for most environmental problems within OECD countries and beyond, including what is potentially the most important environmental problem – climate change. Much of the rationale behind the rise of incentive-based instruments – reduced costs, dynamic efficiency, effectiveness and equity – were equally addressed in the Kyoto Protocol's flexibility provisions, notably the flexible mechanisms, which foresee an international emissions trading (IET) regime. The Kyoto Protocol's flexible mechanisms in return have generated strong interest in the EU for emissions trading at both the EU and member state level (see Haites and Mullins, 2001; IEA, 2001). However, there is still relative uncertainty regarding the international climate change regime, despite the recent Marrakesh Accords.[9] As a result, companies, member states and the EU alike focus on domestic and EU-wide trading schemes.

Emissions trading within the EU does not fall under the provisions of the Kyoto Protocol as there is no exchange of permits between parties. Consequently, emissions trading within the EU is a domestic action – policies and measures in Kyoto Protocol language – and legally speaking unaffected by the rules on IET, although a lack of compatibility between domestic ET and IET will most probably affect the efficiency of both in a negative way. ET within the EU is also domestic action as a result of the EU's joint commitment (the 'EU bubble') on the basis of Article 4 of the Kyoto Protocol. See Figure 1.2.

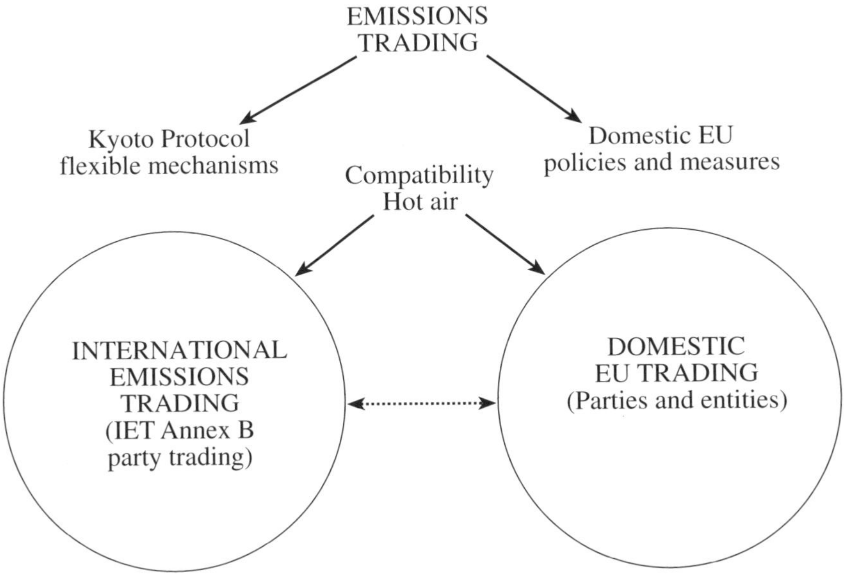

Figure 1.2 International emissions trading versus emissions trading

Economic benefits

Marginal abatement cost differences in Europe are considerably lower than at a global scale,[10] and therefore compliance cost reduction as a result of tradable permit schemes as well. Nevertheless, efficiency gains of an EU-wide trading scheme are still considerable. The European Commission's own calculations assume the economic benefits of additional EU-wide trading by the energy supply and energy-intensive industries – covering approximately 45 per cent of total CO_2 emissions – in addition to national trading of the magnitude of one-fifth of total compliance costs (CEC, 2000). The Institute for Prospective Technological Studies, using Poles world energy model, concludes that emissions trading among all sectors in the EU would reduce compliance costs by 25 per cent (IPTS, 2000). The National Technical University of Athens estimates that GDP losses with EU-wide trading among all sectors can be reduced by 30 per cent (E3M Lab, 2000). For a complete overview, taking into the account the Marrakesh Accords, see Chapter 3.

Effectiveness

Emissions trading is also favoured for reasons of effectiveness, notably cap-and-trade schemes. The overall cap on pollutants ensures that participants in a trading scheme do not emit more than is foreseen through the initial allocation of permits. Two qualifications need to be made, however.

Effectiveness is not necessarily ensured in the case of rate-based credit schemes (see below) and in cases where compliance is either not ensured or non-compliance does not lead to mitigation effects in what follows. That is typically the case where non-compliance leads to financial penalties lower than the actual carbon price, and therefore polluters can buy themselves out of mitigation actions.

3.1 Tradable Permit Schemes: Allowance Trading as Dominant Model

Allowance or cap-and-trade emissions trading schemes[11] have become the dominant model for emissions trading (see OECD, 1992; Klaassen, 1996; McLean, 1996; OECD, 1997b; Norregaard and Reppelin-Hill, 2000). The Kyoto Protocol setting absolute emissions limits has reinforced this. Cap-and-trade schemes with their absolute caps appear as the most appropriate approach for dealing with absolute Kyoto ceilings. Furthermore, the most important application of an ET scheme has arguably been the tradable permit system in the USA that regulates SO_2 emissions, the primary precursor of acid rain (see Joskow et al., 1998). Not only is it the best-researched ET programme, but it also marks a transition from a baseline-and-credit to a cap-and-trade programme (see Klaassen and Nentjes, 1997a) for reasons of greater environmental certainty and lower transaction costs (see Hahn and Hester, 1989). The US SO_2 programme is recent enough – the first phase started in 1995 followed by the second, initiated in 2000 – in order to make it a natural first reference point for ET analysis.

Allowance programmes or cap-and-trade schemes establish an allowable level of pollution permits and allocate them among the participants in the trading scheme in the form of permits, which can be freely traded among participants. Governments set an environmental goal in quantified terms and polluters are required to hold a permit or certificate for each unit of pollution they emit. The polluter has the choice to either abate pollution or buy permits to cover its existing pollution level. The premise on which trading is based is that different emitters incur different abatement costs. Low-cost emitters can cut emissions below the target and sell the difference to high-cost emitters, if the permit price makes this attractive. Such a system provides incentives for over-achievement by providing financial compensation for reduction of pollution levels beyond the ceiling introduced by the government, provided the permit price is right. In theory and in a first-best world, tradable permits have the same economic efficiency benefits as taxes. While in cap-and-trade schemes, the quantity of pollution is fixed, resulting in a price emerging for the trade of permits, which represent the restricted commodity, tax systems charge prices for pollution, which results in the demanded reduction of the polluting commodity.

Most controversial is the allocation of permits that hold property rights (see Tietenberg, 2001; Burtraw et al., 2001). In theory, allocation can be through free distribution (grandfathering) or via sales (auctioning) by the government or by a mixture of both. Industrial acceptability of cap-and-trade programmes is higher in the case of grandfathering. Auctioning has the same effect as taxation, however, with the tax rate set by the market. To avoid extra costs due to auctioning, revenues can be recycled (see Fischer et al., 1998), thereby in theory allowing for the so-called 'double dividend',[12] analogous to green taxation. Recycling advocated by economists to make auctioning budgetary-neutral is generally contested by industry on grounds of 'additional government intervention', meaning the actual process of redistributing receipts from the auction. Irrespective of what method is chosen (grandfathering or auctioning), a number of permits must be withheld to accommodate new entrants. Joskow and Schmalensee (1998) argue that freely allocated allowances offer much more potential to address distributional effects of permit schemes. A study of emissions trading of particulates in Santiago de Chile found another potentially important benefit of grandfathering. Registration and control of emissions from point sources permitted the identification of previously undisclosed sources of pollution (Convery, 2001).

There is a choice between upstream and downstream systems. An upstream scheme aiming at fuel producers and importers (for example extraction, production, processing, transportation) of energy has effectiveness advantages in that it can cover nearly all emissions from fossil fuels, including those from households and transport.[13] A downstream approach targets fuel users, such as power generators, large industrial users, commercial facilities and even household and car owners (Table 1.4). In order to keep transaction costs low, downstream schemes tend to limit their coverage to the major industrial sectors such as electricity, steel, chemicals and cement, with additional instruments such as taxation and regulation being applied for the other sectors.

Upstream and downstream schemes have different initial effects on equity, mainly due to the fact that allowances are owned by different groups. Due to their comprehensiveness and their relative lack of flexibility when compared with downstream schemes, upstream programmes have proved to be difficult to implement due to political opposition, at least in the initial phase (see Fischer et al., 1998).[14]

3.2 Baseline-and-credit Programmes Revisited

Initial emissions trading programmes originated from the US Environmental Protection Agency (EPA); however, they were based on (baseline-and-) credit programmes. Under these programmes, credits are *created*, that is, 'earned'

Table 1.4 Comparison of design options

	Upstream system	Downstream system	Hybrid system
Description	Fuel producers surrender allowances for carbon processed, transported or sold	Large fuel users surrender allowances for carbon in fuel consumed; small users covered by Policies and Measures (PAMs)	Large fuel users surrender allowances for carbon consumed; other energy use is covered upstream
Point of imposition	Refineries, coal preparation plants, some coal mines, natural gas pipelines, natural gas processing plants, fossil fuel importers	Trading system covers major industrial sectors (electricity, steel, chemicals, cement, others)	All those listed for both upstream and downstream systems
Options for expansion	Covers nearly all fossil fuels. Could expand to include non-CO_2 gases	Other industrial sectors can be added over time, so could land use, other gases; high transaction costs if expanded to cover small sources	Same as upstream
Options for excluded sources	Sequestration; other gases could be brought in on a project basis or covered by PAMs	Sequestration; other gases could be brought in on a project basis, but difficult to include small CO_2	Same as upstream

Allocation	Auction, grandfathering or combination	Auction, grandfathering or combination	Auction, grandfathering or combination
Other implications	Requires close tracking of international energy trade		Additional administrative complexity involved in coordinating upstream and downstream components
Advantages	Comprehensive coverage of energy-related emissions improves economic efficiency, environmental effectiveness; may be possible to phase in one member state at a time	Can be tailored around existing measures, phased over time; more attractive to companies and can improve efficiency of existing PAMs	Similar to upstream system in terms of efficiency, environmental effectiveness
Disadvantages	An EC-wide scheme means member states cannot tailor it, unfamiliar instrument	Low coverage of trading system reduces economic efficiency; uncertainty regarding environmental effectiveness	Like an upstream system, cannot be crafted or phased in at Community level

Source: FIELD (2000), p. 24.

when a source reduces emissions below the level required by existing, source-specific limits based on either an NEA or regulation. Credits are tradable to enable the same or another firm to meet its control target. In baseline-and-credit programmes, sometimes referred to as credit-based or resource-based schemes or simply credit trading,[15] the certificate is only needed for the units of pollution firms want to sell. There is no need to hold a certificate for the units of pollution that are allowed under source-specific limits. A major difference between allowance and credit trading is that in the case of the former a participant holds property rights over all allowances (that is, the total number of allocated allowances to a firm). In credit trading, property rights extend only to the 'earned' credits, which polluters can obtain by over-achieving the emissions reduction targets. Targets can be expressed in absolute terms, that is, x tons of pollution, or in relative terms, x per cent energy efficiency improvement (see Tables 1.5 and 1.6). Credit trading based on specific targets is also referred to as rate-based credit trading. If targets are expressed in absolute terms, baseline-and-credit schemes are very similar to cap-and-trade schemes in terms of effectiveness, although they give rise to more transaction costs, as will be discussed later. If targets are expressed in relative terms – sometimes referred to as specific targets – the amount of

Table 1.5 Examples of ET in companies with a specific and an absolute target

Example 1: A fictitious company and an emissions target in absolute value			
Year	n	n+1	n+2
Emissions target in tonnes of carbon equivalent (TCE)	1000	900	800
Tonnes of carbon equivalent actually emitted	1000	800	875
Δ/in emissions target	0	100	(75)
Emissions credits obtained (or debited) in TCE	–	100	(75)

Example 2: A presumed company and an emissions target in specific value			
Year	n	n+1	n+2
Specific emissions target in TCE of product	1	0.9	0.8
Actual production in tonnes of product	1000	800	1100
Carbon actually emitted/tonne of product	1	0.88	0.87
Δ/in specific emissions target	0	0.12	(0.07)
Emissions credits obtained (or debited) in TCE	–	96	(77)

Source: EpE/MIES (2000), p. 14.

Table1.6 The operation of credit trading: an illustration

Definition of aggregate target An aggregate performance rule is defined for a sector, or group of sectors, reflecting the desired policy objective (that is, an absolute limit, a specific target, or a mix of both). There is no initial allocation of permits to firms (or auctioning of permits) as in the case of allowance trading.
Breakdown of target to firm/sector level Individual performance rules are defined for each firm in the sector(s), such that these are consistent with the aggregate performance rule – that is, if each firm satisfies its own performance rule, then the aggregate rule is also guaranteed to be satisfied.
How to earn credits If a firm beats its performance rule, then it is allowed to generate performance credits which it can sell to other firms covered by the scheme. The purchasing firm can then use these credits towards satisfying its performance rule. However, net of all transactions, each firm must satisfy its own performance rule for the period.
Flexibility All firms may be subject to the same performance rule (for example all firms must achieve the same specific energy consumption). Alternatively, it is possible to differentiate the performance rules of firms, or groups of firms, by using performance adjustment factors. (For example, in the case of an aggregate target for specific energy consumption for the chemicals sector, it would be possible to differentiate between firms producing basic chemicals and those manufacturing pharmaceuticals. The adjustment factors allow the distributional impact to be varied, while ensuring that the aggregate performance rule is not compromised.)
Fungibility Firms will be able to sell credits to or purchase credits from firms in any of the three sectors (absolute, relative, mixed), and this trading should ensure that the aggregate performance rule is satisfied at the lowest total cost.

Source: Salmons (2000).

pollutants emitted is de-coupled from economic growth, hence potentially threatening an absolute pollution target. This point will be discussed in Section 4. See also Chapter 3.

There have been several significant applications of credit emissions trading programmes, mainly in the USA, under the US EPA,[16] which will be reviewed below. They include, for example, the early US EPA Emissions Trading Programmes, the phase-down of leaded gasoline and US heavy duty motor vehicle engine emissions trading.

More recently, in Europe baseline-and-credit schemes have aroused interest again in the context of their practical application to existing NEAs with specific targets (Salmons, 2000; EpE MIES, 2000; DEFRA, 2001). In North America, a number of proposals on voluntary early GHG emission reductions ('early action schemes') use baseline-and-credit mechanisms[17] (Rolfe et al., 1999). Finally, the AIJ (activities implemented jointly) under the UNFCCC and the two Kyoto project mechanisms, CDM and JI, are credit schemes,[18] having rekindled the research on baselines and practical aspects of their implementation such as transaction costs (see Dudek and Wiener, 1996; OECD, 1999a; 1999b and Stavins, 2001 for an overview). The US EPA, which generally prefers cap-and-trade programmes for reasons of effectiveness and efficiency (for example they are less complex), recognizes that, notably in the context of early action schemes and ongoing voluntary programmes, credit schemes could be used (Kruger, 2000).

Early EPA trading programmes

The early EPA ET programmes began as step-by-step experiments pushed by local requirements for flexibility. The aim was to introduce some flexibility into a command-and-control approach rather than establishing a trading regime. Begun in 1974, ET was part of the Clean Air Act's programme for improving local air quality by addressing VOCs, CO, SO_2, particulates and NO_x. Firms that reduced below the required levels received credits to offset emissions elsewhere. Companies were allowed to use concepts such as 'netting' or 'bubbles', and later 'banking' and 'offsets'.

Netting concerned modifications of old installations and investment in new ones. Under the old ET schemes, requirements became more stringent in the case of modifications. Under netting provisions, firms could apply the old emissions standards to the old installations as long this could be compensated by overall reductions calculated for the new installation. Not really a market instrument, netting was often used between 1974 and 1984. It had high transaction costs since, with each netting, the regulatory regime needed to be modified. *Bubbles* allow firms to calculate the emissions of several installations at once instead of submitting each specific source to one emissions target. The environmental effect is the same, while firms reduce their abatement costs. About 130–150 bubbles were agreed at the US federal and sub-federal level. Bubbles remained controversial due to the fact that they were linked to non-attainment areas ('hot spots'[19]). Estimated compliance cost savings from bubble programmes exceeded US$430 million. *Banking* provides the possibility to store earned credits for future use, and allows for future internal expansions or sale. Only five projects, including 18 firms, were approved. The main reason for lack of enthusiasm was that banking was unattractive in non-attainment areas. Public opposition created a climate of regulatory uncertainty. *Offsets* (introduced in 1977) go further by allowing firms to trade emissions credits. Firms that wish to establish new sources in areas that are not in compliance with ambient standards (so-called non-attainment areas) must offset their emissions by reducing existing emissions in the same area by reducing internal emissions or through trade.

Although the EPA codified these approaches, they have not been widely used. Nevertheless, estimated cost savings were significant and are expected to be somewhere between US$5 and 12 billion over the years (Stavins, 2001).

Lead trading

The US lead treading was more successful both in terms of environmental effectiveness and cost-effectiveness (Kerr and Maré, 1997). The purpose of the programme, launched in 1980, was to allow gasoline refiners greater flexibility in meeting emissions standards at times when the lead content was reduced to 10 per cent of its previous level.[20] In 1982, EPA authorized inter-refinery trade of credits that could be earned if refiners produced petrol with lower lead content than was mandated by law. The purpose was to reduce the financial burden on smaller refineries, which were believed to have significant higher compliance costs. Subsequently, banking was introduced in 1986. The programme was terminated in 1987, when the lead phase-down ended. Cost savings compared to alternative programmes without lead banking were estimated at 20 per cent or US$250 million per annum.

Heavy-duty motor vehicle emissions

Another example of a credit trading programme (of a voluntary nature) with mandatory performance standards is the averaging,[21] banking and trading (ABT) provisions of the emissions standards for a variety of pollutants caused by heavy-duty truck and bus engines. The programme established by the US EPA foresees that avoided emissions as a result of better than legally mandated performance standards of engines can be credited for ABT provisions. The objective is to allow, as of 2004, making credits fungible with other trading programmes (Stavins, 2001), thereby combining mobile with stationary sources and providing incentives for improved fuel efficiency.

Main Results

While it is acknowledged that the early ET programmes of the EPA have not been very successful or even failures (Godard, 2000), the gasoline lead trading approach was successful in terms of both effectiveness and efficiency (Kerr and Maré, 1997). There is no explicit evaluation of the heavy-duty motor vehicle scheme, but the extension of the ABT provisions seems to suggest that EPA and manufacturers are satisfied with such a programme (Haites and Hornung, 1998). In order to evaluate the future opportunities of such schemes it is instructive to look at what went wrong with the EPA early ET programmes and to identify the criteria for success and failure.

Klaassen and Nentjes (1997a) have identified three main threads of market imperfections responsible for failure: (1) Regulatory restrictions; (2) uncertainty about status of property rights; and (3) high transaction costs. Regulatory restrictions were manifold and stem mainly from the fact the programmes were implemented incrementally and drafted as a complement to existing command-and-control regulation. The logic was not so much to implement market-based solutions but to allow for decentralized measures to displace the established regulatory constraints on pollution emissions geographically or over time. For example, demand for permits was limited by a number of 'hybrid elements' such as that new sources still had to meet tight emissions standards, emissions could be increased in attainment areas, existing sources did not need to reduce, and too stringent baseline setting did not generate credits. Regulatory restrictions can be blamed to some extent for a lack of familiarity with this, then new, instrument. This does not, however, explain everything. Godard (2000) blames fastidious and bureaucratic implementation and reciprocal mistrust between business and the administration for part of the failure. More fundamentally, overall, the volume of trade tends to be lower in credit trading than it is for allowance trading. One reason is that in the latter property rights are better defined. Credit-based programmes require the computation of controversial baselines in order to establish a reference point for future reductions, similar to CDM and JI (see Dudek and Wiener, 1996). There is risk of ending up with permanent administrative oversight and/or cumbersome certification procedures. This may also lead to government pre-approval of trades, which have particularly hampered the earlier EPA trading schemes (Hahn and Hester, 1989). This shortcoming is particularly felt in the case of large stationary sources with complex installations. In the case of type-approval (for example engines, appliances), certification and administrative oversight are likely to be far lighter (Haites and Hornung, 1998). This might explain the success of the heavy-duty vehicle emissions programme.

The earlier EPA programmes have suffered markedly from high transaction costs in two areas: finding a trading partner and obtaining approval from the authorities. While the first category can be remedied, among other means, by the existence of brokers and market intermediaries (Kerr and Maré, 1997), the latter kind of transaction cost is a mixture of regulatory failure and the complexities of credit-based programmes. One shortcoming of credit-based trading has been uncertainty due to baseline setting and certification, leading subsequently to higher transaction costs and thereby discouraging trading (Stavins, 1995),[22] which have particularly hampered the earlier EPA ET programmes (Hahn and Hester, 1989). Stavins (2001) concludes from this to use absolute baselines, that is, absolute targets rather than rate-based targets to reduce high transaction costs and argues for avoiding hybrid programmes – cap-and-trade combined with voluntary 'opt-in provisions', which create incentives for paper trades between the regulated and non-regulated sector and thereby push up the total emissions cap. Whether one needs to go so far remains unclear, although from this analysis it appears that transaction costs are indeed the crucial shortcoming of credit trading programmes.

3.3 European Emissions Trading Initiatives

In the aftermath of the Kyoto Protocol a number of ET initiatives were launched at both member state and EU level.[23] They include a cap-and-trade scheme for tradable CO_2 emissions permits in the power sector in Denmark, which began on 1 January 2001, the UK scheme (DEFRA, 2001) to start in 2002 and the proposed Norwegian[24] cap-and-trade programme, covering approximately 80 per cent of all emissions (Hasselknippe and Høibye, 2001). Further schemes are discussed in Sweden (for example Haites and Mullins, 2001)[25] and France, where the Entreprises pour l'Environnement (EpE), a grouping of major French companies, has developed a credit scheme (EpE MIES, 2000). In addition, companies such as BP and Shell have launched internal cap-and-trade schemes.[26] The European power sector has embarked on a virtual scheme called GETS (PricewaterhouseCoopers, 2000). Both Vattenfall through its subsidiary HEW in Germany and the Nordic power group Fortum have made two transatlantic CO_2 trades as part of a learning-by-doing process with Transalta and Edmonton Power Corporation (EPCOR), respectively. At the same time the European Commission proposed a directive in October 2001 (CEC, 2001a) to create an EU-wide CO_2 cap-and-trade programme.

The proposed schemes (Table 1.7) vary and include pure cap-and-trade schemes (Denmark, Norway, Sweden), hybrid programmes such as the ones from the UK, which combine elements of cap-and-trade and credit trading with both absolute and relative targets, and France, which is a pure credit scheme also combining absolute and relative targets. BP and Shell have both opted for a cap-and-trade model. Except for the Swedish and to some extent

Table 1.7 Main features of European emissions trading schemes (adopted or under discussion)

	Allowance trading	Credit trading	Upstream	Downstream	Cross-border	Domestic
Danish power sector	✓ G			✓		✓
UK emissions trading group	✓ G	✓		✓		✓
French EpE	✓	✓		✓		✓
Norwegian scheme	✓ G + A		✓			✓
Sweden	✓ A		✓	✓		✓
BP target & trade	✓ G			✓	✓	
Shell STEPS	✓ G			✓	✓	
EU proposal	✓			✓	✓	

Notes:
G = Grandfathering of allowances.
A = Auctioning of allowance.

Source: Ellerman (2000), Hasselknippe and Høibye (2001), Haites and Mullins (2001), CEC (2001a).

the Norwegian proposed schemes, allowances are meant to be grandfathered, although with some element of benchmarking. Except for Norway and Sweden, schemes are downstream plus policies and measures programmes. There is also a variety in terms of national and cross-border schemes. Company schemes (BP, Shell, GETS) as well as the EU model typically operate cross-border while those schemes developed by industry and governments (for example Denmark, UK, France, Sweden, France) remain domestic. To avoid distortions to competition notably in the EU internal market – for a preliminary analysis see Egenhofer and Mullins (2000), Legge (2001) – and to reap economic benefits of a larger scheme, the European Commission has proposed a EU-wide scheme to run as of 2005 (CEC, 2001a). The EU proposal is basically a cap-and-trade scheme, although it accommodates baseline-and-credit elements.

The challenge that most of these programmes face is that the economic literature agrees that tradable permit systems should ideally be allowance-based (that is, cap-and-trade), upstream, allowances should be auctioned and the auctioned proceeds recycled (see Fischer et al., 1998), while the 'real world' gives a different picture. Proposed programmes (Table 1.7) are allowance- or credit-based or hybrids and mainly downstream. The allocation of allowances is mainly performed via grandfathering. Proceeds are not foreseen to be recycled, with the possible exception of Sweden. Most of the proposed schemes represent themselves as policy mixes. The following section analyses motivations and prospects for these policy mixes.

4. PROSPECTS FOR POLICY MIXES: LINKING EMISSIONS TRADING WITH OTHER INSTRUMENTS

While the previous sections have discussed the relative merits of instruments regarding effectiveness, efficiency and equity, the following examines the long-trend shift towards linking instruments. There are two distinct levels of linkages, one at the macro- and one at the micro-level. At the macro-level the issue is one of allocating the most appropriate instrument to a given environmental problem, sometimes referred to as 'horses for courses'. For an example of such an approach, see Appendix 2. At the micro-level the issue is to integrate different instruments (across sectors and across pollution sources), to combine the advantages of instruments *vis-à-vis* different – sometimes conflicting – policy objectives such as efficiency, effectiveness, equity or acceptability.[27] Ultimately, the objective is to develop policy mixes which optimize incentives in a broad sense to mitigate pollution ('optimal policy mix').

We will concentrate here at the micro-level and on policy mixes combining emissions trading with voluntary agreements in the form of negotiated agreements, regulation and taxation. As will be shown, combination in the form of a policy mix can be a means to deal with effectiveness, efficiency and equity.

4.1 Emissions Trading and Negotiated Environmental Agreements

Industrial associations within the EU have expressed their strong preference for using negotiated environmental agreements (NEAs) as the 'instrument of first choice' in the context of climate change,[28] while pursuing the option to use the Kyoto mechanisms in parallel. For a definition and further analysis see Chapter 2. Industry has further stated that it would like to continue to use the traditional targets in the form of relative improvements or so-called specific targets. This position largely reflects the current situation that by far the majority of NEAs are based on specific targets (Börkey and Lévêque 2000; ten Brink and Morère, 2000). This is contrast with the absolute greenhouse gas (GHG) emissions limits the Kyoto Protocol has set, which are independent of exogenous factors such as economic growth.

Specific targets of NEAs at first sight appear to be incompatible both with the absolute national targets and the allowance-based emissions trading (CCAP, 1999; Zapfel, 1999; FIELD, 2000) that have been proposed by the European Commission (CEC, 2001a). See also Chapter 3. In analysing the case of the European chemicals industry, Zapfel (1999) argues that NEAs with relative targets 'will result most likely in absolute emissions increases rather than decreases'. The reason for this is the expected output growth, which is unlikely to be offset by carbon efficiency improvements. Zapfel insists that it is not possible to 'square the circle' to develop a policy mix that provides for ET and relative targets. The implicit conclusion is that industry and policy makers must choose between an agreement with relative targets and no trade, and trading under an allowance-based scheme (which might eventually include an agreement with an absolute target).

A number of approaches (see Appendix 1) offer different practical answers to this dilemma. The scheme of the UK Emissions Trading Group and now proposed by the Department of the Environment, Transport and the Regions of the United Kingdom (DEFRA, 2001) distinguishes between a so-called 'absolute' sector (with absolute targets) and a 'unit' sector (sectors with agreements on specific targets) to bring as many sectors as possible within the trading scheme. The heavy energy users can gain a rebate up to 80 per cent of their climate change levy if they engage in bilateral agreements that meet the agreed benchmarks. (See also Chapters 2 and 3.) The combination of absolute and unit sector is presented as a transition period with the stated objective to see as many companies as possible move to an absolute cap. The concept worked out by Entreprises pour l'Environnement (EpE) attempts to reconcile NEAs based on absolute and specific targets with an emissions trading scheme (EpE/MIES, 2000). It is a combination of baseline-and-credit scheme (partly rate-based) with NEAs. Trading participants can earn credits if they over-comply with their targets (expressed in absolute terms or in energy efficiency per unit). The system is flexible in principle to allow companies to move from relative to absolute targets. A similar approach is currently being explored in Germany in a comprehensive study on the compatibility of the German Industry Climate Agreement (*Selbstverpflichtung*) and emissions trading. The authors of the study have indicated that the scope for ensuring compatibility of the two instruments is limited and that the best avenue might become combining the Agreement with a baseline-and-credit scheme (RWI, forthcoming).

Effectiveness

In industry's view, allocating absolute GHG emissions caps to sectors constitutes a limit on economic growth and could imply as a consequence a loss of market share for European producers. This market share would be captured by non-Annex 1 producers with lower energy efficiency and, hence, higher global GHG emissions and a worsened atmosphere (that is, carbon leakage). For governments, once having accepted absolute targets under the Kyoto Protocol, relative industry targets create a different problem. Governments are confronted with uncertainty if the contribution of some sectors depends on *ex-ante* uncertain economic growth[29] (Table 1.8). In the case of non-compliance, governments face the unpleasant choice of targeting the relevant sectors at short notice – close to the end of the first commitment period – with the risk of inflexible and ineffective policies or targeting other sectors for additional contributions beyond their initial targets. Alternatively, and from an effectiveness point of view, a less favourable reaction would be to accept possible penalties related to non-compliance. Another alternative would be to obtain credits from CDM or JI projects, a route that is currently being explored in Germany.

The scheme proposed by the UK Emissions Trading Group (DEFRA, 2001) separates the sectors with absolute (absolute sector) from those with relative targets (relative sector) by a 'gateway'. (For a short description, see Appendix 1, and Chapters 2 and 3.) With trading within the absolute sector unrestricted, trade between the 'absolute' and 'unit' sectors is only accepted as long as there is no net flow from the 'unit' to the 'absolute' sector. In addition, the absolute sector will be able to participate in international emissions trading (IET), whereas the unit sector will not. The 'gateway' (when closed) is the insurance to maintain the integrity of the cap over the absolute sector emissions. The gateway effectively ensures that the unit market can participate fully in trading via the gateway only if the marginal abatement cost in the absolute sector remains lower than in the unit sector. Whether this

Table 1.8 Advantages and disadvantages of absolute versus specific targets

	Absolute targets	Specific targets
Advantages	• Authorities can match more easily with absolute target stemming from the Kyoto Protocol • Lend a final character to the negotiation between the company and the authorities[a] • Fit in well with activities whose volume is highly predictable in the medium term	• Explicitly based on a dynamic of efficiency gains[b] • Enable the company to develop economically and allow new markets to be tapped without a financial penalty • Do not encourage relocation • Resolve the problem of factory closures or new entrants
Disadvantages	• Do not include any explicit reference to technical progress[c] • Cannot readily be adapted to fluctuations in the business cycle or in prices (for example prices of metals or chemical products) or to the opening up of new markets or to new applications, except by buying permits whose price in the medium term is not well known • Offer a greater incentive to relocate • Introduce greater complexity than specific targets as regards the management of new entrants and factory closures	• Make it more difficult for the authorities to predict the overall level of the country's emissions • Require that certain types of production be incorporated by defining a significant indicator[d] • Could induce a company that emits less than its target but is greatly increasing its production to reap an added profit by selling any credits obtained

Notes:

[a] Leaving the company fully responsible for the choice of the means it adopts.

[b] Allow a more technical discussion to be conducted.

[c] Any reduction achieved might be the result of production cutbacks.

[d] For example, a GHG emissions target could not be set per kg of product manufactured for each special chemical product, which could seriously complicate any technical discussion.

Source: EpE/MIES (2000).

assumption is realistic remains to be examined. The opposite, that is, a net inflow of permits from the unit sector, would prompt the closing of the gateway. As a result the gateway separates the absolute from the unit sector. At the same time, it appears that the gateway increases the transaction costs significantly. In the GETS 2 simulation, undertaken by electricity companies (see Appendix 1), a gateway similar to the one suggested by the UK was successfully tested, however with monitoring costs 'far superior to those required in a standard system' (PricewaterhouseCoopers, 2000).

The French scheme proposed by EpE is a baseline-and-credit programme using both absolute and relative targets (EpE/MIES, 2000), which both have their respective advantages (Table 1.8). EpE states that there is no theoretical argument that either absolute or relative targets are more difficult to achieve, although absolute targets provide a higher degree of certainty to meet the Kyoto targets. In reality the presumed certainty of allowance programmes might be misleading. The industry sector in the EU accounts currently for less than 20 per cent of total CO_2 emissions and this figure is falling. In addition, there is a significantly higher margin of error in monitoring and measuring in the sectors covering the remaining emitters, notably transport, households and agriculture, which are far less well monitored (and accounted for) than industry. Uncertainty for governments might rather come from these sectors than from industry, irrespective of absolute or relative targets. This can be illustrated by an example based (but modified) on Nollet (2000). If overall country emissions are 200 million tonnes (mt) of CO_2 equivalent (CO_2e), industry emissions are typically 50 mt. Under an NEA target of 25 mt CO_2e with all companies being in compliance with specific target but with 2 per cent higher output level than assumed, the additional emissions level will be 0.5 mt over the forecast. The example would, however, look different with a less ambitious target. Assuming that the target is 20 per cent reduction rather than 50 per cent as in the first example, then the additional emissions under a 2 per cent additional output growth would be 0.8 mt CO_2e. If output growth is 5 per cent, then the figure would be 2 mt CO_2e, or 1 per cent of the countries' total emissions. The situation looks different, however, if the electricity and heat supply sector is also included in a trading scheme. This would bring the total of covered emissions to around one half of total GHG emissions. In our example, industry and electricity/heat supply cover 100 mt CO_2e. With an ambitious target of 50 per cent and 2 per cent additional output growth, the extra emissions would be 1 mt CO_2e. The corresponding figures for a 20 per cent target and 5 per cent additional output growth would be 4 mt CO_2e – a significant figure. The experience with NEAs can offer valuable lessons. As a result of uncertainty about absolute emissions, most NEAs have updatable targets (Chapter 2).

In conclusion, it is true that relative targets expose government to a risk of non-compliance as a result of higher than projected output growth of a sector. The exact risk depends on the share of total GHG emissions and of the sector(s) concerned, and the level of ambition regarding the target. The bigger the share and the less ambitious the target, the higher the risk for governments. The risk of non-compliance might be managed by close monitoring and eventual purchase of credits via CDM, JI or IET or on the EU permit market.

Efficiency

The main questions about efficiency arise from transaction costs, (inefficient) design and the clarity of property rights. They are discussed in turn.

Transaction costs In market economies, transaction costs are ubiquitous. They arise whenever a transaction is made. Transaction costs inhibit market activity and consequently limit the gains from trade. If the transaction costs inhibit trade in permit trading schemes, the advantages of these will not materialize. As emissions trading is a relatively new instrument, considerable attention has been given to transaction costs. See also Chapter 3 and Chapter 2 for NEAs.

The literature (see Stavins, 1995; Dudek and Wiener, 1996; Kerr and Maré, 1997) distinguishes between a number of different transaction costs. First, the buyer and seller must find each other (search costs), communicate, and exchange information. They must bargain and make the deal, which could also involve, for instance, the drawing up of a contract (negotiation costs). Dudek and Wiener (1996) have grouped these as costs of market transactions. Further, trades must be permitted by a regulatory authority (approval costs). This will require independent monitoring and verification (monitoring costs), and there must be a way of enforcing compliance (enforcement costs), although it is likely that the government will bear this cost (Stavins, 1995). Further transaction costs may arise if the trading involves asymmetric or incomplete information (information costs). Trading may also involve the release of confidential information, making firms reluctant to trade except internally (Kerr and Maré, 1997; Convery, 2001). There are also opportunity costs arising from choosing one kind of emissions trading regime over another more flexible or cost-effective one (Dudek and Wiener, 1996). The risk of failure of the transaction, due for instance to engineering failure or a change in political environment, may lead to insurance costs, either in the form of insurance policies or through an investment decision, such as diversifying a portfolio against the risk of failure in any one enterprise. For an overview see Table 1.9.

Transaction costs are usually absolute: for large economic entities with economies of scale, transaction costs may constitute a small part of the

Table 1.9 Variables affecting transaction costs borne by market participants

Category of transaction cost	Description	Variables affecting costs
Search costs	Seller and buyer must find each other and communicate and exchange information	• Liquidity • Transparency Can greatly be enhanced by market intermediaries, trade exchanges and 'optimal' definition of markets (for example no ceilings)
Negotiation costs	Seller and buyer must bargain and draw up contract	• Transparency • Volume of trades ('economies of scale')
Approval costs	Trades need to be supervised by governments	• Complexity of government regulation • Clarity of property rights
Monitoring costs	Necessity for independent monitoring and verification	• Complexity of government regulation • Clarity of property rights Does not differ from other instruments, such as voluntary instruments, taxation or regulation
Enforcement costs	In case of non-compliance, governments need to enforce it	• Simplicity/complexity of compliance system • Clarity of property rights
Information costs	Lack of, asymmetric or release of, confidential information	• Individual firm response
Opportunity costs	Arising from choosing one kind of ET scheme over another	• Carefulness of design of scheme
Insurance costs	Against technical or political risk	• Quality of installation • Predictability of government policy • Clarity of property rights and obligation

Source: Stavins (1995), Dudek and Wiener (1996), Kerr and Maré (1997), Stavins (2001).

decision whether to participate, but for small companies the same transaction costs may be prohibitively expensive. On the other hand, participation of small economic entities is preferable for efficiency (for example increase of liquidity) and effectiveness reasons. Results of a simulation study for a GHG emissions trading scheme in the German State of Hesse suggest, however, that transaction costs depend less on the size of the company than on characteristics of the emissions source (Hessen, 2001).

In the US permit market for lead emissions, for instance, the efficiency losses from transaction costs amounted to 10 per cent. According to Kerr and Maré (1997), the lead trading programme was 'very close to a textbook tradable permit market' in that most of the traders are large sophisticated agents such as large refineries and oil companies, but even in this market some participants faced higher than average transaction costs, leading to efficiency losses. This loss would be exacerbated '[i]f the market was composed of small, unsophisticated players, or players with few existing connections among themselves' (Kerr and Maré, 1997). Dudek and Wiener (1996) suggest numerous concrete proposals to address transaction costs in international emissions trading such as Kyoto trading, and identify the clarity of property rights and obligations as the crucial determinant for transaction costs. A whole set of transaction costs is related to flaws in design when setting up emissions trading schemes, which were reported in Section 3.2 and are discussed below.

Nevertheless, as Stavins (1995) notes, even a complicated trading system with high transaction costs and consequently lower than expected levels of trade will most likely be more cost-effective than a conventional command-and-control approach (see also Chapter 4). A trading system in which no trading takes place will probably be less costly than, for instance, a technology standard, 'because a trading system provides flexibility to firms regarding their chosen means of control'. Choosing a trading system with high transaction costs obviously bears an opportunity cost of the more flexible system that was not chosen.

The implication is that, although tradable permit systems are demonstrably more cost-efficient than command-and-control systems, transaction costs can greatly affect the relative cost-effectiveness of differently designed systems. The devil is in the detail, since careful attention to the details of design of specific systems will reduce the risk of overselling these policy ideas and help create systems that can be implemented in the real world. As Convery (2001) points out, 'economists tend to pay relatively little attention to institutional design – and associated legal and administrative frameworks – but they are central considerations if emissions trading is to be successfully mobilised'.

Inefficient design The economic benefit of cap-and-trade emissions trading lies in the fact that it produces the least-cost emissions reductions by placing the same price on emissions. It imposes a cost at the margin as each additional unit of emissions requires to surrender an allowance. The advantage is that technology standards are avoided, leaving participants in trading schemes free to apply whatever strategies they feel are most economic. Rate-based credit trading, which typically are based on performance standards, however, may have the same effect as direct regulation and may force firms, at least to some extent, to apply expensive technology to comply with a standard in order to gain eligibility for participation in a trading scheme. In principle, this can be offset by applying concepts such as netting, bubbles, averaging, banking and borrowing, as the early US emissions trading schemes have done. As we have shown, this makes trading schemes more complicated, and therefore increases the transaction costs (see Klaassen and Nentjes, 1997b; Godard, 2000), as was discussed in the previous sections.

Inevitably, with the complexity of a scheme, bargaining between industry and government on which performance standard to adopt and which flexibility provisions to accept becomes more difficult and less predictive. Suspicion by the public may increase, as is sometimes the case with NEAs, which face a similar bargaining process (see Chapter 2). This might increase administrative oversight and, more generally, the risk of inefficient economic solution, for example hybrid solutions using at the same time market mechanisms and command-and-control elements, as was the case in EPA early trading schemes (see Klaassen and Nentjes, 1997a). It might also create political uncertainty (see Godard, 2000) and eventually 'regulatory creep', meaning continuous changes of the regulatory environment. Such bargaining is largely absent in cap-and-trade programmes as well as credit trading on the basis of absolute targets, except when deciding on overall number of allowances, which is based on an implicit performance standard coupled with output growth assumptions. Once the target is set, the politically most contentious issue is over and, with it, the risk of mistrust leading to 'double regulation'.

Uncertainty of property rights Credit schemes do not 'allocate' permits *ex ante* as in allowance schemes, but rather *ex post* and continuously. While companies that have 'allocated' permits can be sure to 'own' them, companies participating in credit schemes need to 'earn' them first to acquire property rights. There remains uncertainty due to baseline setting and certification. For an analysis of risks with baseline setting for JI/CDM see Chapter 3. The most critical and politically sensitive issue in emissions trading is the allocation of 'allowance' (in cap-and-trade) and 'credit' (in baseline-and-credit), the transfer of property rights from the government to the firm. While in allowances trading schemes, political controversy usually comes to a halt once the allowances are allocated and when the scheme starts, in credit schemes political controversy may continue, increasing the risk of governments changing the rules, as witnessed in the early US EPA schemes. Whether this will be an issue in the future remains to be seen. Increasing experience with NEAs and JI/CDM, where crediting is comparable with credit schemes, might reduce risk of changing rules on allocation in credit schemes.

Conclusions: credit trading as a suitable tool for a policy mix?

The preliminary analysis in Table 1.10, comparing the major sources of transaction costs for cap-and-trade and baseline-and-credit schemes, points out that credit-based schemes have a number of disadvantages, leading potentially to higher transaction costs. On the other hand, credit trading programmes have the advantage of combining NEAs and emissions trading and can even be a means to combine emissions trading with rate-based NEAs, as is suggested in approaches such as from Haites and Hornung (1998), Albrecht (1999), EpE/MIES (2000), DEFRA (2001) or Salmons (2000). As a result, credit schemes may play a considerable role as a bridge between NEAs and emissions trading during a transition period. The exact role in which and frequency with which credit schemes will be applied will, however, depend on the outcome of a more in-depth analysis of transaction costs. After all, it was transaction costs, which in the end have made the US EPA change over from earlier credit-based schemes to cap-and-trade programmes, that by now have become the 'traditional trading scheme'.

In addition, credit trading does lend itself to the 'customizing' of tradable permits beyond large stationary sources. Credit schemes could be extended to products, they could provide additional incentives to manufacturers of goods to improve the environmental performance of their products, such as cars or white goods. Car manufacturers that produce engines that go beyond current fuel efficiency standards may obtain tradable credits based on the avoided emissions estimated on the average life cycle of a car (see Haites and Hornung, 1998; Albrecht, 1999). Crediting would be relatively simply since engines are identical and certified as part of the type-approval procedure. The avoided emissions would then simply be multiplied by the assumed distance driven during the car's lifetime to establish the credited amount. Such a system provides incentives to stimulate cross-sectoral energy efficiency investments, which are largely absent in a cap-and-trade system. Manufacturers would be rewarded for developing and producing more energy-efficient products. The strength of incentive would not depend on – at least not directly – government regulation or subsidy, but on the carbon price as expressed in the permit price, which the manufacturer could trade. Examples of such schemes include the previously mentioned US EPA heavy-duty motor vehicle engine emissions trading. Another example is a scheme in Illinois that provides

Table 1.10 Occurrence of transaction costs for cap-and-trade and rate-based credit trading systems

Category	Cap-and-trade	Credit trading	Comments
Search costs	Low	High	Depends on liquidity and transparency of markets. Theory and empirical findings suggest that cap-and-trade schemes have more liquidity and therefore tend to increase market transparency
Negotiation costs	Low	High	Depends on volume of trade and market transparency. For the same reasons as for search costs, cap-and-trade schemes appear to bear lower transaction costs
Monitoring costs	Moderate	Moderate	Depends on the actual design, which is neutral as to the two different systems
Enforcement costs	Low	High	Depends on the design of compliance system and the clarity of property rights. While the design of the compliance system is neutral as to the two different systems, property rights are more clearly defined in cap-and-trade systems
Information costs	Moderate	Moderate	Depends on the firm's response and is therefore neutral as to the two different systems
Insurance costs	Moderate	High	Depends on the quality of the installation (which is neutral as to the two different systems) and the predictability of government policy. Empirical evidence suggests that credit trading is exposed to a higher regulatory risk due to the fact that property rights are allocated *ex post* and continuously.
Opportunity costs	Moderate	Moderate	Depends on design of scheme and therefore is neutral as to the two different systems.

Source: Stavins (1995), Dudek and Wiener (1996), Kerr and Maré (1997), Stavins (2001).

credits for car scrapping, already initiated in 1993. A similar programme exists in Louisiana. In both cases, credits can be traded with stationary sources. Finally, the US state of Georgia allows vehicle fleet operators to earn credits for vehicles that over-comply with federal clean-fuelled fleet regulations. For an overview see Stavins (2001) and for a further development of the concept see Naimon and Knopman (1999).[30]

4.2 Regulation and Emissions Trading

In principle, technology standards are incompatible with allowance trading. Prescribing the application of specific technologies would reduce the cost-effectiveness of an allowance trading programme. The reason is that technology standards require specific reductions at plant level while tradable permits schemes aim at least-cost reductions. When an allowance trading programme is introduced, instead of issuing a permit, companies or sectors would get an overall allowance of GHG emissions, thereby making a permit linked to a technology standard obsolete. Given that GHG emissions in most cases are not yet regulated, in practical terms there are no technology standards for such gases that might reduce the cost-efficiency of GHG allowance trading (FIELD, 2000). However, typically firms are subject to regulation for a number of other emissions than greenhouse gases. Within the EU particular industrial activities must obtain a permit that includes emissions limits for a number of pollutants including SO_2, CO or volatile organic compounds (VOCs) and others, as laid down in the so-called IPPC Directive.[31] Although the IPPC Directive does not cover explicitly any of the six GHGs included in the Kyoto Protocol, it calls for taking into consideration energy efficiency when determining the technology standard. Companies would be subject to double regulation, once through a technology standard – energy efficiency requirements – and once through the overall ceiling of allowances.

In principle, energy efficiency standards are incompatible with an allowance trading scheme for the same reasons as other regulations. Emissions trading seeks least-cost abatement rather than reductions from specific sources. As a consequence, in the case of the introduction of allowance trading, energy efficiency standards would need to be abolished to avoid double regulation. Efficiency standards may have important secondary benefits, which have to be weighed against the potential negative effects of 'double regulation'.[32] Such a secondary benefit applies especially to households and the commercial sector. Traditionally, energy efficiency has been impeded by a number of market barriers such as lack of information by the consumers, lack of capital and split incentives (landlord/tenant relationship[33]), and its overall small potential reduces incentives. This is the area where economic analysis typically acknowledges a role for government among others in the form of regulation. As a result energy efficiency standards have tackled these barriers and might therefore have to be maintained (CCAP, 1999) to ensure overall energy efficiency policy for the sectors concerned, although important methodological and empirical problems on the effectiveness of energy efficiency programmes still exist (see Clinch and Healy, 2001). A downstream allowance trading programme is likely to address this dilemma by excluding those parts that suffer from market barriers to improve energy efficiency from emissions trading or any other market-based instrument. An upstream programme, as proposed in Norway, would be better suited to tackle this dilemma since the point of imposition is not at the consumption but the production level.

On a more practical level, energy efficiency standards could be combined with rate-based credit trading, as has been explained in the conclusions of the previous section.

4.3 Taxes and Emissions Trading

Economic theory is inconclusive on whether to prefer taxes or allowance trading in a second-best world. Whether taxes or trading is the preferred instrument depends on a case-by-case analysis of welfare effects and political acceptability. There is another strong link between non-environmentally motivated taxation and emissions trading, somewhat neglected. Both will be discussed in the following.

Price versus quantity policies

The difference between taxation and allowance trading is that price controls – through taxation – fix the marginal costs of compliance and lead to an uncertain level of compliance. Quantity controls – via allowance trading – fix the level of compliance but result in uncertain marginal costs. This fundamental difference under uncertainty leads to different welfare outcomes intensively analysed by economists (Weitzman, 1974; Pizer, 1997). Pizer finds in his application to climate change policy that an optimal hybrid policy[34] based on initial allocation with additional permits available at a fixed price leads to welfare benefits only slightly higher than an optimal tax policy. However, in the case of a suboptimal hybrid policy with stringent targets and higher trigger prices, the hybrid system yields better welfare gains as a 'straightforward permit system with the same target'. Such hybrid schemes are also favoured by Hourcade and Fortin (2000), who plead for internationally tradable emissions permits with price floors, respectively ceilings, as a safeguard against hot air or too high costs and to increase acceptability. Bernard and Vielle (2000) find in their simulations of the Kyoto Protocol in various configurations that permits are more effective but that the spread is

Table 1.11 Summary of the basic criteria for the choice of policy instruments

Criteria	Tax versus tradable permit
Information intensity	*First-best scenario* • Emissions taxes and tradable permits are equally preferred under a first-best scenario (when information on marginal damages (MD) and marginal abatement costs (MAC) is available) • Under uncertainty about MD, both are equally inefficient • Under uncertainty about MACs, efficiency losses are minimized with an emissions tax if the MD curve is relatively flatter than the MAC curve and vice versa *Second-best scenario* • Tradable permits are preferred if regulator wants to achieve an absolute target
Efficiency in non-competitive market	• When output market is non-competitive, neither instrument is efficient, but efficiency losses are lower under permit scheme with heterogeneous firms (with different abatement technologies) • When permit market is non-competitive, permit programme is inefficient
Monitoring/ enforcement	• Tax is preferable since requiring monitoring of emissions only, while a permit system requires the monitoring of trades in addition
Flexibility in the face of change	• Permits are preferred since once established such systems maintain the desired level of emissions. A tax needs to be adjusted in the face of change of the ambient air quality
Dynamic incentives	• Both are equivalent in terms of encouraging new *technologies*, except if permit market does not function • In terms of *industry*, grandfathering may create a barrier to entry
Revenue-raising	• If permits are auctioned, both systems are equivalent. If permits are grandfathered, tax has higher potential
Distribution	• If permits are auctioned, both systems have the same effect • If permits are grandfathered, new entrants are discriminated against
Competitiveness	• Tradable permits are preferred since they make the cost of environmental protection less visible

Sources: Norregaard and Reppelin-Hill (2000); Stavins (2001).

small, and recommend the creation of an international permit system as well as national taxes or subsidies to address equity concerns. Norregaard and Reppelin-Hill (2000), in examining the relative merits of taxes and tradable permits, argue that on balance neither seems preferable to the other.

Thus the use of either price (tax) or quantity (allowance trading) instrument depends on the outcome of the analysis which leads to the best welfare outcome and political acceptability.[35] More recently the earlier idea of hybrid programmes has been revisited. Hybrid systems are a safety valve in the light of high uncertainty – as in the case of climate change – against the risk of ever-rising permit prices (Kolstad and Toman, 2000). Such policies set the target via an initial distribution of allowances, but allow additional permits to be purchased at a fixed price (McKibbin and Wilcoxen, 1997) or set floor and ceiling prices for permits to reduce the impact of 'hot air', respectively to avoid excessive costs (Hourcade and Fortin, 2000). Such systems work like a permit system by fixing the level of emissions as long as the marginal cost, reflected in the permit price, lies below that ceiling. They work like tax schemes by fixing marginal costs through the fixed price ceiling when the marginal costs hit that ceiling. In theory, by encompassing both tax and permit mechanisms, the hybrid policy should always perform as well as either of its components, depending on whether it works as a tax or permit scheme. One could argue that hybrid schemes are a result of a lack of decisiveness and, indeed, the rationale for the proposed hybrid policies focuses not so much on efficiency considerations but on targets and timetables and uncertainty regarding possible costs related to such targets. Since hybrid policies have not yet been implemented, 'real-world' experience is still missing. Pizer (1997) maintains on the basis of a modelling exercise that suboptimal hybrid policies based on stringent targets and high trigger price can generate better welfare than straight permit systems and this thus makes them a possible alternative to either straight tax or permit policy.

Taxation as a stick

There is an additional, sometimes neglected, aspect of taxation. Taxation can be a stick to 'entice' firms to opt for tradable permit schemes, which they might otherwise not prefer (FIELD, 2000; Hasselknippe and Høibye, 2001). This is for example explicitly expressed in the UK trading scheme (DEFRA, 2001), which allows companies to enter into negotiated agreements while at the same time making them participants of the trading scheme. One of the incentives for companies in the UK is that they avoid the climate change levy. Similar approaches are pursued in Sweden and Norway, where the incentive to join cap-and-trade schemes is the possibility of exemptions from environmental taxes. Traditionally, companies tend to shy away – at least in the first instance – from accepting an absolute cap, often a precondition for participating in emissions trading.

Tax exemptions

EU-specifically, there is another complication between taxation and tradable permits within the EU's internal market. In downstream tradable permits systems, companies that opt in such a scheme are typically exempted from taxation, as is the case in the UK's climate change levy. Exemptions from taxes for companies have traditionally caused concerns about distortions to competition within the EU's internal market, at least as long as participation in a trading scheme is not open to all to the same extent. In the light of the emergence of emissions trading schemes, at least within the EU, the relationship between taxation and tradable permits might need to be reconsidered with regard to distortion of competition between firms in different member states and between different sectors which directly compete with each other (for example steel and cement on the construction market.)

Capital gains taxes and environmental incentives

Little work on taxation of permits has been done. This is surprising given that incentives to use tradable permit schemes can change considerably depending on the tax schemes applied. One of the consequences of tradable permits, both credit and allowance schemes, is the change of property rights. Permits are thus potentially taxable. The fact that permits are potential assets as well as inputs to production poses an additional challenge in choosing the proper tax treatment and achieving efficiency in the permit market (see Fischer, 2001).

Most interesting in this respect is banking. As Fischer et al. (1998) demonstrate, there are 'lock-in' and 'reverse lock-in' effects due to taxation of capital gains. 'Lock-in' effects describe a situation where a tax on capital gains induces a firm, instead of selling (and paying a tax on capital gains) and investing the proceeds, to hold the permit in expectation of further capital gains, and to be taxed only when sold. The permit holder might keep the permit longer than economically efficient. As a result, lock-in effects may stimulate excessive banking, which, from an effectiveness point of view, might be positive given that banking is crucial for effectiveness through over-compliance (see Tietenberg et al., 1998; Burtraw and Mansur, 1999; Norregaard and Reppelin-Hill, 2000). Reverse lock-in effects appear when permit purchases cannot be deducted as inputs until use, which might deter purchase of bankable permits in advance. Deferral of deductibility is necessary as a disincentive to tax avoidance. For example, taxes could be saved by purchasing and deducting permits, while the proceeds of revenues from the sales could be booked in years of less good performance. As Dutschke and

Michaelowa (1998) argue, immediate deductibility creates an incentive for profitable firms to buy permits, while the opposite is true for firms with poor results. So far unanswered is the question whether this kind of tax arbitrage will undermine the efficiency and effectiveness of tradable permit schemes (see Fischer et al., 1998). To date, recent proposals for tradable permit schemes (EpE/MIES, 2000; DEFRA, 2001) pay scant attention to capital gains taxation in the context of design options for emissions trading.

5. CONCLUSIONS: ENSURING COMPATIBILITY OF INSTRUMENTS TO ACHIEVE THE OPTIMAL POLICY MIX

In the previous sections we described the long-trend shift from traditional command-and-control instruments towards incentive-based instruments in the European Union. This shift has been influenced by a diversity of motivations including economic (for example cost-effectiveness), sociopolitical (for example political acceptability), environmental (for example implementation and enforcement) or legal (for example competencies). This shift of government policies is also reflected in a shift of position by the regulated firms. While firms' strategies under command-and-control regimes consisted of cost-avoidance strategies, the shift towards incentive-based instruments is mirrored by companies exploiting the environment in the form of markets for new products to gain a competitive edge. Nowhere is this better demonstrated than in climate change, where companies such as BP and Shell have launched cap-and-trade emissions trading schemes based on ambitious targets. The rationale for this is partly to prepare the firms for an increasingly carbon-constrained world, that is, first-mover advantage in a broad sense. This includes, for example, seeking least-cost reductions and providing stimuli to research and technological development (Appendix 1).

The long-trend shift is advancing very unevenly among countries, depending *inter alia* on the political conditions, the legal system, national preferences, the nature of the environmental problem, different environmental priorities and many others. For example, while tradable permits have played a minor role in the EU, voluntary initiatives and negotiated agreements have flourished. In the USA, it has been the opposite.

At the same time as we have seen a shift towards incentive-based instruments, we have increasingly witnessed a shift towards combining instruments. Increasingly environmental policy integrates instruments into policy mixes. There are two reasons for this. New environmental legislation has to build on existing realities. For example, introducing cap-and-trade emissions trading schemes into the 'real world' has to accommodate existing instruments such as NEAs. Integration has the further advantage of being able to capture the different advantages of the various instruments. Hence policy mixes are designed to 'optimize' the incentives to mitigate pollution. Another driver for integration at the micro-level is the newly emerging climate policies, for example within the framework of the Kyoto Protocol. The Kyoto Protocol establishes three genuine market-based instruments for international action (that is, IET, JI and CDM), but leaves contracting parties free to apply whatever instrument best suits domestic abatement. As a consequence, governments have started to think about combining the Kyoto mechanisms or, more generically, international emissions trading and project mechanisms with domestic policy approaches such as NEAs, taxation, domestic emissions trading or regulation. The variety of policy instruments applied among the different EU member states is formidable. Germany, for example, relies on a mixture of a voluntary approach in the form of a unilateral commitment by industry and taxation and regulation for the remainder of the economy. Regarding the use of the flexible mechanisms, there seems to be a preference for the project mechanisms. The Netherlands' preferred instrument is also NEAs, mainly rate-based. While initially there was little interest in emissions trading, this has changed rapidly (Nentjes, 2001). The UK from the outset attempts to combine emissions trading, both allowance and baseline-and-credit trading (based on absolute as well as relative targets), taxation and negotiated agreements. All these approaches have in common that they present an attempt to combine instruments – in fact to develop a policy mix based on a variety of instruments (see Figure 1.3).

As this policy mix becomes more pronounced, the intensity of balancing different – often conflicting – objectives increases. If under a command-and-control approach the main factor that mattered was environmental effectiveness (and thereby choosing the appropriate technology), when using incentive-based instruments efficiency and equity concerns move up the agenda. This is self-evident since, as we have seen, the rise of incentive-based instruments is due to increasing interest in efficiency. This in return has triggered the interest in equity questions. All three elements are equally important for political acceptability both with the public and the 'regulated' firms. One of the lessons of early EPA credit trading schemes was that public acceptability of emissions trading schemes is important. Regarding efficiency, we have learned that simplicity matters and high transaction costs will lead to failure. As a result, the US EPA has adopted allowance trading, which combined to a much better extent effectiveness, efficiency and, after a period of transition, public and stakeholder acceptance.

Within the context of climate change policies, there are a number of issues related to policy mixes that will play a major role in the future.

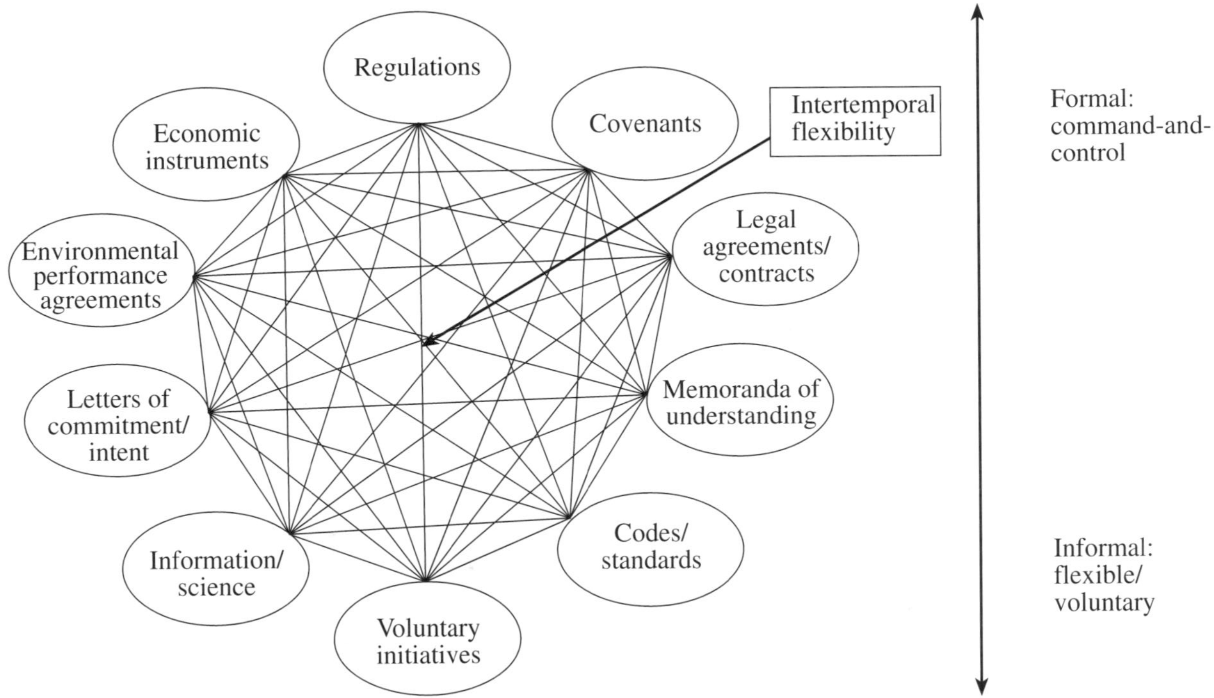

Source: Adapted from Yachnin et al. (2000).

Figure 1.3 The optimal policy mix

The Fallacy of Allowance versus Credit Trading

Allowance trading programmes have the double advantage over credit trading schemes that they ensure environmental effectiveness and tend to have fewer transaction costs. These were the two main reasons why the US EPA, which implemented most actual emissions trading schemes, has moved over time from credit trading to cap-and-trade schemes, which by now can be said to be the 'traditional' approach. Within the context of the Kyoto Protocol, which sets absolute emissions limits, allowance trading is at first sight the most suitable approach of the different models of emissions trading.

This contrasts with the situation in Europe, where numerous voluntary or negotiated agreements between government and industry have set relative or specific targets. This situation, along with proposals in the USA and Canada for voluntary early action credit trading schemes,[36] has rekindled interest in credit-based schemes, based both on absolute and relative targets, the latter being referred to as rate-based credit trading. The main advantage of credit trading is that it can be combined at the same time with policies based on absolute and relative performance, such as in NEAs and a mix thereof, which is the reality in Europe. Only rate-based credit trading programmes are compatible with the 'European model of voluntary approaches' (Börkey and Lévêque, 1998). Among others, Canada, France and the UK have started to experiment with policy mixes blending cap-and-trade emissions trading with baseline-and-credit schemes (both based on absolute and relative targets), NEAs and even taxation. The UK proposal for emissions trading is particularly interesting since it attempts to combine all instruments, including the different forms of emissions trading, NEAs and taxation. See also Chapters 2 and 3. Also it would appear that for Germany baseline-and-credit schemes would be a suitable policy mix for linking domestic policies (for example NEAs) to EU or international emissions trading.

Baseline-and-credit programmes have the further advantage that they might find wider acceptance among stakeholders. Unlike permits in allowance schemes that are allocated, most often for free, credits must be earned by better than mandated behaviour and are awarded by governments *ex post*. In this way, credit schemes can overcome popular criticism that allowance trading gives participants a 'licence to pollute' for free. Advocates of credit trading also hold that it avoids the contentious issue of initially allocating allowances and may therefore be more equitable than cap-and-trade systems with grandfathering, which in a modified form with benchmarking is the most common way of allocating allowances.

So far, as Convery (2001) notes, 'there is relatively little discussion of the equity issues involved in how quotas are allocated, but these considerations are likely to dominate the debate as allocation decisions that directly affect households are made'. Again, cap-and-trade and baseline-and-credit programmes have different effects on equity. Equity considerations can be addressed either *ex ante* (for example as part of the allocation process) or *ex post* via redistributive policies. An example for the former would be to allocate allowances to end users instead of to large producers and importers, as is generally expected, and as a result to keep transaction costs low. An example of the latter would be to tax those that have received the allowances for free and use the proceeds as subsidies for the end users, thus increasing the transaction costs. A hybrid system could consist of auctioning with a price cap, recycling the proceeds to households (Kopp et al., 1999). The choice of the policy mix might be strongly affected if equity considerations are included to a far greater extent than is currently the case.

Credit trading has two major disadvantages, which we have analysed in some detail. Rate-based credit trading, as any other rate-based policy, bears the risk that the participating sector emits more than the assumed target due to output increases. Rate-based policies are essentially de-coupled from economic growth. Within the context of the Kyoto Protocol, where parties have to obey absolute emissions limits, government might want to avoid such risks. As we have shown, the risk depends on what share of the overall emissions is covered by the ET scheme. Typically, downstream credit-based schemes will apply to industry, which accounts for around 20–25 per cent of overall emissions on average in OECD countries. The risk also depends on how ambitious the target is. The more ambitious the target, the less weighty is the potential increase due to output growth, since the overall increase of emissions is a function of the output increase and decrease in emissions per output, based on the specific target. Overshooting due to output increase could also be addressed by the purchase of credits from JI and CDM or by permits on the international emissions trading market. Irrespective of the international permit markets, European companies could also buy within the EU, once domestic or EU-wide emissions trading schemes are in operation.

Recent initiatives to develop rate-based or mixed credit schemes have not yet addressed in detail the issue of transaction costs, which in the past have led to the rise of allowance trading schemes. It seems that some of the main pitfalls, such as failure to trade and government pre-approval of trades, can be addressed within the institutional design, including implementation. For an overview see Tables 1.9 and 1.10. So far, no comprehensive study of how to avoid earlier implementation problems when designing greenhouse gas credit schemes, especially if rate-based, has been undertaken.

Blending NEAs and ET

Irrespective of whether NEAs and ET will become compatible in the first commitment period, NEAs can still function as a starting-point and learning process in the design and implementation of emissions trading. Several clear linkages are emerging.

First, environmental targets are negotiated between government and industry in an NEA. In addition, an emissions trading scheme enhances the efficiency by allowing least-cost abatement. Trading and banking possibilities on top of the NEA give participants incentives to abate beyond the target. This approach has been chosen in the UK (DEFRA, 2001) and France (EpE/MIES, 2000). A more theoretical approach has been developed by Albrecht and François (forthcoming).

Second, experience from negotiating and setting targets (that is, benchmarking) in NEAs will provide the methodologies and data upon which to base an absolute ceiling needed for a cap-and-trade programme. The most advanced countries in this field are the Netherlands, Germany, Finland and Denmark, with a long experience and a sophisticated institutional framework. Chapter 2 provides evidence.

Third, accepting a cap in the framework of an NEA can play an important transitional role in the period with more severe carbon constraints, such as the envisaged first commitment period of the Kyoto Protocol. Similarly, the NEA could be a transitional instrument to allow for a soft transition from command-and-control to emissions trading (see Torvanger and Skodvin, forthcoming).

While the first two linkages are examples of a policy mix whereby instruments are integrated, the last linkage merely represents the transition from one instrument to another.

The Role of Regulation in Environmental Policy Mixes with ET

The rationale for using market-based instruments, including tradable permits, is their economic advantages over command-and-control approaches. See also Chapter 4. The reason is that market-based instruments equalize the incremental amounts that firms spend on pollution control. As a result technology standards and tradable permit programmes are mutually exclusive. This raises few problems for GHG emissions control, which so far is not regulated at all. As a result policy makers have the freedom to design policies without the constraints of prior policies. A number of qualifications need to be made, however. First, firms are typically subject to regulation for a number of other emissions than GHGs. As a result, constraints through regulation of other than GHGs may impose technology choices which reduce the efficiency of the tradable permit scheme for GHG. As is the case for the EU's

IPPC Directive, in taking energy efficiency into consideration when regulating a firm's emissions other than GHGs may subject firms to double regulation. The firm is regulated once by a – albeit indirect – technology standard, that is, energy efficiency standard, and once through the overall ceiling of allowances. Second, market-based instruments, including tradable permit schemes, do not address 'hot spots'. In our case, this can be neglected since climate change policies address GHG concentration in the atmosphere and not localized pollution. Third, tradable permit schemes cannot always function as substitutes for energy performance standards. Energy performance standards address, in addition to pollution, market barriers, notably in the domestic and commercial sector. Such barriers typically include lack of consumer information, lack of capital for small-scale projects due to transaction costs, split incentives and others. They are likely not to be addressed by tradable permit schemes.

Energy efficiency standards, if applied to products rather than processes, could, however, be combined with rate-based credit trading and thereby provide an additional incentive for manufacturers to go beyond the mandated standard if this can be done at a cost lower than the permit price. This could typically work where transaction costs are low and in areas such as engine technology, appliances and the like, where credits can easily be calculated. Such schemes have been operated in the USA.

Another possible practical linkage between regulation and emissions trading could be for government to use stringent regulation as a 'stick' to make firms accept emissions trading (cap-and-trade) more easily by analogy with taxation (see below). So far, we have not been able to find concrete examples where such an approach has been taken.

Adapting Quantity Policies to 'Real-world Solutions'

The debate on policy mixes has shed new light on the debate on 'quantities versus prices'. Rather than defining the circumstances under which one instrument is preferable[37] to another (for an overview see Table 1.11), we have identified the linkages and how they can eventually be combined. In the context of climate change policy, the idea of hybrid systems has been revisited. These hybrid systems set a target via an initial distribution of allowances but create, in the form of floor and ceiling prices for additional permits, a safety valve against ever-rising permit prices to increase efficiency and acceptability (Hourcade and Fortin, 2000).[38] Whether such schemes could work in practice remains to be seen since hybrid programmes have not been implemented in the past.

Efficiency and effectiveness of tradable permits are influenced by capital gains taxes. Under both allowance and credit trading schemes, permits or credits constitute assets. Choosing proper tax treatment is one element of achieving efficiency in the permit market. There is a potential trade-off between efficiency in the market and equity considerations. Taxation of permits might become a convenient tool to address equity issues *ex post*. There is a need for further analysis in this somewhat neglected field.

On a practical level, taxation is increasingly used as a stick to persuade industry to accept cap-and-trade emissions trading programmes. Typically firms accepting an absolute cap and participation in cap-and-trade programmes are exempted from environmental or carbon taxes, as evidenced in the UK, where companies agreeing to participate in the ET scheme are exempted up to 80 per cent of the UK climate change levy. Similar approaches have been adopted in Sweden and Norway (for further details see Appendix 1). A threat of a carbon tax has also facilitated participation of German industry in the industry-led NEA. Further evidence of synergies between taxes and other instruments are reported by the European Environment Agency (EEA, 2000). In two cases in Denmark (collection of nickel-cadmium batteries; PVC and phthalates) a tax was introduced to complement an NEA which did not achieve its objectives. In the Belgian case taxes were used as an enforcement mechanism. For a number of products firms only have to pay environmental taxes if they fail to meet certain targets in recycling. Since in practice very few taxes are paid, it can be assumed that environmental targets of NEAs were met. Further evidence of the synergies between NEAs and taxation is provided in Chapter 2.

NOTES

1. Article 5 of the EC Treaty (current – Amsterdam – version).
2. Emission standards and BAT standards are often the same in practice.
3. While the original version of Treaty (Treaty of Rome in 1957) made no reference to the environment, it was the Single European Act in 1987 which provided a legal base for the environment, mainly based on regulatory approaches. Subsequent Treaty revisions in 1992 (Maastricht Treaty) and 1997 (Amsterdam Treaty) called for integration of environmental policy in other policy areas.
4. See OECD (2000).
5. For a full overview of the political economy see Oates and Portney (2001).
6. The relative absence of EU 'regulation' in the fields of energy and households may seem surprising but can be explained by a lack of EU competencies.
7. Database on environmental taxes in the European Union member states, plus Norway and Switzerland: http://www.europa.eu.int/comm/environment/enveco/env_database/database.htm
8. During the recent revision of the EU Treaty, finalized in Nice (France) in December 2000, EU member states decided to maintain the *status quo*.
9. The Marrakesh Accords are available to download from the UNFCCC website, http://www.unfccc.int/cop7/documents/accords_draft.pdf
10. In the extreme case, under full global IET in Annex 1 countries, for example, ABARE (1999) has concluded that IET could reduce the Kyoto Protocol compliance costs up to four-fifths. Whether this will indeed be the case will depend on the exact rules for IET

depending, *inter alia*, on the scope for trade (parties and/or entities, global trade), access to 'hot air', inclusion of sinks or eventually the possibility to gain access to non-Annex 1 emission reductions. Although such a scenario is unlikely, efficiency gains are still considered to be considerable. In its Third Assessment Report, the International Panel for Climate Change (IPCC, 2001) reports that the majority of studies show that full emissions trading between industrialized countries would halve the cost of compliance from 0.2–2 per cent to 0.1–1 per cent of GDP by 2010. For an overview see OECD (1998a, 1998b); Weyant (1999) and Hourcade and Fortin (2000). For more recent analysis after the pullout of the USA see Böhringer (2001) and Buchner et al. (2001). For a more complete overview and analysis, see Chapter 3 by Janssen.

11. The two terms will be used interchangeably.
12. On the 'double dividend', see, for example, Carraro and Siniscalco (1996).
13. On ways to include transport into an allowance programme, see Ayres (1997) for theoretical aspects and, for example, CCAP (2000) and Bone et al. (2001) for practical aspects. The proposed approaches fall short of including incentives to reduce emissions at product level (that is, cars, trucks). This idea will be further developed below.
14. The introduction of an upstream scheme in the EU is further hampered by EU law. The EC Treaty, Article 175, paragraph 2, foresees unanimity in the EU Council of Ministers for 'measures significantly affecting a Member State's choice between different energy sources and the general structure of its energy supply', therefore providing each member state with a veto. Given the effect of an upstream scheme on the energy mix of member states, which vary dramatically, there are few chances for an adoption of such a scheme at EU level.
15. The three terms will be used fully interchangeably.
16. For a complete overview see, for example, Norregaard and Reppelin-Hill (2000), Godard (2000) and Stavins (2001).
17. For example Chafee Bill; Lazio Bill; Canadian Early Emission Reduction Program (CEERP); Proposal by the Center for Clean Air Policy; CAST Program (Coalition for the Advancement of Sustainable Technology), USA.
18. Strictly speaking, AIJ is not an emissions credit system because projects are pilot projects and no credits are given. Nevertheless AIJ could be the precursor of future emissions credit programmes on the basis of international agreements.
19. Hot spots are areas of very high concentrations of pollutants. Tradable permits facilitate the creation of hot spots if, as can be the case, they allow clustering of emissions in vulnerable areas.
20. The EU applied a tax differentiation scheme, which also proved to be effective.
21. Averaging means credit offsets for engines manufactured during the same year whose emissions are better than the performance standard.
22. In the case of rate-based credit trading, the additional issue of environmental effectiveness has played a role, which was addressed above.
23. For a complete overview, see Ellerman (2000), Hasselknippe and Høibye (2001) and Haites and Mullins (2001); for a more detailed description, Appendix 1; and for more analysis, see Chapter 3.
24. Norway is as a consequence of the European Economic Area (EEA) agreement fully part of the EU internal market, including judicial review by the European Court of Justice.
25. The Swedish proposal was planned to be put before parliament in 2000. Because of a lack of agreement at COP6, the parliament reading was proposed.
26. For further details of the BP and Shell schemes, see Appendix 1.
27. While one instrument, for example cap-and-trade emissions trading, has advantages in terms of effectiveness and efficiency, it has shortcomings regarding acceptability (by industry, by the public at large) and equity.
28. See CEFIC (1997, chemicals industry), CEMBUREAU (1998, cement industry) and the European-level industry association grouped in UNICE (2000).
29. In addition to the risk that industry fails to achieve its relative target to which it has agreed in the NEA.
30. As part of the current policy review of climate change it appears that the US government

is studying a credit-based incentive structure to improve the fuel economy of motor vehicles (*Financial Times*, 19 December 2001, p. 4).

31. Council Directive 96/389/EC of 24 September 1996 concerning integration pollution and prevention control (the so-called IPPC Directive). The substances covered are listed in the annex of the Directive.
32. The economic literature is still undecided on whether historical energy efficiency improvements in the long term are caused by 'price' or 'regulation' (see Villa, 2000).
33. By the landlord/tenant relationship we mean the split incentives to invest in energy savings equipment between the landlord and the tenant. The tenant, who benefits from better energy savings equipment, has no incentive to invest, since he will leave the apartment. The landlord, who needs to do the investment, has no direct benefit, other than an uncertain possibility of demanding higher rent. As a result policy has reacted by regulation, that is, by forcing landlords to install metering and energy savings equipment.
34. According to Stavins (2001), no such hybrid policy has been developed in reality, however.
35. Stavins (2001) concludes that 'policy instruments that appear impeccable from the vantage point of research institutions, but consistently prove infeasible in real-world political institutions, can hardly be considered "optimal"'.
36. Early action schemes by definition work before the first Kyoto commitment period, when there are no firm carbon constraints.
37. Both tradable permits and taxation reduce compliance costs if compared to command-and-control measures. Which one is preferable has to be judged on a case-by-case basis.
38. Ellerman (2001) reports that extreme price hikes are possible. Within the framework of the Californian RECLAIM NO_x permit trading programme, prices rose to over $60.000 in the third quarter of 2000 as a result of the electricity crisis after averaging around $500 throughout the 1990s. This has prompted the government to set a price cap at $15.000.

REFERENCES

ABARE (1999), *Economic impacts of the Kyoto protocol*, Canberra: Australian Bureau of Agricultural and Resource Economics.

Albrecht, J. (1999), 'Making emissions trading more effective: integrating tradable certificates', paper presented at the Second EFIEA Policy Workshop – Integrating Climate Policies in the European Environment, Milan, 4–6 March 1999 (cited from Convery, 2001).

Albrecht, J. and D. François (forthcoming), 'Negotiated Environmental Agreements and CO_2 Emissions Trading', in P. ten Brink (ed.), *Voluntary Environmental Agreements: Process, Practice and Future Use*, Sheffield, UK: Greenleaf Publishing.

Ayres, R.U. (1997), 'Environmental Market Failures: Are there any Local Market-based Corrective Mechanisms for Global Problems', *Mitigation and Adaptation Strategies, for Global Change*, **1**(3).

Bernard, A. and M. Vielle (2000), 'Comment allouer un coût global d'environnement entre pays: permis négociables vs taxes ou permis négociables et taxes?', *Economie Internationale*, **82**(2), 103–5.

Böhringer, C. (2001), 'Climate Politics From Kyoto to Bonn: From Little to Nothing?', ZEW Discussion Paper No. 01–49 (ZEW: Zentrum für Europäische Wirtschaftsforschung GmbH).

Bone, H. et al. (2001), *Flexible Instrumente der Klimapolitik im Verkehrsbereich. Exploratory Study*, Commissioned by the Ministry of Environment and Transport, Baden-Württemberg.

Börkey, P. and F. Lévêque (1998), 'Voluntary approaches for environmental protection in the European Union', ENV/EPOC/GEEI(98)29/FINAL, Paris: OECD.

Börkey, P. and F. Lévêque (2000), 'Voluntary approaches for environmental protection in the European Union – A survey', *European Environment*, **10**, 35–54.

Bretteville, C. and G. B. Søfting (2000), 'From Taxes to Permits? The Norwegian Climate Change Debate', CICERO Report 2000:6. Oslo: CICERO.

Buchner, B., C. Carraro and I. Cersosimo (2001), 'On the Consequences of the U.S. Withdrawal from the Kyoto/Bonn Protocol', Fondazione Eni Enrico Mattei, draft paper.

Burtraw, D. and E. Mansur (1999), 'The Effects of Trading and Banking in the SO_2 Allowance Market', Discussion Paper 99-25. Washington, DC: Resources for the Future.

Burtraw, D., K. Palmer, R. Bharvirkar and A. Paul (2001), 'The Effect of Allowance Allocation on the Cost of Carbon Emission Trading', Discussion Paper 01-30. Washington, DC: Resources for the Future.

Carraro C. and F. Lévêque (eds) (1999), *Voluntary Approaches in Environmental Policy*, Dordrecht: Kluwer.

Carraro, C. and D. Siniscalco (eds) (1996), *Environmental Fiscal Reform and Unemployment*, Series Economics, Energy, Environment, Dordrecht: Kluwer.

CCAP (1999), *Design of a Practical Approach to Greenhouse Gas Emissions Trading Combined with Policies and Measures in the EC*, Washington, DC: Center for Clean Air Policy, study prepared for the European Commission, November.

CCAP (2000), *Transportation and Domestic Greenhouse Gas Emissions Trading*, Washington, DC: Center for Clean Air Policy.

CEC (1992), *Towards Sustainability. Fifth Environmental Action Programme*, Brussels–Luxembourg: Office for Official Publications of the European Communities.

CEC (1996), *Environmental Agreements. Communication from the Commission to the Council and the European Parliament*, European Commission, COM (96) 561 final.

CEC (2000), *Green Paper on Greenhouse Gas Emissions Trading within the EU*, COM (2000) 87 final of 8 March.

CEC (2001a), *Establishing a Framework for Greenhouse Gas Emissions Trading within the European Community* (Commission Proposal for EU Directive on Greenhouse Gas Emissions Trading), COM(2001)581 of 23 October 2001 http://www.europa.eu.int/comm/environment/climat/emission.htm

CEC (2001b), *European Climate Change Programme*, Final Report (www.europa.eu.int/comm.environment/climat/eccp.htm)

CEFIC (1997), *Voluntary energy efficiency programme – VEEP 2005*, Brochure by the European Chemical Industry Council (CEFIC). Brussels.

CEMBUREAU (1998), *Climate Change, Cement and the EU*, Brochure by the Association of the European Cement Industry (CEMBUREAU). Brussels.

Clinch, P. (2000), 'Environmental Policy Reform in the EU', in J. Pelkmans and G. Galli (eds), *Regulatory Reform and Competitiveness in Europe*, Volume 2. Cheltenham, UK and Northampton, MA: Edward Elgar.

Clinch, P. and J. Healy (2001), 'Cost–benefit analysis of domestic energy efficiency', *Energy Policy*, **29**, 113–24.

Convery, F. (2001), 'Emissions Trading and Environmental Policy in Europe', paper presented at the pre-summit conference Knowledge and Learning for a Sustainable Society (Climate and Global Justice Session), Göteborg University, Sweden 12–14 June; www.emissionstradingnetwork.com (June 2001).

Delbeke, J. (1996), 'The Future Use of Economic Instruments in the European Union', in CEPS, *The Rise of Economic Instruments in EU Environmental Policy: Why and How Fast?*, CEPS Business Policy Report No. 1, Brussels: Centre for European Policy Studies.
Dinan, D. (1999), *Ever Closer Union*, 2nd edn, Basingstoke: Macmillan.
DEFRA (2001), *UK Emissions Trading Scheme*. (See: http://www.defra.gov.uk/environment/climatechange/trading/index.htm)
Dudek, D. and J.B. Wiener (1996), *Joint Implementation, Transaction Costs, and Climate Change*. Paris OECD.
Dutschke, M. and A. Michaelowa (1998), *Der Handel mit Emissionsrechten für Treibhausgase*, HWWA-Report 187. Hamburg: Institut für Wirtschaftsforschung.
EEA (1996), *Environmental Taxes*, Environmental Issues Series No. 1, Vol.1, Copenhagen: EEA.
EEA (1997), *Environmental Agreements*, Environmental Issues Series No. 3, Vol.1, Copenhagen: EEA.
EEA (1999), *Environment in the European Union at the Turn of the Century*, Luxembourg: Office for Official Publications of the European Communities.
EEA (2000), *Environmental Taxes: Recent Developments in Tools for Integration*, Environmental Issues Series No. 18, Luxembourg: Office for Official Publications of the European Communities.
Egenhofer, C. and F. Mullins (2000), 'The Role of Emissions Trading in EU Climate Change Policy', CEPS Working Party Report. Brussels: CEPS, September.
Ekins, P. (1996), *Environmental Taxes and Charges – National Experiences and Plans*, Dublin: European Foundation for the Improvement of Living and Working Conditions.
Ellerman, D. (2000), *Tradable Permits for Greenhouse Gas Emissions: A primer with particular reference to Europe*. MIT Joint Program on the Science an Policy of Global Change, report No. 69, November.
Ellerman, D. (2001), *Presentation to the European Research Network on Emissions Trading*, Paris: CIRED, 14 May.
Epe/MIES (2000), 'Implementing an emissions credits trading system in France to optimize industry's contribution to reducing greenhouse gases', MIES (Interministerial task force on climate change–Industry working group. 31.3.00 Final) http://www.epe.asso.fr
E3M Lab (2000), The Economic Effects of EU-Wide Industry-Level Emissions Trading to Reduce Greenhouse Gases. Results from PRIMES Energy Systems Model (E3M Lab, Prof. P. Capros, Dr L. Mantzos, Institute of Communication and Computer Systems of National Technical University of Athens). May 2000. http://europa.eu.int/comm/environment/enveco/studies2.htm
FIELD (2000), 'Designing Options for Implementing an Emissions Trading Regime for Greenhouse Gases in the EC', London: Foundation for International Environmental Law and Development, Study prepared for the European Commission, 22 February 2000.
Fischer, C. et al. (1998), *Using Emissions Trading to Regulate U.S. Greenhouse Gas Emissions: An Overview of Policy Design and Implementation Issues*, Discussion Paper 98–40, Washington, DC: Resources for the Future.
Fischer, C. (2001), *Multinational Taxation and International Emissions Trading*, Discussion Paper 01–18, Washington, DC: Resources for the Future.
Godard, O. (2000), 'L'expérience américaine des permis négociables', *Economie Internationale*, **82**(2), 13–43.

Hahn, R. and G.L. Hester (1989), 'Marketable Permits: Lessons for Theory and Practice?', *Ecology Law Quarterly*, **16**(2).
Haites, E. and F. Mullins (2001), 'Linking Domestic and Industry Greenhouse Gas Emission Trading Systems', paper prepared for EPRI, IEA and International Emissions Trading Association (IETA) http://www.iea.org/envissu/epri.pdf
Haites, E. and R. Hornung (1998), 'Voluntary Credit Trading with Mandatory Performance Standards', prepared for Multistakeholder Expert Group on Domestic Emissions Trading.
Hasselknippe, H and G. Høibye (2001), *Meeting the Kyoto Protocol Commitments*, Oslo: NHO (Confederation of Norwegian Business and Industry).
Hessen (2001), Pilotprojekt 'Planspiel zum CO_2 Emissionshandel. Abschlussbericht – Ergebnisse and Empfehlungen', Hessisches Ministerium für Umwelt, Landwirtschaft und Forsten. http://www.mulf.hessen.de/umwelt/atomaufsicht/_fr_atom.htm
Høibye, G. (2001), 'Participation in the EU Trading Scheme: Norway', presentation at the Centre for European Policy Studies, Brussels: CEPS.
Hourcarde, J.-C. and E. Fortin (2000), 'Impact économique des politiques climatiques: des controverses aux enjeux de coordination', *Economie Internationale*, **82**(2), 45–73.
IEA (2001), *International Emission Trading. From Concept to Reality*, Paris: IEA.
IPTS (2000), 'Preliminary Analysis of the Implementation of an EU-wide Permit Scheme on CO_2 Emissions Abatement Cost', Results from the POLES Model, accessed in September 2000 at: http://europa.eu.int/comm/environment/enveco/studies2.htm
IPCC (2001), 'Climate Change 2001: Mitigation', Contribution of Working Group III to the Third Assessment Report of the Intergovernmental Panel on Climate Change.
Jordan, A. and T. Jeppesen (2000), 'EU Enviromental Policy: Adapting to the Principle of Subsidiarity?', *European Environment*, **10**, 64–74.
Joskow, P. and R. Schmalensee (1998), 'The Political Economy of Market-Based Environmental Policy: The U.S. Acid Rain Program', *The Journal of Law and Economics*, **41**(1), 37–83.
Joskow, P., R. Schmalensee and E. Bailey (1998), 'The Market for Sulphur Dioxide Emissions', *American Economic Review*, **88**(4), 669–85.
Kerr, S. and D. Maré (1997), 'Transaction Costs and Tradable Permit Markets: The United States Lead Phasedown', http://www.arec.umd.edu/skerr/transact.htm, 5 June.
Klaassen, G. (1996), *Acid Rain and Environmental Degradation: The Economics of Emissions Trading*, Cheltenham, UK and Brookfield, USA: Edward Elgar.
Klaassen, G. and A. Nentjes (1997a), 'Sulphur Trading under the 1990 CAAA in the US: An Assessment of First Experiences', *Journal of Institutional and Theoretical Economics*, 153(2), 384–410.
Klaassen, G. and A. Nentjes (1997b), 'Creating Markets for Air Pollution Control in Europe and the USA', *Environmental and Resource Economics*, **10**, 125–46.
Kolstad, C. and M. Toman (2000), 'The Economics of Climate Change Policy', Discussion Paper 00-40, Washington, DC: Resources for the Future.
Kopp, R. et al. (1999), *A Proposal for Credible Early Action in U.S. Climate Change Policy*, Washington, DC: Resources for the Future.
Kruger, J. (2000), US Environmental Protection Agency, personal communication.
Legge, T. (2001), 'Emissions Trading and the European Union Internal Market', in Ministry for the Environment, Nature Conservation and Nuclear Safety, Germany, *Joint Implementation and Emissions Trading as a Chance for Central and Eastern European Countries*, Berlin.

Lévêque, F. (1996), 'The European Fabric of Environmental Regulations', in F. Lévêque (ed.) (1996), *Environmental Policy in Europe*, Cheltenham, UK and Brookfield, USA: Edward Elgar.

Lévêque, F. (ed.) (1996), *Environmental Policy in Europe*, Cheltenham, UK and Brookfield, USA: Edward Elgar.

McKibbin, W. and P. Wilcoxen (1997), 'A Better Way to Slow Climate Change', Brookings Policy Briefs No. 17, Washington, DC: Brookings Institution.

McLean, B. (1996), *Evolution of Marketable Permits: The U.S. Experience with Sulphur Dioxide Allowance Trading*, Washington, DC: US EPA.

Naimon, J. and D.S. Knopman (1999), 'Reframing the Climate Change Debate', Progressive Policy Institute, Policy Report, November.

Nentjes, A. (2001), 'Presentation to the European Research Network on Emissions Trading', Paris: CIRED, 14 May.

Nollet, P. (2000), personal communication.

Norregaard, J. and V. Reppelin-Hill (2000), 'Taxes and Tradable Permits as Instruments for Controlling Pollution: Theory and Practice', IMF Working Paper WP/00/13.

OECD (1992), *Climate Change: Designing a Tradable Permit System*, Paris: OECD.

OECD (1994), *Managing the Environment – The Role of Economic Instruments*, Paris: OECD.

OECD (1996), *Implementation Strategies for Environmental Taxes*, Paris: OECD.

OECD (1997a), *Evaluating Economic Instruments for Environmental Policy*, Paris: OECD.

OECD (1997b), 'International Greenhouse Gas Emissions Trading', Annex I Expert Group on the UNFCCC, Working Paper No. 9.

OECD (1998a), Proceedings of the OECD Workshop on 'Climate Change and Economic Modelling. Background Analysis for the Kyoto Protocol', Paris, 17–18 September.

OECD (1998b), *Economic Modelling of Climate Change*, OECD Workshop Report, Paris: OECD.

OECD (1999a), 'Status of Research on Project Baselines under the UNFCCC and the Kyoto Protocol', OECD Information Paper, ENV/EPOC(99), 19/FINAL.

OECD (1999b), 'Experience with Emissions Baselines under the AIJ Pilot Phase', OECD Information Paper, ENV/EPOC(99), 23/FINAL.

OECD (2000), *World Energy Outlook–1999 Insights: Looking at Energy Subsidies: Getting the Prices Right*, Paris: OECD.

OECD (2001), *Environmentally Related Taxes in OECD Countries: Issues and Strategies*, Paris: OECD.

Oates, W.E. and P.R. Portney (2001), 'The Political Economy of Environmental Policy', Discussion Paper 01-55, Washington DC: Resources for the Future.

Pelkmans, J. and C. Egenhofer (1993), 'Defizite in Politikfeldern der EG-Integration', in C. Jakobeit and A. Yenal (eds), *Gesamteuropa*, Opladen: Leske + Budrich.

Pelkmans, J. (2001), *European Integration*, Second Edition, Edinburgh: Pearson Education.

Pizer, W. (1997), 'Prices vs. Quantities Revisited: The Case of Climate Change', Discussion Paper 98-00, Washington, DC: Resources for the Future.

Potier, M. (1996), 'The Experience of OECD Countries in Their Domestic Use of Economic Instruments for Environmental Management', in CEPS, *The Rise of Economic Instruments in EU Environmental Policy: Why and How Fast?* CEPS Business Policy Report No. 1.

PricewaterhouseCoopers (2000), *GETS 2 Report*, Brussels.

Reinhardt, F.L. (2000), *Down to Earth: Applying Business Principles to Environmental Management*, Boston: Harvard Business School Press.

Rolfe, C., A. Michaelowa and M. Dutschke (1999), 'Closing the Gap: A Comparison of Approaches to Encourage Early Greenhouse Gas Emissions Reductions', HWWA Report, HWWA-Institut für Wirtschaftsforschung-Hamburg.

RWI (forthcoming), 'Möglichkeiten und Rahmenbedingungen einer Verknüpfung der Selbstverpflichtung der deutschen Wirtschaft zur Klimavorsorge mit flexiblen Instrumenten nach dem Kiotoprotokoll', Report by Rheinisch-Westfälisches Institut für Wirtschaftsforschung (RWI) and University of Essen.

Salmons, R. (2000), 'Performance-based Credit Trading: Cost-Efficient Implementation of Rate-Based Environmental Strategies', CSERGE, University College London, draft paper (forthcoming in *Environmental and Resource Economics*, 2002).

Shaw, D. et al. (2000), 'The Application of Subsidiarity in the Making of European Environmental Law', *European Environment*, **10**, 85–95.

Stavins, R. (1995), 'Transaction Costs and Tradable Permits', *Journal of Environmental Economics and Management*, **29**, 133–48.

Stavins, R. (2001), 'Experience with Market-based Environmental Policy Instruments', Discussion paper 01-58, Washinton, DC: Resources for the Future (To be published in K.G. Mäler and J. Vincent (eds), *The Handbook of Environmental Economics*, Chapter 21. Amsterdam: North-Holland/Elsevier Science (forthcoming).)

Storey, M. (1996), 'Demand Side Efficiency: Voluntary Agreements with Industry', Annex 1 Expert Group on UNFCCC, Policies and Measures for Common Action, Working paper 8, OECD.

Sun, J.-M. and J. Pelkmans (1995), 'Liberalisation, Centralisation and EU Telecoms', *The World Economy*, **18**(5), 635–64.

ten Brink, P. and M. Morère (2000), 'Voluntary Initiatives to Address Climate Change in the EU', *European Environment*, No. 10, 178–92.

Tietenberg, T. et al. (1998), *International Rules for Greenhouse Gas Emissions Trading*, Geneva: United Nations/UNCTAD.

Tietenberg, T. (2001), 'The Tradable Permits Approach to Protecting the Commons: What have we Learned?', paper given at the International Workshop on 'Trading Scales: Linking Industry, Local/Regional, National and International Emission Trading Schemes', Fondazione ENI Enrico Mattei (FEEM), Venice, 3 and 4 December.

Torvanger, A. and T. Skodvin (forthcoming), 'Environmental Agreements in Climate Politics', in P. ten Brink (ed.), *Voluntary Environmental Agreements: Process, Practice and Future Use*, Sheffield, UK: Greenleaf Publishing.

UNICE (2000), *EU strategy responding to climate change*, brochure by the Union of Industrial and Employers' Confederation of Europe (UNICE), Brussels.

Weitzman, M.L. (1974), 'Prices vs. quantities', *Review of Economic Studies*, **41**(4), 477–91.

Villa, P. (2000), 'L'intensité énergétique à long terme', *Economie Internationale*, No. **82**(2).

Weyant, J.P. (ed.) (1999), 'The Costs of the Kyoto Protocol: a Multi-Model Evaluation', Special Issue of the Energy Journal, International Association for Energy Economics.

Yachnin, R. et al. (2000), 'The Optimal Policy Mix', The Conference Board of Canada, 297-00, June.

Zapfel, Peter (1999), 'Negotiated Agreements and "Flexible Mechanisms": Building Blocks for Efficient Kyoto Implementation Strategies in the European Union?', in ZEW, Flexible Mechanisms for an Efficient Climate Policy (ZEW – Zentrum für Europäische Wirtschaftsforschung – Centre for European Economic Research).

APPENDIX 1: OVERVIEW OF MAIN EUROPEAN EMISSIONS TRADING INITIATIVES AS PART OF POLICY MIXES

National Emissions Trading Schemes

Within the last two years, a number of EU member states have developed national emissions trading programmes. In what follows, they are briefly described. For a complete overview see Ellerman (2000), Hasselknippe and Høibye (2001) and Haites and Mullins (2001). For a more selective analysis see Chapter 3.

1. The Danish cap-and-trade power sector scheme

The Danish scheme for tradable CO_2 emissions permits (cap-and-trade scheme) is the first of its kind in the European Union and has been in operation since 1 January 2001. It is limited to the electricity sector, which is responsible for about 40 per cent of total Danish emissions. It is limited to companies producing more than 100 000 tonnes of CO_2, that is, 90 per cent of the sector. The cap is initially set at 23 million tonnes (mt) or 70 per cent of emissions during the reference period 1994–98, and will annually be reduced, decreasing to 20 mt in 2003 when the scheme ends. Allocation was made on the basis of grandfathering according to historical criteria and technical criteria that favour more energy-efficient producers. In case of non-compliance, a fine of Dkr 40 (about €5.40) is levied for every excess tonne of CO_2 emitted.

In order to ensure a level playing field for new entrants in the Danish power market, new producers will be allocated allowances when starting up production on the basis of transparent and non-discriminatory criteria. This was required by the European Commission to avoid distortions to competition within the internal market.

The scheme has several objectives. The main goal is to limit CO_2 emissions by setting a cap and to create incentives to produce more environmentally friendly electricity. The programme is also intended to mitigate potentially negative effects on the competitiveness of Danish power companies, which are subject to competition within the Nordic electricity market comprising Denmark, Finland, Norway and Sweden. The Nordic electricity market is the first cross-border electricity market subject to full competition. Early evidence seems to suggest that the scheme is not able to avoid negative effects on the competitive position of Danish electricity producers *vis-à-vis* their competitors from Finland, Norway and Sweden. One of the major driving forces for the scheme is also to prepare the energy industries for the implementation of the Kyoto mechanisms. By running the scheme, both government and industry are preparing to put in place the necessary procedures, methodologies, institutions and internal company structures to deal with international emissions trading. The scheme is therefore also meant to be linked to other national or EU schemes if possible and as they emerge.

Since the scheme is largely intended as 'learning-by-doing' or a pilot programme, it is not yet integrated into a broader policy mix, although elements of this are emerging. Companies below the 100 000 tonnes CO_2 threshold, which are exempted from the scheme, are subject to taxation. However, it is not intended to tax the entire electricity production but only the emissions above the level needed to cover domestic use of electricity. This will limit the net export based on fossil fuels, especially coal, on which Denmark relies heavily. Naturally, this creates incentives to produce electricity from more environmentally friendly sources and will reduce Denmark's dependence on coal in the medium and long run. As a consequence, it reduces the competitiveness of Danish power producers based on generation from coal in the short run. Both the *de facto* price cap on permits due to setting the maximum penalty at €5.40 for the trading sector and the exemption from taxes for the non-trading sector for electricity used by households are means to avoid too steep price rises.

Sources: Hasselknippe and Høibye (2001), Haites and Mullins (2001), Press service of the European Commission.

2. The UK emissions trading scheme

Under the Kyoto Agreement the UK's legally binding target is to reduce their GHG emissions by 12.5 per cent of their 1990 emissions by 2008–10. This corresponds to a 7 per cent reduction (~50 mt CO_2e) in 2010 for a business-as-usual (BAU) approach. However, the UK government has declared a target of 20 per cent reduction by the Kyoto period; this again corresponds to about a 15 per cent reduction in 2010 from a BAU emissions forecast. The UK climate change strategy is based on the combination of a climate change levy (CCL) to be implemented in 2001, and an ET system, which will be complementary to the CCL.

The CCL will be charged on all industrial and commercial users of energy covering industry, including fuel industries, commerce, agriculture, public administration and other services.[1] Electricity produced in high-efficiency combined heat and power plants, and electricity from renewable sources are exempt. Ten of the most energy-intensive British trade associations and their companies are expected to sign negotiated agreements for energy efficiency targets under the CCL. By doing so these sectors will qualify for an 80 per cent discount on the levy. The CCL and its negotiated agreements are expected to reduce British GHG emissions by 5 million tonnes carbon per year by 2010.

Development of the ET scheme was driven to a large extent by business. The initial Emissions Trading Group was set up in June 1999 by the Confederation of British Industry (CBI) and the Advisory Committee on Business and the Environment (ACBE) comprising companies and the government. The scheme would be open to all companies operating in the UK willing to commit themselves to binding carbon or GHG limits agreed by government under its rules. Overseen by an Emissions Trading Authority, three categories of participants would be allowed to trade: (1) Firms that abide by declining annual – absolute – emissions limits set by government (the 'absolute' sector); (2) firms that accept an output-related or specific target (the 'unit' sector); and (3) firms that deliver specific carbon or GHG emissions savings projects. In order to prevent permits from the unit sector swamping the absolute sector in periods when output is relatively high, the scheme attempts to limit sales from the unit sector to the absolute via a 'gateway'. Trading among the unit sector is unrestricted ('cap-and-trade'). Trade between the 'absolute' and 'unit' sectors is accepted as long as there is no net flow from the unit to the absolute sector.

The GETS 2 simulation of a gateway revealed significant transaction costs, questioning whether such a gateway will indeed work in practice. The UK government has also made it clear that in the longer run it would want to see companies accepting an absolute cap under the emissions trading scheme, rather than energy-efficiency (that is, relative) targets.

The UK scheme is a good example of how policy mixes work together to accommodate the 'real world'. The scheme combines emissions trading with the climate levy and NEAs. The main element to link the instruments was the introduction of rate-based credit trading. The gateway was introduced to avoid carbon leakage. This in return seems to increase transaction costs.
http://www.defra.gov.uk/environment/climatechange/trading/index.htm (last visited: 18 December 2001)

3. The French credit trading scheme

The French Kyoto commitment requires a stabilization of the country's 1990 emissions by 2008 to 2010. On a BAU approach, this corresponds to a 12 per cent reduction (~60 mt CO_2e) for the Kyoto period. Recent reports suggest that France might be below the 1990 GHG emission levels already, as a result of including sinks in the emissions account. However, as the use of fossil fuels in the transport, residential and services sectors is expected to increase, additional measures will have to be taken.

The centrepiece of the French climate change strategy was a new energy tax that was announced by the French government in January 2000. The tax system would have addressed a number of sectors, including industry, construction and other sectors. The tax was scheduled to start in 2001, with one tonne of CO_2 equivalent charged at €23–30. This price would rise towards the Kyoto period, and could be around €76 per tonne by 2010, according to official French sources. However, the tax was declared unconstitutional by the Conseil constitutionel on 29 December 2000, leaving the government without its key policy in place. It may be revived at the next budget, depending on the political constellation at the time.

The French government was to alleviate energy-intensive industries from this tax system, in exchange for NEAs for emission reduction. These NEAs would have been negotiated by both companies and sectors of industry. Targets within these NEAs can either be expressed in absolute (cap) or relative (energy efficiency target) terms. Energy can be exonerated from tax when used principally as a raw material and for electricity generation.

Within this context, a grouping of major French companies (Entreprises pour l'Environnement – EpE) together with government has attempted to develop an ET scheme that can be combined with NEAs and ultimately make taxation obsolete (EpE/MIES, 2000). As industry's contribution to the French National Plan for tackling climate change, the plan focuses on large industrial emitters. The basic approach is to negotiate a target expressed in absolute or relative terms at the level of each industry in an NEA. The sector target would then apply to companies belonging to this sector. Each sector would then be assigned a decreasing target based on a company's current situation up to the target value in 2008–12, whether it is an absolute or a specific value. The setting of the target would combine grandfathering with other criteria, including benchmarking whenever available.

NEAs would be linked with a credit-based emissions trading scheme; that is, companies that do better than their target can sell the accumulated credits. Those who do not meet their targets can buy credits. The overall targets for the sectors or industry as a whole can be met with more flexibility. This credit-based reward system would avoid the problems associated with allowance allocation and public opposition to ET due to it being seen as a 'licence to pollute'. This complaint was particularly strong in France, making the introduction of a cap-and-trade system impossible.

The proposal of EpE is basically a means to increase flexibility gained from NEAs further by adding emissions trading without jeopardizing environmental effectiveness. Since most NEAs were made on the basis of relative targets, the EpE proposal chooses a credit-based scheme. The scheme is also an attempt by industry to avoid taxation. Over time, it can be expected that more and more companies will accept absolute caps after they gain more experience with GHG abatement and get a better database. The 'carrot' to do so could be the potentially lower transaction costs of cap-and-trade programmes.

http://www.ceps.be/Research/WorKyoto Protocolarty/cop6/March%2016/Boyd.htm (last visited 18 December 2001).

4. The Norwegian upstream scheme

In 1990 the total Norwegian emissions of greenhouse gases was given as 52 mt CO_2e. Norway is allowed a 1 per cent increase from the 1990 levels under the Kyoto Protocol, corresponding to a 17.4 per cent reduction (~14.3 mt CO_2e) for a BAU emissions forecast.

The first discussions took place with the Norwegian Commission on Emissions Trading on the effects of the Kyoto Protocol on Norwegian climate policies in general and business and industry in particular to ensure that Norway would meet their Kyoto commitments in a cost- and environmentally efficient way. Following the Commission's recommendation, the government published a White Paper in June 2001 which lays out the basic proposal.

As already proposed in the Commission on Emissions Trading, the proposed scheme would cover all GHGs and as many sectors as possible, as well as carbon sinks. The units of trade should be standardized for full exchangeability. Using units of 1 tonne CO_2 equivalent (UNFCCC definitions) ensures that the global warming potential of emissions is accounted for. Permits should not be restricted to certain years, but traded freely after the introduction of an ET system (that is, full banking).

To achieve a large coverage, the scheme would include large industrial sources and covers the wholesale (that is, wholesalers, importers) level fossil fuels for heating and transport fuels. This amounts to a coverage of approximately 80 per cent of all GHGs. Although becoming mandatory in 2008, the scheme would be voluntary before then. The incentive to join is the possibility of exemption from the Norwegian CO_2 tax.[2] This may also be one of the reasons why industry is fully supportive of the scheme (see Bretteville and Søfting, 2000).

The Norwegian approach is different from that of the UK, which – as we have seen – combines different instruments, in that the objective in Norway is to base climate change policy on a comprehensive upstream emissions trading scheme that would cover 80 per cent of all emissions. However, it is similar to that of the UK and other schemes in using taxation as a stick. If passed as currently designed, companies that participate would not be subject to a CO_2 tax.

Sources: Høibye (2001), Haites and Mullins (2001).

5. The Swedish proposal for a hybrid upstream/downstream allowance programme

The total Swedish GHG emissions for 2008 are expected to be around 74 mt of CO_2e. This must be reduced by 8.3 per cent (6.1 mt CO_2e) under the Swedish commitments for the Kyoto period. This corresponds to a 4 per cent increase from 1990 levels under the EU burden-sharing agreement.

The objective of the Swedish proposal was to bring a domestic programme into line with the potential EU trading scheme (CEC, 2001a). By including emissions of the sectors potentially covered by the EU programme, the Swedish programme would only cover 30 per cent of 1998 emissions. Therefore it was suggested to include upstream at the wholesale level of fossil fuels for transport and households, bringing total coverage to 80 per cent of 1998 GHG emissions. The scheme would start with CO_2 and only include the other GHGs in 2008. Banking into the first commitment period would be allowed. Sinks should not be included, but reconsidered in 2008.

The Swedish currently have a CO_2 tax that would be replaced by the introduction of an ET scheme. This CO_2 tax should be phased out by 2005, and completely replaced by the ET scheme. It is recommended that industry not currently taxed on CO_2 emissions should be left out of an ET scheme to ensure the ability for these industries to compete internationally. Although this approach is almost certain to be contested by the European Commission on grounds that it might cause distortions to competition in the internal market, the Swedish approach is another example of how taxes are used as a stick to entice industry to accept absolute targets that are necessary for a cap-and-trade scheme.

There is a specific twist, however, in Sweden. Rather than grandfathering allowances, as done in most of the cases, Sweden opts for auctioning. Currently, the government thinks that quotas should be auctioned off according to the 'polluter pays' principle. The highest bid will receive the permit, and this supposedly leads to equal rules for all participants in the scheme. New entries are included in the scheme on the same basis as existing industry. In the Swedish case, nevertheless, it is reckoned that companies still had a net benefit. The government assumes that the auction of permits would generate less income for the government that the currently high CO_2 taxes in Sweden. This raises the question whether industry indeed will feel threatened by the taxation stick (CO_2) to accept the carrot (cap-and-trade ET based on auctioning). A further twist might be introduced by the worries of the Ministry of Finance about lower revenues. There is some talk that the Ministry might slow down the implementation of the ET scheme.

Sources: Hasselknippe and Høibye (2001); Haites and Mullins (2001).

Company Schemes

In addition to governments experimenting with domestic emissions trading schemes, a number of companies or sectors have developed emissions trading schemes.

6. BP Group initiative on greenhouse gas targets and emissions trading

In May 1997, the BP Group announced that it was taking precautionary action to control GHG emissions. The announcement was followed in September 1998 by a commitment to reduce emissions by 10 per cent from a 1990 baseline over the period to 2010. Finally, in January 2000, the then BP Amoco launched a group-wide emissions trading system to help meet this target. This internal trading system followed a successful pilot programme that ran from September 1998 to December 1999. BP has worked closely with the US group Environmental Defense in developing this system.

Emissions trading Group-wide emissions trading is a central tool to deliver GHG reductions both transparently and in the most cost-effective way possible. This system was extended to cover the former ARCO business units from the start of 2001.

A fundamental part of this approach has been to put in place rigorous protocols for monitoring, reporting and verifying GHG emissions throughout the Group. Another objective is to share experience that is gained. For example, BP contributes to the industry-wide emissions trading initiative that is under development in the UK.

The 10 per cent reduction commitment compares with an overall Kyoto obligation for developed countries of 5.2 per cent from the 1990 baseline by 2008–12, and covers all BP Group operations worldwide. This commitment is ambitious in that it will require an absolute reduction of about 40 per cent by 2010 from the trend of GHG emissions, taking into account the Group's projected growth. This is the equivalent of Austria's Kyoto target.

The BP approach turns the climate change target into a business target for each of the Group's business units. The emissions trading mechanism enables business units to incorporate the financial impact of GHG management into routine project planning. The motivational effect of these factors ensures that a business focus is applied to the GHG issue. According to BP, this approach 'focuses creative energy in ways that help identify unexpected economic opportunities'. Achieving the carbon target is an integral part of the performance contracts and bonus system for managers.

Technology development and diffusion An internal network called the 'Green Operations Network' (Green Ops) supports business unit efforts to identify and implement the most effective and economic GHG reductions by sharing best practice and undertaking R&D. In this way, options to make cost-effective reductions in emissions are made available throughout the Group. The network comprises around 800 members and operates a 'question and answer' forum for exchange of information and ideas across the Group.

Examples of technology improvements that are promoted by GreenOps include divided wall distillation, co-generation, and use of low-grade heat. Reductions are captured on a project database and used to refine the 'cost of abatement', information that is vital for the business units to be able to build GHG management into their business planning models.

Project teams are established within GreenOps to fill technology gaps. Over the next few years there will be increasing emphasis on new technologies (for example CO_2 separation from flue gases) and BP continues to develop less carbon-intensive energy sources with its solar business and its interest in other emerging technologies (for example hydrogen fuel).
http://www.bpamoco.com/downloads/340/ghg_emissions_trading_in_bp_may2001.pdf
http://www.bpamoco.com/downloads/446/GO_page1.pdf (last visited 18 December 2001).

7. The Shell STEPS scheme

Launched in January 2000 at 22 production units located exclusively in Annex 1 countries, the scheme accounts for 30 per cent of the Group's CO_2 emissions (110 mt). It forms part of the commitment of Shell companies to reduce GHG emissions by 10 per cent in 2002 compared to 1990 and by 5.2 per cent in 2010 compared with 1990. Permits were grandfathered, however, with 5 per cent being withheld for auctioning. The system is managed via a website by the Group's European Trading Manager. To ensure compliance, STEPS foresees penalties for those who do not reach the targets, in addition to the fact that the emissions situation will influence the yearly capital allocation among the different businesses and firms.

Shell's overall objective was to gain experience in how best to cope with the Kyoto target. This covers issues such as

- reducing emissions,
- introducing reporting measures,
- influencing business decisions,
- reducing customers' emissions,
- getting acquainted with the Kyoto mechanisms and finally
- increasing the public's awareness of climate change.

www.shell.com/steps (last visited 18 December 2001).

NOTES

1. The CCL will be charged at the following rates: electricity 0.65 US cents (0.43 pence)/kWh, coal and natural gas 0.23 US cents (0.15 p)/kWh, and liquified petroleum gas 0.11 US cents (0.07 p)/kWh.
2. For example tax rates (all in € per tonne of CO_2): gasoline (39), gas and oil on the continental shelf (39/34); light fuel oil (23); heavy fuel oil (19).

APPENDIX 2: A CASE STUDY OF A POLICY MIX: EU CLIMATE CHANGE POLICY

Area	Policy mix options [available for <€20 (€20 – €50)]
Energy: [349–376 (309–318) mt/year]	Consequences of **liberalization** of energy markets: [129–156 (185–194) mt/year] • *Fuel switching*: 88 (25) mt/year • *Fuels conversion efficiency*: 0 (100) mt/year • *Energy services*: 40–55 (7) mt/year
	Technical standards/regulation • reduction of *methane emissions*: 34 mt/year
	Consumer information • energy *labelling*: 10 mt/year
	Subsidies (coupled with regulation) • Promotion of *renewables*: 126 (74) mt/year • Promotion of CHP: 1–12 (53–64) mt/year • CO_2 *sequestration*: 0 (50) mt/year
	Industrial policy measures • Facilitating *turnover of capital stock*: 60 mt/year
Transport: [57–203 (8) mt/year]	**Fiscal measures** • *Infrastructure, use and charging*: 40–60 mt/year • *Road taxes*: 17 mt/year
	Transport fuel efficiency standards: 0 (40) mt/year
	Consumer information • awareness raising: 50 mt/year
	Negotiated Agreement (NEA) • Agreement with car industry on CO_2 *emissions* • Agreement with car industry on *Light Commercial Vehicles*: 0 (5–10) mt/year
	Engine technology: 30 mt/year
Env. performance of product/ processes: [45–175 (57–62) mt/year]	**Technical standards/regulation** • *Energy efficiency standards in buildings*: 50–65 (8) mt/year
	Green public procurement: 40–65 (8) mt/year
	Audit schemes, best practice: 40–55 (4) mt/year
	EU recommendation/guidelines for member state supporting action: 25 mt/year

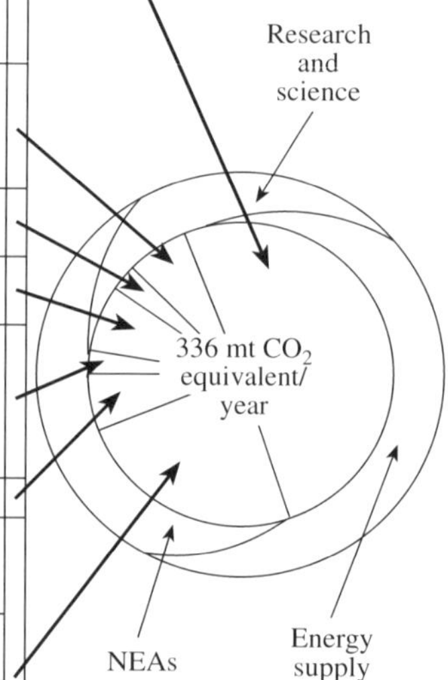

Source: CEC (2001b).

Figure 1A.1 EU climate change policy: individual policies and measures, and overall policy mix

2. Negotiated agreements and climate change mitigation

Patrick ten Brink, Marina Morère and Jane Wallace-Jones

1. INTRODUCTION

Aim of the Chapter

The aim of this chapter is to present state-of-the-art insights into the actual and potential role of negotiated environmental agreements (NEAs) for climate change mitigation – whether to help meet the Kyoto targets or simply to meet national commitments independent of the Kyoto Protocol.[1] It therefore spells out the key issues regarding the instrument's use, impacts and context, and also methodologies for its assessment. The chapter will therefore give details of where the tool has been applied, and will also include an analysis of the incentives for adopting an agreement, and incentives for action within the agreement; this includes a brief review of the basic risks and rewards.

Regarding the impacts, the chapter presents the available data on emissions reductions and also evaluates whether agreements offer sufficiently significant impacts to be considered a serious instrument to address climate change. Furthermore, it spells out the methods for the assessment of agreements and the limitations of these methods.

The chapter also assesses the linkage of an NEA to other instruments in the policy instrument portfolio, and briefly addresses the question of whether NEAs can be stand-alone effective policy instruments, or whether their effectiveness can only be a result of a linkage to other policy tools within a portfolio of instruments that governments use to address climate change.

Definition of NEAs

When people speak of voluntary environmental agreements or initiatives, many different names and tools come to mind: codes of conduct and responsible care programmes, voluntary measures such as self-declarations or commitments, implementation of (accredited) environmental management systems (such as ISO14001 or EMAS – Eco-Management and Audit Scheme), voluntary auditing, eco-labelling, voluntary environmental reporting, green purchasing and ethical investment, public voluntary programmes, technology support programmes, multilateral environmental agreements (MEAs), gentleman's agreements, covenants, and negotiated environmental agreements.

In turn these are categorized under different classifications of governance: co-regulation (Lévêque, 1996), joint environmental policy making (JEP, 1998), and cooperative environmental governance (Glasbergen, 1998; see also UNEP, 1994, 1997, OECD, 1998, UNEP/US EPA, 1997, and ten Brink, 2002 for discussions of different types of EAs).

This chapter's focus is on negotiated environmental agreements (NEAs), and those self-commitments and public voluntary programmes that contain similar elements – stakeholder involvement, objective setting, implementation – or insights of particular value to NEAs. This therefore includes the last three groups in the recent OECD classification (OECD, 1999), which distinguishes between instruments according to the parties involved:

1. Unilateral commitments made by polluters
2. Agreements achieved through direct bargaining between polluters and those polluted
3. Environmental agreements (NEAs) negotiated between industry and public authorities
4. Voluntary programmes developed by public authorities (for example environmental agencies) in which individual firms are invited to participate.

This chapter focuses explicitly on NEAs, which can be more fully defined as those commitments undertaken by firms and sectors that are the result of negotiation with public authorities and/or are explicitly recognized by the authorities. Traditionally they have been launched by (central and local) government or by industry.

NEAs containing targets fall into two broad categories. *Implementation NEAs* address existing and explicit objectives in national strategies or plans, while *target-setting NEAs* specify their own targets, though sometimes helping to implement a predefined objective (see EEA, 1997). Others (Bastmeijer, 1996 and Ost 1994 in Sairinen, 2000) broaden the definition of NEAs across three functions:

- *Bridging function* – where the NEA forms an intermediate step en route to legislation, and is replaced once legislation is in place (the Portuguese Pulp Paper agreement (EEA, 1997) can be seen as an example of this type of agreement, as can the Finnish Packaging Waste Agreement, which was launched in advance of EU membership (Sairinen, 2000)).

- *Supporting function* – where NEAs support the implementation of requirements under legislation (hence the implementation agreement noted above) or related to existing legislation (for example more demanding targets than related standards within legislation).
- *Independent function* – where the NEA's objective is not related to existing or future expected regulation, but where the existence of the NEA makes legislation unnecessary – therefore either in place of legislation or truly additional in cases where no legislation would be passed.

At this stage, most environmental agreements for climate change perform an independent function, with the exception of those that concern energy efficiency, which may have bridging or supporting functions.

Independent of whether they are target setting or implementation agreements, or whether they have bridging, supporting or independent functions, it remains possible to present a common set of basic 'building blocks' of NEAs. Figure 2.1 illustrates such 'building blocks' of an NEA (agreement, monitoring and reporting) and the linkages between these.

A further distinction in the types of NEAs can also be usefully made: some focus on emissions from processes and others focus on product-related emissions. This can have important implications for the design of the agreement and what targets can be set, and at what level agreements can be set. Within the EU context, national level agreements have focused on both process and product-related emissions, while EU-wide agreements have focused to date only on emissions from products.

Overview of Use of NEAs for Climate Change

In the early 1990s, many EU countries adopted NEAs as part of their national policy for CO_2 reduction and/or for improving energy efficiency. Since the UNFCCC was agreed in Rio, the use of NEAs for energy efficiency has increased (Chidiak, 1998; Storey, 1997). However, none of these NEAs was designed in view of binding targets. Examples of these pre-Kyoto voluntary agreements include:

- 'Energy Conservation Agreements' established in 1992 and 1997 in Finland (Sairinen, 2000);
- the 'Long Term Energy Efficiency Agreements' (1992) in the Netherlands (see Dutch Ministry of Economic Affairs, 1998);
- 'Eko-Energi Programme' (1994) in Sweden;

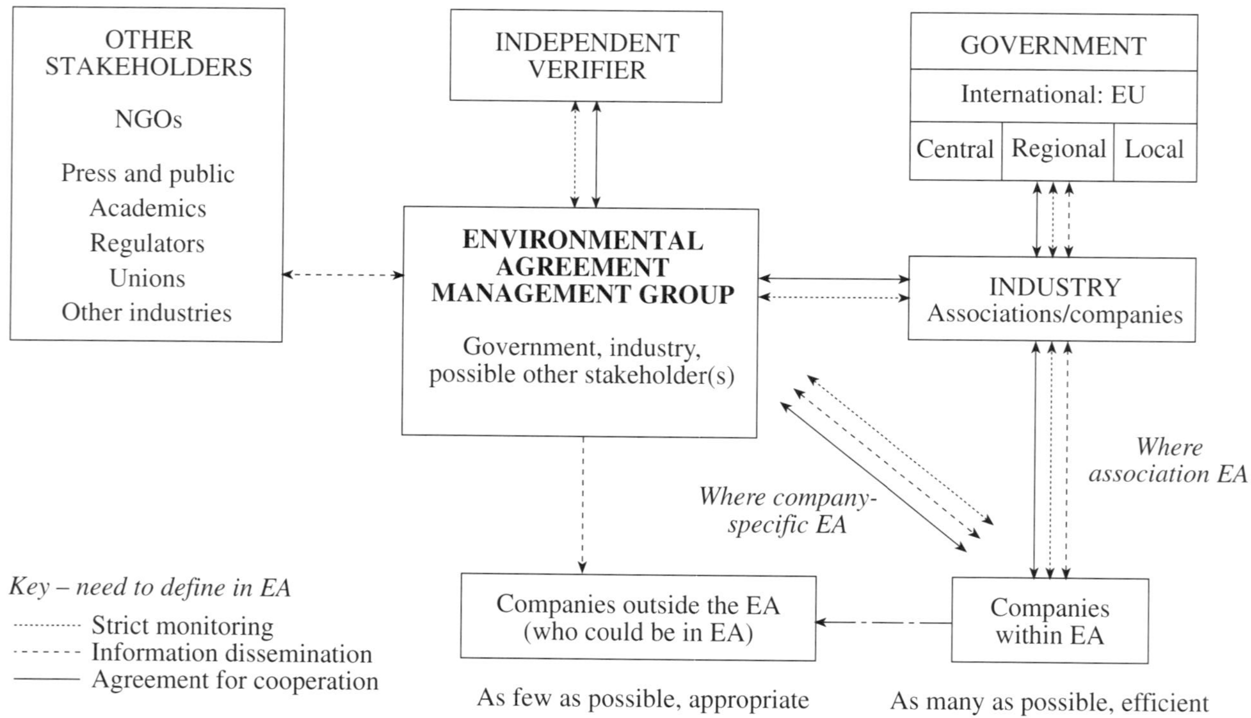

Source: ten Brink (2002).

Figure 2.1 NEA stakeholder involvement

- the 'Declaration by German Industry on Global Warming' established in 1995, then amended in 1996 and 2000 in Germany (see BDI, 1995, 1996a, 1996b, 1999 and Eichhammer and Jochem, 1998 and see note for 2000 commitment[2]);
- the agreements on energy efficiency improvements in the UK (see DETR web page[3] and Salmons, 2000);
- the 'Agreements on CO_2 Reduction and Energy Efficiency' (1995) in France;
- the Danish 'Agreements on CO_2 Emission Abatement' (1996–97) in Denmark (see Millock, 1999);
- the US 'Climate Challenge Program' and other American initiatives: Green Lights, Energy Star, Climate Wise (see ten Brink, 2002);
- the Japanese 'Keidanren Agreement' (see ECOTEC, 1999, and Keidanren web page[4]);
- the Australian 'GHG Challenge Programme' (see AGO web page); and
- the New Zealand 'Climate Challenge Programme' and industry voluntary agreements.[5]

The European experience is illustrated by Chidiak (1998), ten Brink et al. (1999) and ten Brink (2002).

In contrast, since the Kyoto Protocol, agreements are being defined with consideration for a priori targets. Among these new initiatives are:

- Switzerland's 'Action Program Energy 2000';
- Italy's 'Climate Pact' between government, industry and NGOs (November 1998), which specifies that NEAs are to be the favoured policy tool to address climate change;
- the UK's climate change levy (CCL), which defines a group of NEAs (known as climate change agreements, or 'umbrella agreements') as complementary measures. The umbrella agremeents, approved in 2001,[6] have been designed as a complement to the recently established CCL which is to be used in conjunction with the domestic tradable permit system. This policy mix represents the main tool that will be used in the UK to meet the Kyoto commitment. There are now over 40 such NEAs in the UK;[7]
- the Dutch benchmarking agreements;[8]
- the European Commission recognized the ACEA/JAMA/KAMA commitment – the agreement with automotive industries on specific CO_2 emissions from new cars (1999 and 2000 – the Japanese and Korean associations signed up in 2000). (See CEC, 1999 for Recommendation 1999/125/EC for ACEA.)

In addition to these agreements, a number of companies have also launched purely self-commitments (no negotiations and no discussions). Two key examples are Shell and BP. Shell promised to cut greenhouse gases (GHGs) by at least 10 per cent from 1990 levels by 2002 (ENDS Daily, 16/10/1998). In addition, these two companies have also launched voluntary emissions trading initiatives in order to understand the likely mechanism of this particular Kyoto instrument (Torvanger and Skodvin, 2002).

It is clear that there are some additional trends in the use and design of NEAs. First, not only is the number of NEAs increasing (and particularly for climate change and energy-related issues), but also the sophistication of the agreements is developing. NEAs are also increasingly being used as an instrument integrated within a portfolio mix, with linkages to taxes and to emissions trading and Kyoto mechanisms beginning to be seen. Furthermore, as noted by Convery and Lévêque (2001), there appears to be an evolution from a situation in which the targets (if any) are set under the leadership of firms, to one in which government agencies are the predominant shaper of the targets to be achieved by the voluntary agreement. Furthermore, voluntary agreements have become more demanding over time in terms of quantitative targets, delivery and reporting mechanisms – not just for next-generation NEAs, but also as part of the process of development of existing NEAs. These issues are explored within the selection of case studies and discussed in more detail throughout this chapter.

Selection of Case Studies

To obtain the necessary insights from practice, this chapter focuses in depth on five case studies, and makes reference to other practice where pertinent. The five case studies chosen for study are:

- the 'Declaration by German Industry on Global Warming' established in 1995, then amended in 1996 in Germany, and again in 2000 – of interest given its broad coverage of economic sectors, its role as a process and the improvements over time;
- the Danish agreements on CO_2 emissions abatement – of specific interest as linked to the use of a tax);
- the UK climate change agreements – notable as they are linked to the UK climate change levy and a key initiative within the UK;
- the Dutch LTAs – of specific interest given the importance of the LTAs as vanguard agreements and also of interest given the link to operation permits;
- the Italian agreements – of interest as they have been elected, in spite of the lack of historical experience in their use, as a favoured instru-

ment for meeting national CO_2 reduction commitments. We focus mainly on the Pact for Energy and the Environment and the more recent ENEL agreement, drawing on elements in the Montedison agreement at times.

Reference to other practice will be made where relevant.

Structure of the Chapter

Sections 2 and 3 focus on the incentives for participation and the risks and rewards for the various stakeholders. This is important for policy makers to assess *ex ante* whether an NEA is appropriate given their situation, and if so, what measures need to be put in place to minimize the risks, maximize the rewards and hence help ensure that the agreement fulfils its potential. Section 2 presents the theoretical aspects and Section 3 the empirical results.

Section 4 presents the evidence regarding the actual impact of the agreements for climate change, looking into the evidence of energy savings and reductions in emissions of greenhouse gases. Section 5 looks at the design of the instrument, and notably how the instrument is linked to other instruments and policies. Section 6 discusses evaluation methods and associated good practice guides, leading to Section 7, which presents a summary of the evaluation of NEAs for climate change and preliminary conclusions.

2. INCENTIVES FOR PARTICIPATION AND RISKS AND REWARDS: THEORETICAL ISSUES

This section begins with a brief review of commonly cited arguments for and against negotiated environmental agreements as environmental policy instruments and then, drawing from theoretical and empirical literature, describes in more detail the types of risks and rewards these instruments *potentially* present to key stakeholders: government, industry and NGOs.

The section closes with an examination of the types of incentives used in the case study NEAs, both those incentives which, first, bring about participation in an agreement and, second, those that encourage the implementation of the agreement.

Interest in the Instrument

Environmental agreements and in particular negotiated environmental agreements are increasingly being looked to as a possibly appropriate instrument to help address environmental problems covering a broad range of pollutants and natural resources. The interest has developed in the context of increasing concern that command-and-control legislation and regulation can be too burdensome, and that use of economic instruments, such as environmental taxes, can be too costly, and thus politically and socially unacceptable, for the industries in this rapidly globalizing market. NEAs were seen as potentially being an instrument that could be finely tuned, quick to set up, and able to achieve environmental objectives at lower costs.

The motivations were not only to avoid arguably cumbersome and slow-to-develop legislation or costly taxes, but also reflected a number of other issues:

- the concept of shared responsibility, and the related concept of shared uncertainty, for the environmental problem, building on industry knowledge of what can best be done to address the problem that relates to industrial activities;
- the concept of stakeholder involvement – as it is the stakeholders who know best how they affect or are affected by the environmental issue. This relates to the issue of equity as affected parties should arguably have some means, beyond recourse to law courts, to influence activities that concern them;
- the principle that a problem should be solved at the level at which it can be most effectively addressed – local initiatives are (often) more appropriate than centrally coordinated ones;
- the more ideological interest of reducing the size of the state – this has led to an interest in this instrument, as less regulation and enforcement input would arguably be necessary for the public sector to ensure that the same environmental objective was reached, where the private sector assumes a share of the responsibility for regulation;
- the interest in learning and capacity building, as cooperation can lead to much greater awareness within industry of the measures that could valuably be implemented, could engage employee motivation, and lead to a much increased knowledge on what can be achieved by the regulatory authorities, as well as build formal and informal networks across regulating bodies, environmental institutes, industry and unions, and NGOs.

Generic Arguments for and against NEAs

It is not possible to have a single list of arguments for or against EAs that is applicable to all EAs uniformly. Whether the generic arguments for and against NEAs, as noted in Table 2.1 (for the specific EA of negotiated environmental agreements), are applicable depends on the situation for each EA itself, the context within which it is negotiated and implemented, and the

Table 2.1 Generic arguments for and against negotiated agreements

For	Against
• Flexible, and can be tailored to the particular circumstances for each stakeholder and also to a specific location; can build on business's particular knowledge of its realistic capacity to reduce GHG emissions, and respond to particular requirements of a locality (and hence bridge the gap between national legislation and local needs)	• Can be too flexible if not defined properly and not stringent enough as a result. This may lead to lower effect and effectiveness than otherwise possible with other instruments. This is a case of regulatory capture, where government is persuaded into accepting a potentially second- or third-best instrument
• Participatory, involving those affected by the agreement in negotiations and implementation (third-tier verifier, public)	• Can be exclusionary to the extent that not all affected parties might be represented in negotiations and some valid concerns might not be reflected in the NEA[a]
• Potentially easier to achieve agreement and quicker to implement than alternative instruments	• Negotiations can take time. Agreements might be used as a means of buying time through forestalling regulation
• Low-cost implementation; for companies NEAs can represent significantly lower costs than taxes	• Can be costly to implement: transaction costs and monitoring activities can be resource-heavy (especially for NGOs and internally for a sector with many companies)
• Cost-effective, as the industry can choose the initiatives reflecting the internal knowledge of the hierarchy of cost of initiatives; encourage innovation through cooperation between firms	• NEAs, where they are chosen as a substitute for environmental taxes, can imply that the polluter pays principle (PPP) is not upheld, as polluters do not pay for the remaining pollution

Table 2.1 continued

For	Against
• Encourage dynamic efficiency, through setting renewed/renewable targets without requiring an explicit set of measures	• Can limit the pace of technological change/innovation to rates which industry is comfortable with – depending on the stringency and enforcement of the targets and whether alternative instruments have been substituted. This might only lead to static efficiency gains if targets are not set appropriately and are not revisable
• Brings industry on board, and can encourage greater appreciation of environmental issues. If appropriately designed, clarifies and formalizes responsibilities, rights and roles	• Need to monitor compliance, which can be resource-intensive (especially if contracting an independent verifier). There also needs to be a continued interest and pressure from public authorities and NGOs to ensure that the NEA can develop and improve to allow targets to be met and additional potentials to be realized
• Greater stability provided through long-term planning and objectives set internally rather than surprising external development	• May encourage free-riding – depending on institutional context, NEA design, the level of incentives/enforcement and the size and structure of industry
• Contributes to the development of regulatory capacity through the increased awareness of impacts, measures to address impacts, and through the development of formal and informal networks available to the regulatory agencies	• Can lead to 'non-level playing field' for signatories and non-signatories. This can be problematic where (a) some parties are excluded from being signatories to agreement that benefit signatories and (b) some parties are not included where

Table 2.1 concluded

For	Against
• This can compensate for limited capacity, and also allow regulatory cost savings	signatories face restrictions, potentially compromising the sustainability of the agreement
• Can be a useful tool for conflict resolution, and can address issues that are not covered by existing regulation or regulatory powers • Can also help implement the precautionary principle, where there is a lack of scientific evidence required to allow taxes or regulation to get passed	• Can lead to reduced quality of regulatory control, where a move towards co-regulation leads to reduced regulation by the state and where the increased regulation by the private sector is inadequate. Another case of regulatory capture
• Accelerated behaviour change towards sustainable development practice, building on 'shared responsibility' at all levels of industry/economy	• Does not raise revenue that could be valuable for other environmental initiatives

Note: [a] For example where no NGO or local community is involved in consultation and local community concerns are possibly ignored.

Source: ten Brink (2002) adapted from ECOTEC (1999).

initiatives launched (in the design and during implementation and update of the EA) to ensure that the positive aspects are realized and the negative aspects avoided. Each EA is deemed to evolve over time and experience is gained by the players, thus rendering these arguments useful for appraising the potential value of a EA, but not fully applicable to a process.

Generally speaking, however, to the extent that NEAs are negotiated, designed and implemented according to good practice guidelines, the arguments against will be minimized, though in many cases these arguments can still be critical, and the NEA might not, in some cases, be the optimal instrument. The final choice depends on the exact nature of the environmental objective, the nature of the industry which would be a partner to the NEA, the NEA itself, and the institutional and policy instrument context. This requires a case-by-case discussion, which is also seen in the context of the individual players – government bodies, industry and non-governmental organizations (NGOs). It is indeed beneficial for the future launch of an NEA for each stakeholder to assess *ex ante* the risks and rewards of participating in the NEA. It will enable each of them to raise potential problems/solutions during the negotiations, which might prevent the design and launch of an ineffective NEA.

Risks and Rewards of Choosing NEAs

Risks and rewards to government

Rewards The key reward of NEAs to government is that, if the NEAs work, industry can progressively introduce its own initiatives, building on the particular strengths and opportunities within the industry. To ensure that industry does progressively introduce its own initiatives, it is key that the design and implementation of the NEA ensures that roles and responsibilities are adopted at each relevant part of the industry, thus ensuring that the particular strengths and opportunities are realized. This requires appropriate communication strategies and responsibility allocations, as well as initiatives to encourage the 'greening of people' (see Wehrmeyer, 1997, Strachen and Clark, 1997 and Legerwood, 1997).

A further reward to (some) governments is the value of the NEAs as a process supporting the institution building of the ministry and other public agencies involved, and the increased understanding and trust that can be developed through potential cooperation on the agreement. Institution building can be optimized by involving local authorities and local NGOs. Extensive participation in EAs can lead to a broadening and deepening of the regulatory capacity through the involvement in and access to formal and informal networks of experts dealing with the issues.

Risks The key risk is that of regulatory capture by industry – whereby the industry implements a NEA which represents little more than business-as-usual (BAU), and at the same time persuades the government to drop otherwise more effective (from a CO_2 reduction perspective) alternative instruments. As industry is considered to be one of the most powerful interest groups because of its high incentives to influence, its low organizational costs and its high financial resources relative to other interested parties, an increased likelihood of regulatory capture through the use of NEAs has been one of the aspects most often considered (OECD, 1999), at least from a theoretical perspective.

The risk is clearly dependent on the negotiating skill of the parties concerned, the information asymmetry that might exist between industry and the government negotiation party, the presence/absence of a watchdog (NGOs, academics, press), the organization of government agents involved, as this can favour or limit collusion between industry and agent interests, and the importance of obtaining an agreement. This risk is real, but even with a single NEA it need not be permanent, if appropriate mechanisms to monitor, review and update the EA are in place. Furthermore, with appropriate preparation before negotiations, the risk of information asymmetry can be at least partially addressed.

It is therefore important that there remains an option for EAs to be updated and developed as the results become clear, and also as the understanding of the practical measures to reduce emissions deepens, and the cost implications become clearer.

A good example of this issue is the German 1995 climate change agreement, whereby the government effectively agreed to drop the plans for a carbon-energy tax (and waste heat ordinance) on the grounds that a set of associations would implement an NEA that, in actual fact, did not guarantee significant reductions in emissions. Following significant public criticism and pressure, the agreement was reviewed and updated to include more stringent targets, and a monitoring and independent verification system which would help ensure that significant CO_2 emission could be achieved.

Other risks include the risk of protracted and complex negotiations leading to high transaction costs, possibly a risk of reduced compliance with legislation, reduced revenue where NEAs replace taxes and charges, and a potential conflict with the polluter pays principle (PPP).

Risks and rewards to industry

Rewards There are many rewards for industry in implementing NEAs. There are clearly public image benefits, generally across all NEAs, though as the public becomes more discerning, and industry more environmental friendly, some of these offer weaker 'green' credentials than others. There are certainly some geographic and cultural differences here, and in industries where there are few environmental reports, there are still some first-mover advantages to be had in producing such a report. The public communication of stakeholder initiatives is key to ensuring that these benefits are realized.

Probably the biggest reward is the potential for reduced regulatory burden and reduced tax burden – where NEAs are chosen instead of new regulation or new taxes. Industry thus has the benefit of being able, through its own commitments, to influence government policy, in particular, by pre-empting specific public regulation (see Maxwell et al., 1998 and Brau and Carraro, 2001). These initiatives should be applauded to the extent that the commitments truly reflect good environmental practice, and truly offer at least an equal (from a results point of view) to the alternative regulation or tax. The main method to ensure that there is, and will remain, a reduced regulatory or tax burden is to develop and implement an NEA that is credible and produces results at least comparable to the alternative instruments.

For individual firms, NEAs can bring significant input savings and sales increases if they are effective enough. These may come from product differentiation on environmental quality and the communication of this to consumers via advertising and labelling (see Brau and Carraro, 2001). Either these goods can be sold at higher prices or the firm is rewarded by an increase in market share due to consumer preferences for greener products. The role of information is clearly crucial to obtaining these benefits.

Improved trust with government and government agencies are other key rewards, sometimes leading to easier licensing and fewer inspections, which can translate effectively into a reduction in administrative costs for firms.

Risks The main perceived risk for industry – if the NEA involves cooperation with government – is that the increased knowledge of the government of the true level of environmental performance might lead to the imposition of sanctions, especially where non-compliance with legislation is not atypical of the companies in the sector concerned. This can be addressed by granting a moratorium, or derogation period, on compliance for a period of time during the NEA (this has been the case in Mexico and in Portugal).

A further key concern is that of loss of potentially commercial sensitive information to potential competitors, or indeed investors. This can be addressed through designing appropriate monitoring and communications mechanisms, with a clear distinction between internal and publicly available information (as done in Japan – Keidanren – and Australia, where only public profiles are publicly available), and confidentiality clauses clearly input on internal information. This should only be done to the extent that truly sensitive information is protected; otherwise it can compromise the transparency and, ultimately, the credibility of the NEA.

The most widely discussed risk to industry and to the economy as a whole is that of free-riders – where certain companies make the environmental investments allowing the association or sector targets to be met, while others make no such efforts and rely on the initiatives of the pro-active group within the industry. The risk of this depends on how many players there are in the industry, whether there is an association NEA or a company-level NEA, whether the association has a large representation (to address free-riders external to the NEA), and whether the association has significant influence on its members (to address free-riders internal to the NEA), and finally, whether appropriate monitoring and sanctions are in place. The level of risk also depends on the nature of the NEA, as the incentives to free-ride are significantly lower (often zero) in 'win–win' NEAs – that is, where involvement leads to economic gain as well as environmental improvements. For stricter NEAs, which imply true incremental initiatives, the risk is larger. Free-riding can be reduced if clear incentives exist, compliance systems and sanctions for non-compliance are clear and effective, and in some cases if individual targets are in place and individual responsibilities made clear and communicated.

Risks and rewards to non-governmental organizations (NGOs)

Rewards The greatest reward of NEAs to NGOs is that if NEAs are effective, they can be a valuable instrument to influence industry's environmental behaviour towards sustainable development practices. An NEA can also help with the institution building of the NGO, given that they may achieve a better understanding of the industry's priorities and abilities to address the environmental aspects and impacts of their activities. Furthermore, the NEAs themselves often contain significant information dissemination and awareness initiatives, complementing and supporting the mission of the NGOs. Involvement of the NGO in an NEA, and the consequent building of trust between the other stakeholders and the NGOs,[9] can support this process. This is of great importance in developing countries.

Risks The key risk in participating in an NEA is that of loss of reputation and perception of independence of an NGO, and consequently effectiveness of NGOs generally – where they are associated with an NEA that proves to be ineffective. A further risk is that the involvement of the NGO in the NEA might lend 'green' credibility to a company that might not merit this label;[10] this is especially so for the free-riders within an NEA. Finally, the NEA can be quite a costly initiative to be involved in, and consume a substantial amount of the NGO's resources without necessarily offering commensurate returns in environmental improvements. Early clarification of roles and responsibilities, burden sharing and a phasing of collaboration can minimize this risk; clearly an NGO leaving an NEA can cancel the initial public image of NGO involvement.

Another risk to NGOs posed by many NEAs to date has been the exclusive manner in which negotiations are carried out. Traditional regulation is subject to the rules of democracy and thus under public and political control (Mortensen, 2000), while one of the features which is said to render NEAs attractive is the speed with which they can be concluded, a feature which is obviously influenced by the number of interested parties at the negotiating table. In fact, negotiations are often seen to take place between government agencies and industry, a tendency which runs counter to the general increased emphasis on public participation and transparency, excluding parliament and environmental organizations – which are often considered to represent the public interest and are less vulnerable to the problems associated with capture (Krarup, 2001). NGOs may also be excluded (by their own wishes) as a result of the high transactions costs associated with negotiations (see later section on transaction costs).

It is clear that a type of democratic legitimization is needed and this may be a parliamentary validation (see Mortensen, 2000) such as is used in the Netherlands – which, however, does not guarantee any involvement from the public or stakeholders. Both Mortensen (2000) and Barth and Dette (2001) suggest that instead of negotiating with third parties, a hearing process, thus still keeping the number of negotiators to a minimum, could be used and possibly foreseen in a limited legal framework of basic rules for NEAs. Bailey (2002) suggests that the use of a prior notification built on the practice of environmental impact assessments (EIAs) could help offer a cost-limited route for participation and hence democratic legitimization. Other public participation and conflict-reducing tools such as consultation periods (for drafts of the agreement) and NEA public notifications could also be useful in this respect, and thus somewhat resolve the issue of potential NGO exclusion and costs associated with negotiating. In each of the cases of public discussion of a draft agreement, there will be a clear need to ensure that no undue difficulties and delays to the negotiations occur. It is therefore important to ensure that the design and implementation of these mechanisms is right.

The Economic Analysis of Voluntary Agreements

The economic analysis of voluntary agreements – based on principles of industrial economics, political economy, law and economics – provides some insights into the rationale for participation by industry. Unlike traditional instruments such as environmental taxes and permits, voluntary approaches by industry were devised by practitioners and not by theory (see Lévêque, 1999). Hence the economic analysis of voluntary approaches is based on an *ex-post* analysis of the instrument, using concepts from different economic theory areas. Two major elements in the use of voluntary approaches include the issue of regulatory gains and capture and the issue of free-riding behaviour among the participants. This section briefly presents the main results concerning these two issues, as well as touching on the issues of transaction costs and competition, which, however, are not commented on in depth in relation to the case studies due to a lack of relevant data.

Negotiated agreements are often based on collective action between firms and public authorities, where the economic actors cooperate to obtain a higher gain as compared to the costs. Indeed, regulation by government to tackle externalities from pollution due to market failure infers an abatement cost on the individual polluters. Collective action can offer the potential for lower cost, hence higher gains for industry. However, some individual polluters may not see an incentive to participate in the collective action because they can benefit from the gains without contributing. This is the well-known case of *free-riding*. In some cases free-riding is practically irrelevant if the targets are not tight and can be achieved by a only a small set of the players implementing no-regrets actions. When, however, the targets are tight, cooperation becomes a more necessary condition for sharing the gains across polluters, which involves bargaining between the individual participants. If there are legal obstacles to inter-firm cooperation, such as competition law, or to regulatory agencies to negotiate/sign an agreement with the private sector, then negotiated agreements might not take place. But where NEAs can take place, it is worth exploring what theory has to say about their nature and use. Below we discuss in more detail the important issues of: regulatory gains and capture; free-riding; transaction costs; and competition aspects.

Regulatory Gains and Capture

Why would a polluter participate in 'voluntary' collective action to reduce pollution beyond compliance to regulations in place in one country? This voluntary action would create an additional cost burden for the firm, including environmental management, capital investment to make the production process cleaner, human resources for monitoring, and so on. This is reduced by the efficiency gains of the agreements (for example cost-effective investments with real rates of return). Any net costs will increase the production cost of the firm, hence the price of the products or lower profits (and the converse is true if economic gains are larger than costs). Economic theory says that a firm would then participate if, and only if, the firm gains equal or higher benefits from voluntary collective action (see Lévêque, 1999) than under existing regulation and instruments or under alternative new regulations and instruments.

Indeed, a potential gain from voluntary collective action can be derived from the avoidance of regulation and the associated costs. NEAs can, in principle, reduce transaction and compliance costs to industry. Maxwell et al. (1998) propose four strategies used by industry to influence the regulators, and hence obtain what theory calls 'regulatory gains':

- pre-empting tougher government regulations or regulatory instruments (for the former, standards; and for the latter, taxes);
- weakening forthcoming regulations if full pre-emption is impossible (for example weakening regulation, gaining exemptions to a tax for signatories);

- reducing the extent of monitoring and inspection by regulatory agencies; and
- signalling regulators to persuade them to raise (non-signatory) rivals' costs.

Whatever the true rationale for participation is, a firm or an association will participate in an NEA if the NEA allows them to minimize their abatement costs (or maximize their (net) benefit) relative to the other policy options being considered. This can be the case if the agreed NEA target is lower than what foreseen regulation will require achieving – or allows a lower-cost means of achieving the target. Greater flexibility in ways to internally implement pollution reduction measures to meet the target under an NEA can reduce compliance costs as compared to legislation, which often requires preset solutions. Industry can choose the most cost-effective means to achieve the target. These benefits are hence called regulatory gains, as they are derived from avoiding regulatory action and the associated compliance costs.

Apart from results on benefits to industry, theoretical work on modelling negotiations between government and industry suggests that negotiated environmental agreements offer the potential to increase social welfare in many cases. But in some cases, social welfare can also be reduced and the society at large could be less well off. In particular, social welfare can be reduced if the marginal effectiveness of lobbying by consumer and environmental groups is reduced by industry's lobbying expenditure and if regulators accept weak agreements and therefore do not maximize social welfare (see Maxwell and Lyon, 2001).[11]

The fact that industry is a powerful interest group, and considering the potential for regulatory gains explained above, does raise the question of regulatory capture as industry could indeed use NEAs to 'capture' environmental policy. If industry succeeds in blocking the introduction of more stringent (and realistic) environmental policy instruments by signing an NEA with the government, then environmental policy has been captured by industry. In theory, there is regulatory capture when regulatory costs are equal to zero for industry, for example the NEA target is close or equal to business-as-usual abatement, or where they are less than the expected costs of alternative regulation without compensating social benefits.

Regulatory capture is made more likely if there are existing information asymmetries, as the regulator will find it difficult to ascertain what it is possible for firms to do without incurring significant costs. This may lead to weaker targets being negotiated. There will always, of course, be information asymmetry and always some element of capture, but this is the case with all instruments – NEAs, regulation and taxes. However, the negotiation process within the NEA does offer a particular opportunity for firms to exploit the benefits of information asymmetry, aided by the fact that consumer and environmental interest groups often do not have the resources to counter this threat, given the high transaction costs of involvement.

However, the NEA, if designed appropriately, can lead to real regulatory capacity development as the information asymmetry is eroded. This requires the appropriate design of monitoring, reporting and evaluation/verification systems. In other words, NEAs can be information-revealing instruments, though they will offer different information than taxes, which are a key instrument for revealing information on abatement costs. There is therefore a risk that the information asymmetry will be reduced less by an NEA than by a tax or charge.

Free-riding

Theory suggests that there is an incentive for free-riding in the case of collective action (for example an association agreement with a common association target within the NEA, not supported by complementary individual targets). Here, theory suggests that some members of the association would choose the option of doing nothing or little, while the others' activities lead to the collective target being met. Thus, in an initial agreement there is some implicit assumption of 'burden sharing' to allow the collective target to be met, but with free-riding this burden is shared less equally. In other words, there is collective regulatory gain, and free-riders will seek to maximize their share of this regulatory gain. It should be noted that for some agreements the gains are larger than the burdens, so the issue of burden sharing and free-riding is less important. It could be expected that for some (less demanding) NEAs, those companies that are active will not care about the fact that some companies are free-riders.

Free-riding is not only an issue in the implementation phase of an agreement, but has also been found, at least in theory, to influence the signing of an NEA in the first place. In fact, in their economic analysis, Brau et al. (2000) explore the influence of potential free-riding on concluding a collective agreement. They found that NEAs will not be concluded in a context in which all firms in the sector (independent of whether signatories or not) benefit from the abatement efforts of the signatory firms.[12] More pertinent for future NEA design is their finding that, when assuming that benefits go to the firms who sign the NEA and also partly to the free-riders, the regulator has to impose a minimum participation constraint for the NEA to be signed. In their work Brau et al. found that if the minimum participation constraint is met, all firms have an incentive to sign the NEA. Key issues for free-riding include:

- the incentives for free-riding depend critically on how difficult the collective target will be to reach, as easy targets mean that the body of the collective will not be directly affected by some free-riders, if the non-action by the free-riders has no effect on the collective target being met – if the burden being shared is low, then the concern for a firm not assuming their responsibilities is less;
- the abatement costs, for if these are high, the incentive to free-ride is high;
- the existence or absence of penalties or sanctions – whether internal or external. Internal penalties can be ‘soft’, as in peer pressure, or ‘hard’, if there is an internal accountability rule. External penalties depend on whether the agreement is legally binding, and what the penalties for non-compliance are. These again can be ‘soft’ where no real direct cost implications are to be expected (for example loss of good will, or ‘name and shame’ publicity), or ‘hard’ where significant penalties arise, such as loss of exemption from taxes;
- the likelihood of free-riders being caught. It is clear that in large collectives, there are difficulties in identifying free-riders (given transaction costs and difficulties in obtaining proof, especially in the case of significant information asymmetry). In pure economic terms, the incentive for free-riding then exists if the probability of being caught, times the penalty, is less than the expected regulatory gain. Some firms will be sensitive to the ‘signal’ given by the penalty, independent of the cost, especially where there is a public mention (type of name and shame) of the free-riders.

The risks of free-riding can be minimized by careful design of the NEA, including not just the monitoring, reporting, evaluation and verification procedures, but also the mechanisms in place for accountability and sanctions. There is clearly a balance between what can be done by the regulatory authority and what can best be done by the collective itself. It is valuable to engage collective self-interest in ensuring no free-riders, as this reduces regulatory costs and encourages most sustainable self-regulation. An NEA design option, which avoids the free-rider problem, is that of concluding bilateral agreements between individual firms and the government as opposed to using collective agreements with targets applying to an entire sector or group or firms. The Danish energy efficiency agreements described here are exemplary of this approach. However, the elimination of the free-rider problem and the obvious reduction in transaction costs related to within-sector burden sharing is potentially replaced by that of increased transaction costs for government.

Transaction Costs

Transaction costs refer to those costs associated with the negotiation, burden sharing and administration of NEAs (including monitoring and reporting), and are not to be confused with those costs typically associated with implementation, or rather, meeting the targets of NEAs.

NEAs are often attributed a cost advantage with respect to traditional regulation due to, on the government's part, reduced monitoring and enforcement costs, and on the part of industry, increased flexibility in finding solutions to the environmental problem being tackled. However, as stated by Krarup (2001), the relative cost advantage of agreements can be questioned on several grounds. If the transaction costs of agreements approach the costs related to other types of regulation, the comparative advantage of an agreement diminishes.

This comparative advantage may be seen to decrease if, as is often called for, there is a 'higher degree of structure and a system for monitoring and enforcement by the responsible agencies in place' (Krarup, 2001). Thus, in some cases, environmental effectiveness of an agreement and related transaction costs may prove to be inversely proportionate – in other words the effectiveness decreases as the transaction costs increase. This raises the question as to where the balance should lie.

Krarup further notes that the transaction costs for the agency depend on the type of agreement considered. In the case of agreements with individual firms, the agency has higher transaction costs compared to a situation with branch agreements. This means that the cost burden is shifted from industry to the public sector for agreements with individual firms. However, in both cases there is a trade-off between a more efficient burden sharing among firms and the level of transaction costs.

Finally, it is possible to discern a broad relationship between the number of players in an agreement and the transaction costs: the cost will increase with the number of communication channels and the degree of rivalry. In other words, the cost is lower when we have a large degree of economic concentration and small heterogeneity between companies, that is, when we have a small number of participants and interests involved in the negotiation, which represent a small number of communication channels and of conflict focuses – see Cabugueira (2002).

Competition

Questions raised over the potential impact of voluntary approaches on competition are by no means recent. The 1996 Communication from the Commission on environmental agreements highlights the need to ensure compatibility between environmental agreements and the articles of the EC Treaty regarding the protection of competition within the EU. More recently, a number of theoretical studies have been carried out (including within the European Commission DG Research-funded CAVA project – concerted action on voluntary approaches). These have explored the relationship between voluntary approaches, mainly NEAs, and competition. The question is not confined to theory, though Brau and Carraro (2001) state that there are fewer studies which report on the likely conflict between NEAs and free competition, partly as a result of the fact that many of the *circa* 40 environmental agreements which have been notified to the Commission are still under scrutiny and no official information is available on them.

The salient points of the CAVA project policy brief (Brau and Carraro, 2001) concerning voluntary approaches, competition and market structure are:

- Voluntary approaches, at least according to economic theory, have indirect effects on market structure and competition whether they are principally aimed at improving the environmental reputation of the products sold by participants to the agreement or at pre-empting policy interventions foreseeing regulatory or economic instruments.
- Economic analysis provides support to the intuition that in a few cases voluntary agreements could be adopted with the strategic objective to affect market structure and competition conditions.
- In general, a concentrated market structure has a positive effect on the environmental effectiveness of the actions undertaken within a voluntary initiative. This clearly involves a trade-off between the goal of maintaining competition in the market and the objective of exploiting the well-recognized flexibility that generally characterizes voluntary approaches.

3. INCENTIVES FOR PARTICIPATION, RISKS AND REWARDS: EMPIRICAL ISSUES

Risks and Rewards – Minimizing Risks and Maximizing Rewards: Case Study Summary

Key information on perceived risks and rewards, as well as measures for minimizing such risks and maximizing rewards, is noted for the case studies. The rewards noted below are clearly complemented by the potential reward of the emissions reductions promised to be achieved; this is explored in more detail in Section 4.

- *Italy: risks and rewards to both industry and the government.* A historical lack of trust between public administration and firms could be an obstacle to the correct interpretation of NEAs as instruments, and thus to their consequent implementation.
- In the ENEL agreement, *the main reward* is the internal flexibility with which the group in its entirety can meet its targets (rather than transposition of targets to all ENEL power stations in an equal manner), which is of great importance to the electricity sector with a view to market liberalization. The *reward for the government* is that *circa* 30 per cent of its national CO_2 emission reduction target could be reached through the implementation of this agreement. The significance of this agreement poses *a risk to industry*, however, as it means that should efforts fail to meet what is required, the sector may well be subject to more severe legislation; at no point does the agreement clarify the potential future absence or presence of legislation in this regard. Compliance with command-and-control regulation is very costly for the electricity sector in terms of infrastructure and technological investments.
- *Denmark: rewards to the government and industry* by reducing the possibility of free-riding a priori through the design of the programme: NEAs are mainly bilateral rather than collective.
- *The Netherlands: rewards to the government in the form of industry responsibility* posed by the new Energy Efficiency Benchmarking Covenant in that firms are obliged to do the groundwork both in establishing their efficiency targets, and furthermore, justifying such targets, through a benchmarking exercise which shows that they are indeed among the best in the world. The *risk for the government* is that, in spite of industry efforts, the achievement of such specific targets will not be enough to satisfy the absolute national CO_2 reduction targets.
- *UK: rewards to industry* by signing up to the agreements as all signatories obtain an 80 per cent rebate on the climate change levy. *The risk* to them is that if they fail to meet the targets, they have to pay the tax retrospectively. The *risk to the government* is that the targets are not sufficiently stringent, and therefore the tax rebate is not adequately compensated by performance.
- *Germany: risk of changing position and risk of regulatory capture.* The possible change of government position with a change of government is a risk to business, as there is no guarantee that the agreement will be 'honoured' by a new government. In the German case, the new Social Democrat–Green coalition government introduced an energy tax, despite the fact that the previous government had agreed not to do so. Some might argue that the government did not 'betray the honour' of the agreement, as there had been little proof that industry had truly

made enough progress on emissions reductions. The corollary of this point is that there is a risk to government of 'no action' or 'lip service' by industry, creating effective regulatory capture.

A preliminary set of measures drawn from case studies to minimize risks and maximize rewards is set out below:

- *Denmark: minimize the risk of long and protracted negotiations in the Danish tax agreement package.* Companies can receive tax subsidies during the process of reaching an agreement if they sign a Declaration of Intent. This, however, only holds for a maximum negotiation period of eight months.
- *The Netherlands: minimize the risk of free-riding.* The two-tiered organization in the Dutch LTAs – setting targets at the sector and then at the firm level – and the close collaboration with the competent authorities and the permit system in the latter reduce the risk of free-riding.
- *The Netherlands:* minimize the possibility that efforts in the benchmarking agreement do not suffice with regard to national absolute CO_2 reduction targets by providing for a review of progress in 2004 and promising to take the alternative measures necessary to ensure the respect of national CO_2 emissions reductions targets for the 2008–12 period.
- *Germany: minimize the risk to government of regulatory capture.* In Germany the initial climate change agreement was arguably a weak agreement, and some would see it as constituting regulatory capture, as emissions reductions under the earlier alternative – carbon-energy taxation and waste heat ordinance – was arguably likely to have been greater than under the agreement. However, with the change of government, from the conservative Kohl-led government to the Social Democrat–Green coalition led by Gerhard Schroeder, an energy tax was implemented, and therefore reduced the level of regulatory capture, and increased incentives for industry to implement the agreement.
- *UK: minimize the risk of signature but with non-implementation of the agreement.* The existing retrospective liability clause under the UK climate change agreements stipulates that non-compliance will lead to a loss in the subsequent two years of the climate change levy exemption (80 per cent).
- *The Netherlands: maximize the reward of keeping industry economically intact while pursuing environmental goals.* In the Dutch LTAs, the risk of penalizing economically sensitive industry is minimized, first by involving the sector associations in target setting, thus permit-

ting industry input into this process, and, second, by requiring that the Energy Conservation Plan (ECP) submitted by each company participating in the LTA includes financial information which helps interpret the firms' investment opportunities/limitations in pursuing the environmental goals over the LTA period.

- *Germany: maximize reward to government.* The ability to review and update agreements can produce an important opportunity for the government to demand improved measures by industry. In Germany the 1996 update included new measures for monitoring, and more associations agreed to absolute emissions targets than had hitherto been the case. This arguably improves the likelihood of actual benefits, and certainly improved the government's reputation for policy making and instrument development.
- *Italy: maximize the reward for industry in the form of public image.* The Italian Montedison Agreement is designed as a Group NEA rather than NEAs between the Ministry of the Environment and firms constituting the Group. This gives a unified image of a group strategy. The Group also plans to ensure the credibility of the agreement by including a monitoring and verification facility in the form of an independent third party, as of 2001.

Incentives for Participation and Implementation of NEAs

The design of the NEA offers a key opportunity to ensure that incentives are identified and appropriately integrated into the agreement. The NEA conditions and requirement lead to (a) incentives for participation (depending on how risks and rewards are addressed); and (b) incentives for implementation, to avoid 'lip-service' agreements. The incentives for participation, usually presented as a package, generally include:[13]

- avoidance or reduction of tax liability – as in the UK, Denmark and Germany;
- avoidance of additional legislation, such as in the energy efficiency benchmarking agreement in the Netherlands;
- reduction in administrative costs re permits – as in the LTAs in the Netherlands;
- subsidies for action;
- closer connections to government, with due possibilities to influence future decisions;
- public image benefits of being signatories – as the public might see signing as an indication of commitment.

These incentives for participation are not sufficient to ensure proper implementation and are valuably complemented by additional and explicit incentives for implementation:

- explicit requirements in the agreement for particular action – in the Dutch and Danish cases, energy audits have to be carried out, with action required on specific projects where certain pay-back conditions are demonstrated;
- threat of sanctions – which include retrospective tax liability under the UK climate change agreements and the Danish agreement scheme;
- threat of 'request of special measures' by inspectorates/permitting authorities – this is the case of the Dutch LTA case;
- the monitoring mechanisms in place can encourage action either through internal awareness of benefits of action, or through public pressure due to the reported results (the German agreement, as well as the ACEA/JAMA/KAMA automotive manufacturers' commitment, the Australian Climate Challenge Programme);
- the importance of maintaining trust and cooperation with the regulator – given either the ideological/cultural standpoint, or simply the view of long-term benefits to be gained from this stance (in Germany and the Netherlands);
- the ACEA/JAMA/KAMA commitment (with action by automotive manufacturers focused mainly on technological changes) is linked to market incentives being proposed/implemented by the European Commission – notably labelling and fiscal measures, which will help to encourage the market uptake of lower specific emissions cars. Importantly, these additional measures are not 'within' the commitment, underlining the fact that linked incentives can be internal or external, and deal with the key automotive stakeholders only indirectly through access to the other key stakeholders – the consumers.

Details of incentives for the case studies are presented in Tables 2.2 and 2.3. Although Table 2.2 does not aim to distinguish between incentives for engaging in a new agreement and those for joining an agreement later, this issue has been seen to be pertinent in some case studies. For example, in the case of the German agreement, over time more associations have joined the agreement, underlining the fact that there are two separate issues regarding incentives to participate: one for adopting the 1995 declaration, and another for the additional associations to join during the development of the agreement.

In guidelines regarding incentive compatibility, Convery and Lévêque (2001) point out that this is also of importance for government agents, and not just industry, underlining that 'national government sets out the policy objectives

Table 2.2 Incentives for participation

Case study	Incentives to adopt agreement
Danish NEAs	• Reduction of tax liability • Subsidies of up to 50% of the costs of audit and verification can be granted by the DEA • Companies can receive tax subsidies during the process of reaching an agreement if they sign a Declaration of Intent (maximum negotiation period of eight months)
Dutch LTAs	• Reduction of administrative costs regarding permitting related to energy conservation measures • NOVEM provides *financial support* to stimulate energy conservation. Among the most important are: subsidies for energy programme decree (BSE); tenders for industrial energy conservation (TIEB); subsidy scheme for energy conservation; programme for intersectoral technologies (MINT); tax reduction facility (investment tax reduction scheme) that allows investments in energy-saving equipment to be deducted from profits tax; the scope of the energy investment allowance (EIA) was broadened in 1999
Dutch EEBC	• Not imposing any additional specific measures aimed at further energy conservation or CO_2 reductions • There will be no specific energy tax for the companies, no compulsory ceiling for CO_2 emissions, no additional compulsory energy efficiency or CO_2 targets, no additional conservation commitments and no additional CO_2 or energy requirements • The costs of the commitments undertaken by the Dutch government in relation to JI, CDM and ET will not be borne directly by these companies
German Climate Declaration	• *Avoiding tax liabilities*: The agreement was conditional on the threatened carbon-energy tax and waste heat ordinance being dropped; • It was understood that no other carbon-energy taxes would be implemented in the future (proven not to be the case now)

Table 2.2 concluded

Case study	Incentives to adopt agreement
Italy: Pact for Energy and Environment	• *Avoiding additional legislation.* The CIPE 1998 Act specifies that if NEAs regarding particular issues are not stipulated within a certain period, then the relevant Ministries will legislate • NEA participants are to have priority access to *funding* in applying projects within the NEAs[a]
Italy: ENEL agreement	• Avoidance of regulation which would be costly and damaging to the market structure, particularly in light of liberalization in the sector. Approach allows ENEL to meet targets on the basis of internal efficiency considerations when deciding how to intervene at each of its power stations • The development of a domestic trading scheme which foresees a recognizable credit system
UK	• Signatories to the UK climate change agreements (noted as 'umbrella agreements') obtain an 80% reduction on the climate change levy • Signatories also feel that closer collaboration with the government might help them prepare for (and indeed influence) the eventual UK position in implementing the Kyoto Protocol

Note: [a] This incentive system has also turned out to be seen as a bit of an obstacle to both the negotiation and implementation of NEAs falling within the 'Patto per Energia e Ambiente'. It appears that many potential participants were more reticent about entering into negotiations for NEAs on learning that the funding is given once the projects offering specific opportunities for reducing GHG emissions are identified. As many of the issues to be potentially tackled through NEAs require a concerted 'study' period, it may be that the 'rewards' are perceived as coming long after, and perhaps with some uncertainty, the required input from the parties.

and negotiates the agreement, but implementation – especially as regards enforcement and the free-rider problem – is left to another political jurisdiction, ensure that the latter have some incentive to do so'.

Table 2.3 Incentives for implementation

Case study	Incentives for implementation
Danish NEAs	• *Economic sanctions in the form of retrospective tax liability.* For example, heavy processes without an agreement in the year 2000 pay Dkr 25 per tonne of CO_2 while those with an agreement pay Dkr 3 per tonne CO_2. Liability is determined on the basis of the yearly progress report results seen against the EA objectives. Retrospective liability is only 'called' if there is no successful renegotiation. • *The firm must carry out an energy audit* which identifies the energy efficiency opportunities (more of a requirement than incentive); • *The rules are set a priori* – i.e. the firm must carry out all identified potential improvements in energy efficiency with a payback period of less than 4–6 years • *Monitoring and verification mechanisms. Yearly progress reports* required by the DEA, which keeps a database and thus can track performance. In addition, energy audits are carried out by consultants and then approved of by pre-appointed verifiers • The DEA has already used sanctions when negotiations with a non-conforming firm were found fruitless (Chidiak, 1998)
Dutch LTAs	• *The permitting regime which all firms in the Netherlands are subject to is simplified for the signatories to the* LTA • In the case that firms are expelled from an LTA, the licensing authorities, normally provincial and municipal authorities, can impose *stricter requirements* than the LTA • In order to *participate* in the LTA, individual firms must submit an energy conservation plan (intervention and investment programme) • *NOVEM* provides technical expertise for energy audits and identification of energy improvement opportunities • *Branch organizations* commit under the LTA to diffuse energy improvement opportunities among its members

Table 2.3 *continued*

Case study	Incentives for implementation
Dutch EEBC	• The explicit threat of tighter environmental licence requirements if firms do not respect their commitments. If the competent authority (CA) does not approve the plan, the company may withdraw from the covenant, the CA can tighten the environmental licence requirements, and the government can, if necessary, take other measures
Germany	• The monitoring, reporting and associated public pressure is a key incentive given the institutional culture in Germany; gentlemen's agreements are important within the culture, given a history of consensus and collaboration • The threat of 'loss of reputation' and the prospect of a tax being implemented if the agreement is not successful • There are no additional formal incentives for implementation, and no clear means, beyond information dissemination, by which associations can encourage member companies to act – this is arguably a weakness of the system
Italy: Pact for Energy and Environment	• The annual report on progress in implementing the 'Patto per l'energia e l'ambiente' (CNEL, 1999) states that the NEAs developed to date have *few specific monitoring mechanisms and sanctions in the case of non-compliance*
Montedison & ENEL[a]	• In the agreements between central government and large firms, such as ENEL and Montedison, there are *monitoring and credible verification mechanisms foreseen in the former, and to be improved in 2001 in the latter* • The ENEL agreement states precisely how CO_2 reduction is calculated in relation to the progress of the different initiatives
UK	• The incentives to implement are directly linked to the threat of losing the subsidy of the 80% reduction of the

Table 2.3 concluded

Case study	Incentives for implementation
	climate change levy for all signatories to the climate change agreements. This is a very real incentive, given the scale of the climate change levy • Incentives also exist to 'overcomply' with targets, given potential access to the domestic emissions trading scheme (e.g. where marginal abatement cost lower than the ET price)

Note: [a] These agreements are treated separately from those falling under the Pact for Energy and Environment, mainly because they are between single firms or a group and central government, whilst those of the Pact tend to involve either a sector association or the chain of stakeholders interested in the specific issue which is the object of the agreement. They have also been largely presented to the public as autonomous of the Pact, although the Pact annual report includes details on both these agreements and the third agreement with a large firm, that signed by central government and Pirelli.

Regulatory Capture and Free-riding

Regulatory capture

There does appear to be at least some short-term evidence of regulatory capture. The 1995 German climate change commitment was a very weak commitment, with the associations being able to capitalize on the government interest in having a 'result' at the Berlin COP1. Subsequently, however, some of the weaknesses have been addressed, and the extent of regulatory capture has been reduced. However, there does remain some concern that regulatory capture is still not completely dispelled. Jochem and Eichhammer (1996) made the point that the 2005 interim targets initially agreed were already 80 per cent met at the time of signature due to major restructuring of East German industry. Some of the targets have since been tightened, though there are still some concerns that the targets are not very demanding.

One could also argue that the Italian Climate Change Pact and the ACEA/JAMA/KAMA commitment could be seen as examples of regulatory capture. It is, however, unclear, whether alternative more demanding instruments were realistic alternatives, and hence these cannot be, in a strict sense, regarded as cases of successful regulatory capture – that is, weak instruments accepted when a stronger one was available. Having said that, the fact that no stronger instruments were realistically available suggests that there is a general regulatory capture (where economic concerns predominate over environmental ones), and one that cannot be explicitly attributed to the voluntary instrument.

Free-riding

There is no evidence to date that NEAs have suffered from serious free-riding, though the lack of monitoring and transparency can make such a judgement difficult. However, it is clear that most agreements to date have not had very serious cost implications for the signatories, and therefore there has arguably been little incentive for free-riding, and that incentive has been amply addressed by peer pressure across signatories. For some sectors in some jurisdictions, for example Germany, their protocols seem to allow them to successfully address the free-rider and transaction cost problem, and they find collective agreements culturally and practically congenial (Jeder, 2000 in Convery and Lévêque, 2001). However, one could expect a real problem of free-riders in NEAs, where the stakes are higher. Indeed there is an expectation that there will be a problem of free-riding in the UK climate change agreements, as this concerns very significant rebates (80 per cent) from the climate change levy, and failure to comply with the targets will lead to a loss of the rebate for the subsequent two years. However, it is expected that the mechanisms in place in the climate change agreements will help reduce the likelihood of free-riding. These agreements were scrutinized by the European Commission's Competition Directorate, which looked into the possible problems regarding state aid, given the climate change levy rebate. The European Commission found that the climate change agreements were acceptable.

Transaction Costs

There are few comprehensive assessments of transaction costs, partly as there is a lack of clarity on what should be contained under this title, and partly as time spent by authorities on NEAs is often not totalled and costs apportioned to NEAs. Table 2.4 presents an overview of the types of costs that can make up the transaction cost burden (building on the work of Krarup and Ramesohl, 2000 and 2002). It is generally the case, however, that the transaction costs are often underestimated. Examples of transaction costs include:

- in Denmark, the *expected* extraordinary administrative costs for the governmental agencies because of the agreement scheme amount to €4 million per year (Rigsrevisionen, 1998) (see Krarup and Ramesohl, 2000 and 2002);
- in the Netherlands *the total costs for the industrial energy conservation policy* between 1989 and 1999 amount to €585 million, where approx. €29 million was spent on NOVEM staff (Glasbergen et al., 1997). However, the administrative costs for the LTA alone have not been estimated (see Krarup and Ramesohl, 2000 and 2002);

- in Germany, a rough estimation of the transaction costs suggest that personnel costs of 1–2 person-months (€25 000 Euro) on average can be attributed to the efforts of self-administration by industry per sectoral agreement. Taking the total number of 18 agreements, this would total €450 000, plus approx. €250 000 covering the expenses for the independent monitoring institution and accompanying research. Together with the fairly low costs for government administration, the total administrative costs of the German scheme would not exceed €1 million per year (see Krarup and Ramesohl, 2000 and 2002);
- The Australian Greenhouse Challenge government budget grew from $600 000 in 1995 to $6 million in 1999. However, there was a relaxed reporting requirement for SMEs, which were allowed to report only when changes occurred – hence reducing transaction costs;
- Australian Greenhouse Challenge requires CEO signature, as well as signature by the relevant minister – hence implying some time from top-level staff;
- for the ACEA/JAMA/KAMA commitment, the associations pay for data collection for the first years of the Commitment, and the member states provide data subsequently. While there are no cost estimates for the associations, European Commission, member state and Parliament time (the annual monitoring report goes to the Council and Parliament for consideration every year), the Commission pays at least €50 000 per year for supporting services, excluding their own provision of experts. The data collection and analysis costs for the member states – associated with the implementation of Decision 1753/2000 – is also far from insignificant.

While these costs are usually low, they are often higher than expected given the perception that NEAs are a low-cost measure. In some cases, the high transaction costs prove detrimental to the agreement. For example in an analysis of the American Project XL (see Sexton et al., 2002), some analysts suggest that one reason that the programme failed to fully meet its original goals and objectives was transaction costs – Blackman and Mazurek (1999) note that there were unacceptably high transaction costs brought on by administrative and bureaucratic problems at the EPA.

The negotiation part of the process is also a non-trivial exercise, often consuming significant time and resources. This often leads to NGOs having difficulties regarding involvement. In most cases they are completely excluded, but where they have a chance of being included, the process can be so time-consuming that it becomes hardly affordable to be involved. Very few estimates have been made of negotiations costs. Two examples from the Czech Republic offer some light – the estimated transaction costs for prepa-

Table 2.4 Transaction costs of agreements – possible elements

	Government agency	Industrial associations	Firms	Other: NGO etc.
Preparations	*Ex-ante* assessment of instrument and instrument mix Design of framework conditions of the agreement scheme Gather information about firm conditions Notification of intent to enter negotiations	Gather information about conditions in member firms	Can include: • energy audits • verification • energy management	Preparing position on possible agreement and position paper on appropriate form
Negotiations	Meetings and contact with firms or associations about target setting	Meetings and contact with firms and agency	Meetings and contact with government agency and association	Possible involvement in negotiations, development of position papers
Agreement	Publishing agreement Possible public hearing Possible parliament ratification	Signature of agreement and approval from firms	Possible signature of agreement	Lobbying at public hearing, or in parliament

Administration	Dialogue and guidance of industry Revision of framework conditions for the agreement scheme	Coordination of target achievement by member firms		
Monitoring/ reporting	Developing monitoring form Checking data from industry Own data collection (e.g. member states) Regular reporting Verification	Data collection from member firms Data collation and presentation/reporting	Data collection Self-reporting	Commentary on monitoring/reporting information
Enforcement	Sanctioning in case of non-compliance	Possible internal enforcement to avoid free-riders		Contribution to awareness that enforcement might be appropriate
Evaluation	Evaluations Revision of the scheme in case of problems or unintended effects	Own evaluation of scheme and in some cases member performance	Own evaluation of performance in some cases	

Source: Adapted from Krarup and Ramesohl (2002).

ration, negotiation and conclusion of the agreement for the NEA on detergents was around €12 000–14 000 and for the packaging waste agreement were €15 000–17 000. Clearly, for significant agreements where *ex-ante* assessments and preparations are required, one could easily expect ten times this figure (see Dvorak et al., 2002).

In summary, transaction costs can be higher and more troublesome than expected, and need to be taken into account in the decision as to which instrument to opt for. Importantly, budgets need to be allocated to allow sufficient resources to be available, tenders have to be prepared in advance to ensure that appropriate support, monitoring, and verification experts (often external institutes or consultancies) are available at appropriate times, and practical measures put in place to make the transaction costs cost-effective.

Competition

There is very little empirical evidence of the impact of NEAs on competition or market structures. What can be said from practice includes the following:

- In many cases NEAs allow a way of exempting supposed sensitive industries from costly alternative instruments, notably taxes. The UK and Danish schemes allow for exemption from tax schemes for signatories. This will help the competitiveness of industry. Whether fairly so is a separate question.
- NEAs can make use of competitiveness issues, notably, the product-related NEA (such as agreements on refrigerators and freezers), as it allows market differentiation on environmental credentials. Some have argued that it creates a trade barrier to lower environmental goods (see below), but it has been deemed as proportional to the environmental concern.
- The negotiation position of the NEA can lead to anti-competitive conditions for an agreement. In the UK, discussions on NEAs in one sector led to the sector demanding certain conditions on procurement, which would have resulted in the exclusion of some producers. The government rejected the offer of the NEA and chose an alternative policy instrument instead.
- Competitiveness concerns and concerns of commercial confidentiality have been taken into account in the design of monitoring and reporting schemes. The public cannot access company-specific information on association agreements, while internally relevant information is available. This is the case for the voluntary environmental agreement (VEA) for the pulp paper sector in Portugal (see EEA, 1997) and the cement industry sector within the German self-commitment.

- In many cases concern about adverse effects on competitiveness is a non-issue, as the targets of the NEA are too weak to lead to differential costs for players. Furthermore, many schemes are 'no-regrets' schemes and therefore there is no possible negative impact, though some potential positive impacts.

The most important issue regarding competitiveness is the first point – that NEAs allow exemption from other instruments, and notably taxes. This has been a great attraction of NEAs, and arguably has many merits. Real consideration is, however, needed as to whether the exemptions were too extensive and granted to too many sectors, given the implications for the burden sharing for activities outside of the agreements, and given the possible implication of non-compliance with national emissions reductions targets.

Having said that, it is clear that the use of NEAs must respect the laws regarding trade, competition, state aids, and the internal market. It is worth noting that the UK climate change agreements were under scrutiny to see whether they offered unfair support to the signatory sectors. The final verdict was that the agreements were acceptable and did not contravene any laws. Similarly, the agreement on washing machines was initially regarded as anti-competitive, as the lowest-efficiency machines would no longer be available on the market. But DG Competition found that environmental benefits were there, and that the consumers were saving money, compensating the increased cost. This led to official sanction of the agreement which, in fact, covers 95 per cent of the market.

4. THE IMPACT TO DATE OF THE AGREEMENTS FOR CLIMATE CHANGE TARGETS

Principles

One important aspect for the credibility and functioning of the NEAs is the application of targets. It is important to have quantitative targets covering both milestones (for example intermediate targets giving early warning) and final targets (see Box 2.1 for the conclusions of the European Climate Change Programme (ECCP) sub working group on voluntary agreements with targets for climate change). In a long-term policy vision, initial and intermediate targets may play an important role in 'getting the ball rolling', creating a transition phase where relatively modest targets are set, perhaps by industry, to allow the process to get started and for the actors to develop familiarity with the protocols involved (Convery and Lévêque, 2001), though the key here is that the agreement allows for targets to be revised if it becomes clear that new targets are needed (for example if targets are met half-way through the period). Reviewing targets also has important implications for dynamic efficiency as, for a voluntary agreement to encourage innovation, it must provide for rising standards of performance or for periodic updates; otherwise the code will offer no incentive for continuous improvement and innovation (Moffet and Bregha, 1999).

BOX 2.1 ECCP CONCLUSIONS REGARDING TARGETS

- Concrete absolute targets should be encouraged where possible, but the issue of absolute vs relative targets should be judged on a case-by-case basis.
- Most stakeholders argued that targets should be reviewed and strengthened according to technological development and the need for stronger climate action. It was stressed that any renegotiation of targets should involve all the key stakeholders.
- Most stakeholders agreed that for long-term VAs, interim targets or milestones would be valuable (for example to verify progress), but would not be practical for shorter VAs.
- Binding targets might be required if there is a possibility of linkage to emissions trading (ET) schemes.
- Given the future potential linkage of VAs to emissions trading and the Kyoto mechanisms (JI and CDM), agreements could usefully consider the role of these in meeting targets and the influence on the targets themselves.

Note: While the ECCP used the term VAs, these mainly concerned NEAs.

The role of *quantitative* targets is of particular importance since the underlying aim is to reduce absolute GHG emissions and to help honour the UNFCCC commitments – for those NEAs launched after Kyoto, the Kyoto targets. Ideally the NEA should refer to the national emissions reduction commitments, and set targets which would allow an assessment of the contribution to meeting these commitments. The targets should form a package, noting not just the energy efficiency targets, which are important for industry planning, but also specific CO_2 emissions targets, and ideally also absolute emissions reductions targets – against both a prediction of future emissions and against a reference year. Furthermore, NEAs should all contain interim targets to allow earlier measurement of progress, response to shortfalls in progress, or further target development in cases of easy progress (ten Brink and Morère, 2000 and 2002).

Practice and Lessons

Table 2.5 presents an overview of the use of targets in selected environmental agreements. For several agreements the targets were set during the discussions (target-setting agreements) while others (for example the Dutch LTAs and Denmark) implemented previously negotiated targets.

Some agreements already contain interim targets, with, for example, the ACEA commitment adopting an interim target of 160 g/km to 170g/km for the average new passenger car sold in 2003 (compared to 140g/km in 2008).

The recent ENEL agreement (14 July 2000) in Italy is an example of good practice, a least in regard to what is set out in the agreement. The agreement makes specific reference to government-set Kyoto-related targets for the electricity sector and also makes commitments for ENEL (the leading electricity company) regarding both energy efficiency and absolute CO_2 emissions reduction targets for 2006 and interim targets for 2002, calculated in relation to 1990 data.

Targets are often the main issue to be discussed when an NEA does not bring significant GHG (CO_2) emissions reductions. Absolute targets are not usually favoured by industry, which instead prefers non-emissions reduction targets, such as energy efficiency gains. As already mentioned in the section on risks and rewards, for those countries which promote NEAs with energy efficiency targets with the aim of meeting national absolute CO_2 emissions reductions targets, the approaching emissions reduction commitment target period (2008/12) renders crucial a review of targets and of the entire policy strategy.

Some agreements set targets on an association level, while others choose company-specific targets, which links to the important aspect of collective and individual liability. Collective liability means that the targets as well as the implementation are collective, or set by an association as a whole. In case of non-compliance, this might lead to sanctions against all participating companies independently of individual pollution abatement efforts (OECD, 1998).

Achievements to Date

The agreements can lead to a number of distinct types of benefits:

- actual reduction of (specific or total) energy use;

Table 2.5 The role of targets

Case study	Target type[a]	Level of target	Legally binding	Target details	Interim targets	Updateable targets
Denmark: NEAs	I	Company, mainly	Yes	Energy audits, management and efficiency measures/projects	Yes	Yes
The Netherlands: LTAs	I	Association	Yes, implicitly	20% energy efficiency gain by 2000 (reference point: 1989)	Yes	Yes
The Netherlands: EEBC	TS	Individual firm	Yes	Dutch plants benchmark with the average energy efficiency of the best region in the world or with the best 10% of plants worldwide to determine individual targets	Yes	Yes
Germany: climate declaration	TS	Association	No: moral	20% specific CO_2 overall and/or 20% total reduction (1990 to 2005), with sector-specific targets for each sector (see EEA, 1997). Some targets are absolute, others specific	A few	Yes

Italy: NEAs: The Pact for Energy & the Environment	TS	Association mainly	No	Varies for the different agreements; still unclear for many	Sometimes	Yes
Italy: Montedison	TS	Company	No	Result reduction of CO_2 of *circa* 10 mt/year through the implementation of projects	Yes, not all quantitative	Yes, implicitly
Italy: ENEL	TS/I	Company	No	15 mt reduction of CO_2 in 2002 22 mt reduction of CO_2 in 2006 Specific average CO_2 equivalent emissions of : 0.55 kg CO_2/kWh net (2002) – 13.5% below 1990 levels	Yes	Yes, implicitly
UK: climate change agreements	TS	Association	Legal	Absolute or output-related reduction in CO_2 or energy consumption	Yes	Yes

Note: [a] I: implementation agreement; TS: target-setting agreement.

- actual reduction of (specific or total) GHG emissions and contributions to Kyoto targets;
- learning and awareness of measures within industry;
- development of regulatory capacity through increased learning and informal networks;
- innovation in process or technique;
- development of environmental management systems within industry.

BOX 2.2 LEARNING ASPECTS OF NEGOTIATED ENVIRONMENTAL AGREEMENTS

Some authors (Ramesohl and Kristof, 2002) argue that the emissions reductions benefits from EAs are likely to be small and far less important than the 'learning' benefits of the collaboration. These authors think that the collaboration will lead to the regulator having a much-improved understanding of what measures are actually possible to implement in the practical reality of industrial processes, how best to ensure that these measures are implemented, and how best to link the measures to other instruments. This will lead to improved climate change policy making which will have real benefits, but only in the medium to long term. They also argue that climate change NEAs should be entered into, not to achieve the Kyoto targets, but to ensure that future policy making is improved and long-term benefits are assured.

BOX 2.3 DEVELOPMENT OF REGULATOR CAPACITY

The process of co-regulation and multi-stakeholder involvement can not only lead to an increased awareness of issues (see Box 2.2), but can also develop formal and informal networks which can be used to support the development of policy and also as part of the process of monitoring and implementing the agreement. This is key to ensuring the continued public and political pressure on the agreements, and also key to the ability of the regulatory authorities to respond quickly in negotiations and changing circumstances.

This chapter focuses its assessment on the first two, as these are of greatest interest to policy makers concerned with meeting the national targets to reduce GHG emissions. The learning aspects and development of regulatory capacity are, however, arguably more important over the long term (see Boxes 2.2 and 2.3) as witnessed in Italy where they may prove to be the main impact of the first generation of the Pact-related NEAs due to the prior lack of familiarity with these instruments.

Actual Benefits

Although achievements in energy savings have been realized through NEAs, their contribution to absolute CO_2 reductions from a given reference year (for example 1990) is limited, except where the targets explicitly aim at reducing absolute emissions. Often, the targets are set to increase specific energy efficiency, which means that increased production is considered beyond the scope of the NEA, and no emissions reductions with respect to a given reference year can be guaranteed. Achievements to date are presented in Table 2.6 and where it is too early for information to be available, expected results are noted.

While these results offer some quantitative insights, they are difficult to disaggregate and allocate between the business-as-usual improvements and the direct and incremental effects of an NEA (although Rietbergen et al., 1998 have tried to do this for the Netherlands).

While it is likely that there are some win–win initiatives (combining environmental and economic benefits) through energy savings measures that would not have taken place without the NEA, it is too early to be able to assess whether true incremental cost initiatives have taken place and led to any benefits.

The results do, however, represent a contribution towards meeting climate change targets, but are arguably below the true potential for the NEA instrument, and insufficient to meet the national GHG emissions reduction commitments without other parallel programmes, activities and instruments. Practice to date does not suggest that, even if NEAs were to fulfil their full potential, they could be more than just one instrument in the portfolio needed to meet the emissions reductions targets.

The inability to present more concrete quantitative data and conclusions reflects, in part, the limitations of the existing monitoring and reporting schemes. Again, this underlines the need for targets within national NEAs to be transparently linked to national Kyoto targets, and for monitoring mechanisms to cover the storage of data, reference year emissions inventories, and baseline and policy projections – and all consistent with UNFCCC reporting frameworks.

Table 2.6 Quantitative benefits of NEAs

NEAs and sector	Quantitative benefits: energy use, efficiency improvement and GHG emissions reductions	Comments on NEA contribution
Dutch LTA by chemical industry (base year 1989)	1998: –18.5% (Min of Econ. Affairs, 1999) 1997: –15% in energy use 1996: –12% (combined energy savings) 1998: 17.4% energy efficiency improvement compared to 1989 level, represents total energy saving of 117 PJ and 6.6 mt CO_2 per year.	Only 50% of the achievements are due to the LTAs, other 50% are BAU Within industry performance differs. By 1998, LTAs which had already reached targets are the sectors: sugar refineries, margarine fats oils, coffee roasting, cement and electronics; base metals and chemicals should meet the targets; others, such as light industry, are far from reaching the targets (perhaps reflecting investment cycles)
All-industry LTAs (30 as of December 1998)	24–25% of achieved energy efficiency improvement in industry due to implementation of LTAs (Rietbergen et al., 1998)	
Dutch EEBC	The parties expect a maximum energy saving of 51–124 PJ in 2012. This is consistent with a maximum reduction of CO_2 emissions by 5–10 mt CO_2 in 2012.	Estimates from University of Utrecht report which assumes organic development of energy efficiency in the Netherlands
German climate change declaration	As noted in Krarup and Ramesohl (2002). The degree of sector target achievement in 1997 was between 60% and 160%. Concerning the overall targets for industry, a reduction of absolute CO_2 emissions in the participating sectors of 34.8 mt CO_2 between 1990 (205.1 mt CO_2) and 1997 (170.3 mt CO_2) is indicated, corresponding to a reduction of 17% (RWI, 1999 p.129)	The chemical industry and the iron and steel industry contributed almost two-thirds (21 mt CO_2) to that result

German NEA[a] e.g. cement industry	1987–98: –17% in specific fuel energy consumption 1987: 3500 kJ/kg 1998: 2900 kJ/kg	Reduction represents 75% of the initial target by 2005 Mainly due to efficiency improvements in the Eastern *Länder* Most of progress towards targets was achieved prior to the agreement This raises questions as to the real effectiveness of the NEA
Danish NEA, energy-intensive firms	1996–1999: – 1.4% total energy use as a result of the agreements 1994 Inter-ministerial Committee stated that trade, industry and services were expected to contribute to 4% of total CO_2 reductions (national CO_2 reduction target: 20% reduction in 2005 compared to 1988 levels). The Finansministeriet (1999) evaluation of the green tax system shows that the tax system in trade, industry and services is expected to lead to a 3.8% reduction of CO_2 in 2005. Of this, agreements and investment subsidies are said to contribute 1.8% (corresponds to 1.1 mt CO_2), whilst the other 2% is contributed by increased taxes on energy	Poor improvement by all parties to the agreement in the first generation of NEAs, performance appears to improve. Below (Energistyrelsen, 1999) the estimated savings (based on the general description of agreements, observations in case studies and survey) are given in relation to total CO_2 emissions in the companies which are parties to the agreements Agreement entered 1996–97: estimated impact in 1999: Concrete projects 1.2% Special investigations 0.8 Energy management 0.8 Total impact 2.7% For agreements entered 1996–2000: estimated impact in 2005 is 6.3%, though with a larger share due to energy management, given that fewer projects expected to be identified in second-round audits

Table 2.6 concluded

NEAs and sector	Quantitative benefits: energy use, efficiency improvement and GHG emissions reductions	Comments on NEA contribution
UK recent NEAs	Expected 2 mt C for the first ten climate change agreements together The agreements were launched on 1 April 2001; therefore there are no results	Pre-budget report estimate
Italian Pact for Energy & the Environment	The agreements have just been launched; therefore there are no results	
Montedison	Expected 10 mt/year CO_2 emissions reduction	The yearly progress report relates the individual projects' progress but no breakdown in relation to CO_2 emissions reductions. Improved information expected when Montedison adopt the third-party monitoring system planned for 2001
ENEL	The agreement has just been launched; therefore there are no results – see Table 2.5.	The CIPE 1998 Act stipulates reductions in emissions due to changes in energy use and the use of renewables

Note: [a] Promising results are present for the cement industry, and as noted chemicals iron and steel offered significant emissions reductions savings. See BDI (1999) and RWI (1999) for results on other areas.

Source: Adapted and updated from EEA (1997) and ten Brink et al. (1999).

5. DESIGN OF INSTRUMENTS AND LINKAGE TO PORTFOLIO OF INSTRUMENTS

Choosing NEAs to Address Climate Change

Public sector

The process that leads to government support for the choice of an NEA, instead of, or as a complement to, regulation or market-based instruments, is outlined in Figure 2.2. It highlights the important issues that a government policy maker should consider before engaging in negotiations with a stakeholder group: are traditional instruments inapplicable or insufficient? Does the sector structure lend itself to a collaborative approach? Is there a history of trust between the public sector and the industry or at least a potential for trust and cooperation? Is there political support for a 'shared-responsibility' approach, which in turn reflects culture and credibility issues?

If these conditions are met, and if the economic, political and policy, institutional, legal and cultural contexts are appropriate in each country, then there is an argument for starting discussions on whether an NEA could be an instrument whose development should seriously be considered. If so, a detailed *ex-ante* assessment is needed to explore what the possible targets and timescales should be, who the collaborating parties could be, what the risks and rewards are, what measures could valuably be put in place to reduce the risks and maximize the rewards, and importantly how the NEA fits into the existing portfolio of instruments and how it could be launched together with additional instruments. It is important to clarify before negotiations whether the NEA is going to be the centrepiece instrument, or more of a supporting instrument for a broader policy package. The *ex-ante* assessment could valuably take on board views by experts and stakeholders to ensure that before negotiations start the information asymmetry is minimized.

Ideally, with the *ex-ante* assessment complete, the government should persist in the negotiations until an appropriately formulated and promising environmental agreement is reached, assuming the following conditions are met:

- clear and realistic targets, linked to the national emissions reductions commitments;
- appropriate system for monitoring, reporting and verifying the NEA;
- key stakeholders are involved, and properly informed, at all stages;
- sufficient level of obligation and incentives; and
- sector coverage in terms of total energy consumption and, ideally, GHG emissions, includes the main contributors.

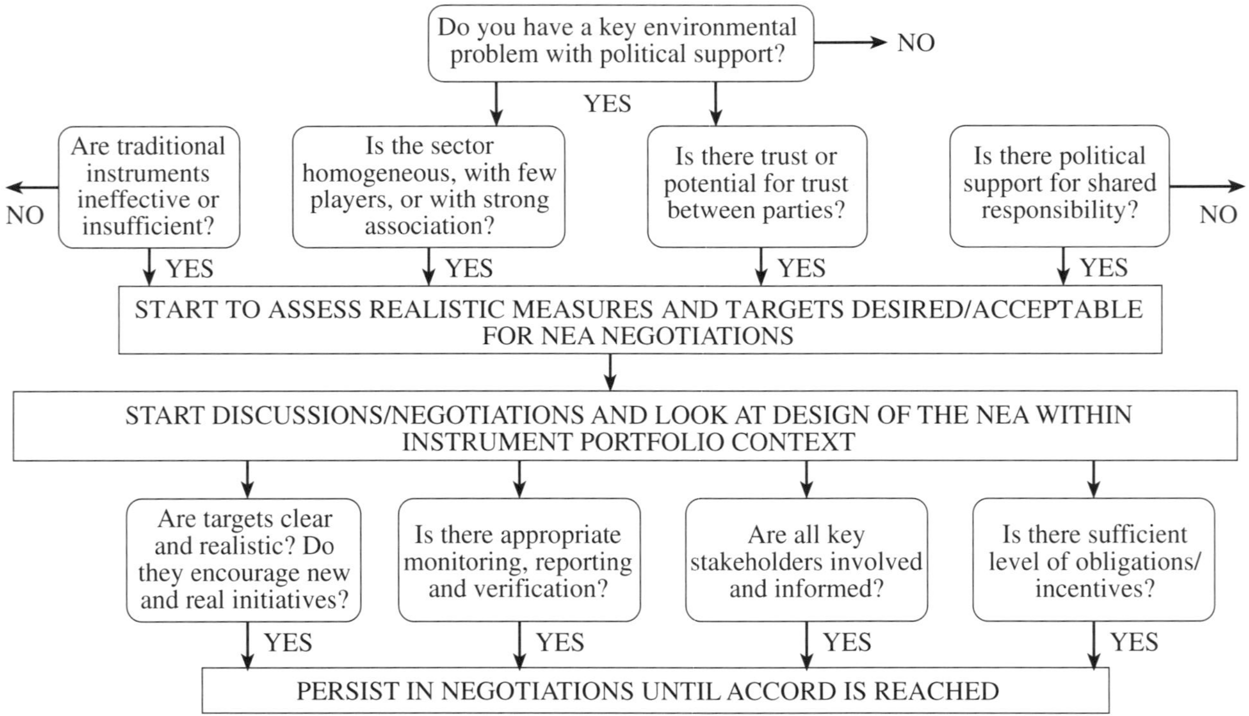

Note: If there is an environmental problem without political support, then the requirements of an NEA will make this instrument unsuitable for addressing the problem.

Source: ten Brink (2002)

Figure 2.2 The process of choosing an NEA: an approach for government

The *ex-ante* assessment of the NEA should then be fine-tuned during the negotiations so that it is clear what implications negotiations agreements will have on the subsequent emissions reductions and hence their contribution to the overall policy goal.

Recent research supports the above, and offers some additional insights to support this process. The NEAPOL[14] project (funded by CEC–DG Research) provides a useful message to policy makers seeking to ensure they have the ideal context for the future use of NEAs. This study examined the performance of 12 negotiated agreements in Europe with the aim of furthering knowledge on the specific characteristics of negotiated agreements and, above all, the factors within their institutional–economic context which influence their performance.

The study argues that the influence of certain socioeconomic aspects (here the study focused on the policy climate (trust, potential trust), the existence of a threat of a sanction ('stick behind the door'), sector homogeneity and representativeness as well as the competitive advantages for firms as a result of consumer pressure) can be outweighed by the influence of other aspects, and that the combined context (all of the four separately studied aspects together) will provide a favourable or unfavourable climate. Thus the socioeconomic context does not need to be favourable in *all* respects; indeed, the results show that:

- the absence of a good policy climate can be outweighed by the use of a strong and credible alternative instrument;
- an alternative instrument on the other hand might not be necessary if the private parties feel that there is a certain degree of market pressure (from consumers or competitors) to succeed in the agreement;
- the disadvantage of a very heterogeneous sector with very different parties can be minimized by the imposition of a strong alternative instrument, making the incentive for these parties the same and stimulating cooperation between them;
- there might be no need for an alternative instrument or market pressure if the private parties and the policy makers already have a very good relationship and if they act as 'gentlemen' towards each other (De Clerq et al., 2001).

Ultimately, what is emerging from this decade of study on NEA design and performance is that there exists a trade-off: voluntary approaches can be a flexible, and therefore cost-effective way of developing environmental policy, but this advantage comes at the expense of public participation and ultimately environmental effectiveness. Conversely if invested with the proper procedural safeguards, voluntary approaches can be an effective policy instrument for achieving ambitious environmental goals, but this effectiveness comes at the expense of flexibility (Convery and Lévêque, 2001).

Private sector

The process for the private sector is somewhat easier, but follows the same main steps. Clearly, industry needs to weigh up the costs and other implications of the alternative instruments being discussed and clarify its preference. It needs to carry out an *ex-ante* assessment to clarify what measures can be implemented at what cost (for example development of the marginal abatement cost curve), and clarify what the burden-sharing implications are for the industry sector and whether the burden-sharing will be accepted by its members. In preparation for negotiations it clearly needs to assess whether the alternative instruments being suggested are realistic alternatives, and explore what areas it can afford to compromise on and which not. This will help inform decisions as to whether the industry should push for an NEA, and if so, what it should negotiate for in terms of, *inter alia*, targets, timetable, measures, reporting.

Linkage to Other Instruments

It is important to consider the question of the objective or policy context before looking at the details on the linkage to instruments, as this offers additional insights into the choice of instrument.

The objective context

The objectives of NEAs can usefully be seen against the background of the threat of implementation of carbon-energy taxes or regulation. In some countries there was insufficient support for these measures, and NEAs were seen as an appropriate, or acceptable, alternative (and indeed sometimes the only available instrument). In other countries, NEAs were complementary instruments, with tax reductions for signatories (for example in Denmark and recently in the UK with the climate change levy). Some examples of objective contexts have included the following:

- In Denmark, the NEA was launched in the context of the national energy plan, along with a tax.
- The Dutch LTAs were chosen to implement the long-term environmental policy plan (NEPP and NEPP+) targets.
- In Germany, the (target-setting) Declaration of Industry was a means to show good will and intention for action at the COP1 in Berlin and to avoid the proposed waste heat ordinance and the discussed and much threatened carbon-energy tax. The choice to accept the original self-

commitment was set against a background of two opposing camps in the German administrations – one group was fighting for an increased use of taxes and charges and another group fighting for a more collaborative approach.

- In Italy, the Climate Pact was launched in November 1998 between the government, industry and NGOs. It follows, and makes direct reference to, the sector-specific CO_2 reduction targets delineated in the CIPE 1998 Act on guidelines for national policies and measures for GHG emissions. The Pact is a 'commitment to commitment' and NEAs are to be the key policy instrument to address climate change in Italy.
- In the UK, a recent (1999) series of climate change agreements (umbrella agreements) complements the new climate change levy (in force since April 2001), which forms the core of climate change policy – together with the domestic emissions trading scheme. Negotiations involved central government and (at first) ten industry associations, with a large number of other sectors being subsequently involved in negotiations.

Linkage of NEAs within the Portfolio of Instruments

The historical debate about 'which instrument is best?' is increasingly being replaced by a more mature debate as to 'which combination or portfolio of instruments is best?', in other words, 'what is the optimal instrument mix?'

Empirical studies so far have concluded that there is no single solution for an effective combination of instruments; the success of each agreement strongly depends on individual contextual parameters. As a general rule it can be concluded that environmental agreements must be accompanied on some level by credible threats that the public authorities do have other means to achieve similar environmental results and are willing to apply these if necessary, as well as by interesting incentives for industry (ELNI, 1998).

This approach to developing instrument mixes has to date been a mixture of simply the result of the winners and losers in often ideological standpoints on particular instruments types – with institutions in some countries historically favouring collaboration (for example the Netherlands) opting for more NEAs, other more sceptical countries adopting command-and-control legislation; in others the use of taxes and charges has found favour. In addition to the ideological influence of preference, the role of pragmatism has also been significant. Pragmatic solutions take into account the perceived strengths and weaknesses of the instrument, note the importance of existing instruments in place, the likely resistance to or interest in the different instruments, the possible timing and costs of the choice, the historical use of the instruments and even the ideological stance with regard to different instrument types.

Table 2.7 NEA linkage to other instruments within the portfolio mix

Case study	Context	Portfolio of instruments and linkage
Danish NEAs	Context of Energy Plan	Linked to carbon-energy tax: signatories partially exempt from tax
Dutch LTAs	NEPP and NEPP+	Linked to permits: facilitated permits for signatories
Dutch EEBC	1997 policy document on the economy and the environment. Action programme: energy conservation 1999–2002	Linked to benchmarking with the best international standard (BIS), which indicates that gap can be bridged through cost-effective measures implemented at the latest by 2005 If it is not possible to reach the BIS in this way by 2008, the company shall take measures to realize the BIS by 2012 or realize a comparable result by other acceptable means, which can include the application of flexible tools such as JI, CDM and ET
German Declaration	UNFCCC commitment, COP1: Berlin	This complements (indirectly) the energy technology support programmes, and is complemented (directly) by the new energy tax. The VDEW (Vereinigung Deutscher Elektrizitätswerke) association was looking into possible links with JI; unclear whether implemented
Italy: Pact for Energy and the Environment	National policies and measures	Linked to funding: Pact NEA projects have priority access to national and regional funds

Italy: Montedison	National strategy for the reduction of GHG emissions	Link to funding from the Ministry of the Environment (MoE) for specific projects Information instruments: MoE to launch promotion campaigns on products developed in the agreement Administrative procedures facilitation: for authorization, e.g. of the re-gasification terminal at Rovigo Support for new laws and norms to favour the entrance of new low-GHG extinguishers, electric vehicles The agreement foresees that 'the parties to the agreement commit to monitoring the evolution of different co-operative instruments for the environment, including the international co-operation mechanisms foreseen by the Kyoto protocol'.
Italy: ENEL	National strategy for the reduction of GHG emissions	Government funding to sustain ENEL commitments to promoting energy efficiency in end use Administrative procedures facilitation Emissions trading: the ENEL group is allowed to use ET on the national level JI and CDM: commitment to sustain the development of JI/CDM and recognize the credits derived from them
UK climate change agreements	Climate change strategy	Linked to climate change levy: signatories exempt from 80% of tax Linkage to domestic emissions trading scheme will be possible, but with 'gateway' restrictions (quantity)

It is not possible to present a step-by-step method for finding the optimal policy mix, but clearly effect, effectiveness, cost, transparency, fairness, openness, accountability, acceptability and workability are important indicators to consider when discussing various combinations of instruments (see also Chapter 1, which discusses the optimal policy mix further). For climate change, the ECCP sub-group on voluntary agreements summarized the view on the optimal policy mix:

> Stakeholders generally agreed that there could be no concrete prescriptions for how VAs should be linked to other instruments. However, it was agreed that it would generally be valuable to carry out an assessment of an 'optimal policy mix' to address climate change and hence assess the role, design needs and instrument links for VAs. This would help to encourage an optimal design for a package of instruments including VAs.

For the case studies explored in this chapter, the linkage to other instruments is presented in Table 2.7. Here we see different choices in different countries, with, *inter alia*, linkage to taxes, to permits, to emissions trading, to subsidies, and to information campaigns. This underlines the common view that there is no single optimal instrument mix, but the choice has to take into account the economic structures, the institutional practices, the legal context, preferences, and the cultural and historical context. In many cases it is easier to 'piggy-back' on existing instruments than redesign the system as a whole, even if a fresh start could be justified.

It is, however, clear that NEAs, when combined with legislation, have features which allow for greater design flexibility and cost-effectiveness and potential administrative cost savings. Regulatory instruments, in turn, can 'safeguard NEAs against their shortcomings, namely low expectations in their environmental targets, weak enforcement provisions and the lack of credible and efficient monitoring and reporting requirements' (OECD, 1999). This suggests, as could be expected, that NEAs should not be regarded as an instrument that can do the job alone.

In fact, in regard to the effectiveness of a background legislative threat in ensuring compliance with agreement obligations, Segerson and Miceli (1999) found that the equilibrium level of abatement is directly related to the magnitude of the threat. For weak threats it was found that negotiations take place but the resulting abatement level is low and can actually be lower than that threatened to be imposed through legislation. Other theoretical studies analysing such policy mixes have recently been written by both Chidiak (1999) and Millock (1999). These theoretical conclusions support the practical policy intuition.

It is clear that the instrument mix needs not only to have the threats in place during negotiations, but that threats are existent and credible during implementation. Ultimately it is up to each country to assess what instrument mix can offer an effective, cost-effective and efficient means of achieving the necessary environmental goal, taking into account the important issues of burden sharing, free-riding, economic sensitivity, commercial sensitivity and the polluter pays principle.

Linkage of NEAs to Emissions Trading and the Kyoto Mechanisms

Currently there is a great debate on the linkage of NEAs to emissions trading (ET) and the Kyoto mechanisms, and the implications of this link for the effectiveness of the different instruments, on design and format, and regarding likely future use. There are four types of options:

1. NEAs are a short-term instrument that will be completely superseded by the use of emissions trading and Kyoto mechanisms, with little connection between the instruments.
2. NEAs can be an intermediary instrument that helps lead to the use of emissions trading and Kyoto mechanisms.
3. NEAs can be a parallel and long-term instrument that has little linkage to the Kyoto mechanisms.
4. NEAs can be a long-term instrument, integrated with emissions trading and the Kyoto mechanisms.

The fourth possibility is perhaps the most interesting, but it is worth exploring the other options in some detail.

1. Short term: no linkage to ET/Kyoto mechanism

This is of little interest to the debate, except to note that it might come about if the effectiveness of this and the next generation of agreements is unsatisfactory and the instrument loses credibility. It might also be a predecessor (as NEAs can, in principle, be launched quickly, in contrast to the ongoing discussions aiming at clarifying the details about ET and Kyoto mechanisms), and offer real benefits, but still be superseded by Kyoto mechanisms.

2. Short term, with implications for Kyoto mechanisms

Where an intermediary option, NEAs can function as a starting-point and learning process for the design and implementation of the Kyoto mechanisms. Hence they play a type of bridging function to future Kyoto mechanism use. For example:

- Parties to the NEA can gain experience in cooperation and shared reductions of CO_2, where the NEA includes collaboration on or investment in projects in JI and CDM countries; the agreement can later be developed into a fully fledged JI scheme or CDM scheme.

- Furthermore, NEA experience can help explore the effective burden sharing (across sectors of the economy) implicit in each NEA, and the internal burden sharing and potential for burden sharing as projections and experience are gained on the measures and GHG abatement costs. In short, experience with an NEA can show 'who does what'; efforts to understand future emissions and costs within an NEA can show 'who can do what'.
- NEAs can also help provide experience with GHG inventories, identification of initiatives to reduce energy use and GHG emissions, clarify monitoring, reporting and verification requirements, and methods and stakeholder involvement. This can facilitate the Kyoto mechanism learning process.
- If the emissions inventories are sufficiently robust, and it is clear what the emissions reductions implications of initiatives are, then this information could support a case for getting credit for early action, or at least not being penalized for early action (this could then link in to the emissions trading scheme – for example help in allocation of permits/credits).

3. Long term: little linkage to Kyoto mechanisms

As the NEAs appear to offer real benefits in terms of learning and capacity building, it is possible that the task of target setting and meeting Kyoto targets is tackled directly by the Kyoto mechanisms, but NEAs can work in parallel on learning and capacity building. Here the NEAs could increase awareness (or mitigation, energy efficiency options) within the economy, learning (climate policy), and regulatory capacity building (through increased communication, and formal and informal networks).

Furthermore, NEAs are used to address many other environmental objectives than just climate change, and even if the Kyoto mechanisms effectively negate any NEA role in meeting the Kyoto targets, NEAs are likely to be in place to deal with, *inter alia*, air emissions, waste, chemicals and by-products.

4. Long term, explicitly linked to emissions trading (ET) and Kyoto mechanisms

NEAs can be developed in a way that explicitly considers the ET Kyoto mechanisms as a means for achieving future NEA targets. In other words, Kyoto mechanisms are a tool within the broader framework of NEAs.

For example, companies might buy permits/credits to meet their NEA targets, or indeed sell permits/credits, if they over-comply (see Box 2.4), though this will only work if the NEA is a legally binding agreement, given the likely cost implications.

The use of NEAs linked to Kyoto targets could therefore help support the liquidity of the domestic emissions trading market, where the NEA does lead to over-compliance, and there is a possibility for the NEA signatories to sell to this national market. This latter point has implications for the design of the NEAs (see Box 2.4 for UK example).

BOX 2.4 PRACTICAL EXPERIENCE OF THE LINKAGE OF NEAS TO KYOTO MECHANISMS

There is so far little empirical evidence of linkage, though some signs of linkage are becoming available. These are noted below.

Linkage to emissions trading

In the UK climate change agreements, there is a potential for purchasing permits where the targets are not met. There is also the potential for the sale of permits to the national scheme, but with some restrictions on the quantity. These restrictions are known as 'gateway restrictions'.

In the Dutch benchmarking covenant (EEBC), there is the potential to purchase emissions permits in 2008, if it becomes clear that the target (for the sector of being in the top 10% of world energy efficiency for the respective sector) will not be achieved.

In the ENEL agreement, a domestic trading scheme for trading between different ENEL units is to be put in practice.

Linkage to JI/CDM

There is little evidence of formal linkage, though it is understood that the German climate change agreement allows consideration for JI projects. Furthermore, there was an intention by one of the agreement associations to seriously consider this option.

The EEBC and the ENEL agreement both foresee the future possibility for commitments to be met through the use of JI and CDM. In the case of EEBC, like ET, this will be addressed in 2008, whilst it is already possible for ENEL to earn 'credits', recognized by the government, through JI/CDM.

Similarly, potential access to emissions trading can help support the agreement of total targets, where previously only specific targets or measures were agreed. This would help the governments in their strategies to meet the Kyoto targets.

In addition, JI or indeed CDM might be a concrete measure for NEAs to meet their own targets, though this would require careful formulation in the NEA itself. Therefore, once again the NEA provides a framework within which the Kyoto mechanisms are a tool. This is a real possibility, when considering that NEAs could have a broader focus than just climate change, and might include air emissions in general, which would respond to many developing-country environmental interests.

These two options can also be viewed as NEAs supporting the uptake of emissions trading and the use of JI or CDM, where the NEAs are defined more narrowly as learning and facilitating tools. It is possible that an NEA across smaller installations or enterprises could highlight potential energy and emissions savings that would not be encouraged by the flexible mechanisms themselves (given transaction costs), and hence bring more emissions credits to the market.

Design implications

The linkage of the instruments – NEA and ET – is likely to have important design implications, including:

- monitoring – the VA monitoring system would need to be compatible with that of ET, and helpfully supported by emissions inventories and baselines;
- targets – VA targets would have to be compatible with ET targets and could usefully be absolute targets to facilitate the linkage;
- compliance – compliance systems would need to be compatible with those of ET;
- provisions – new agreements should make provisions for eventual linkage to ET and contain a review clause to substantiate the linkage in due course.

As noted in the ECCP sub-group on voluntary agreements, these points should be taken into account in the design of new VAs. The design of existing VAs may need to be adjusted if there is to be a future link with ET. Furthermore, the design of VAs should take into account the fact that the sectors and/or companies covered by ET may have a mandatory cap on emissions.

6. EVALUATION METHODS AND GOOD PRACTICE GUIDES

Evaluation Methods: Assessing the Merits of EAs – both *ex ante* and *ex post*

Ideally each EA would be assessed (and indeed assessable) both *ex ante* and *ex post* (and indeed *ex itinere*) – *ex ante* to answer the question of whether the EA is likely to be effective, efficient and credible, under what conditions, and whether it could be an appropriate instrument; and *ex post* to assess whether the instrument was indeed an effective and efficient instrument at the time to address the environmental objectives, given the particular environmental, institutional and economic circumstances (see ten Brink et al., 1999).

The *ex-post* analysis requires consideration of:

- the environmental performance of the participants in the absence of the EA, depending, in part, on a view on the use and effective of alternative instruments;
- whether and which alternative policy instruments might have been used. Commonly EAs are set in a context where alternative instruments are proposed at the same time as the EA negotiation, partly as an incentive to conclude successful negotiations;
- this in turn requires a baseline against which to judge the environmental performance of the EA industry participants. This baseline might be described as business-as-usual (BAU) in the absence of alternative new policy instruments or alternative policy scenarios requiring the identification and assessment of the effect of the likely alternative policy tool(s) (see Box 2.5 and Figure 2.3). This final issue is complicated further by the fact that the choice of EAs is not generally one of a single choice between two alternative instruments, but rather a choice of the appropriate balance within a portfolio of instruments. It is clearly easier (but still difficult) to assess against a BAU baseline.

BOX 2.5 THE *EX-POST* ASSESSMENT FRAMEWORK

When pollution is perceived to be dangerous or creating too much pressure on the environment, a target (T) (or set of targets T_1, T_2, ... , T_n) for a reduction in pollution may be set to be reached in the year $x+t$ (or $x+t_1$, $x+t_2$, ... , $x+t_n$). This target can be a single fixed target, updateable target, or one of a series of targets. And the target can be expressed in terms of percentage or absolute reduction from a given reference year, percentage or absolute reduction from some defined baseline, or reduction in specific emissions (again definable either in percent relative to reference year, or explicitly).

Figure 2.3 presents schematically the evolution of pollution emissions, noting reference year, baseline emissions (here we have presented the level of pollution as decreasing, but it could equally be stabilizing or increasing) and the performance of the industry under the EA.

For the environmental assessment of each EA, the reference situation corresponds to the start of the agreement – in year *x*. The pollution situation (C)[a] is the actual level of pollution at the time of the monitoring and assessment (T_i=1996). The current situation is compared to the various reference values in T_i, given by the points A, B, D, E. The value relative to the reference point can be given in physical tonnes by A–C, or as a percentage reduction from reference year: (A–C)/A. These can be compared to the target, depending on how the target is expressed.

The environmental effectiveness of the EA corresponds to (D–C) (BAU) and to (E–C), assuming that (E) can be defined (policy scenario). Different indicators can be used for the comparison between NEA results and BAU and NEA and alternative policy instruments – for the former a percentage effectiveness value is or given by (D–C)/D and for the latter by (C–E)/E.

Where the policy baseline projection is unknown, the second-best option is to assess the current level of pollution (C) against the level of pollution in the business-as-usual situation (D–C), where the policy preceding the EA is assumed to remain unchanged. Where this baseline is unknown, then one can only use the reference situation (A or A*) to obtain a quantitative measure.

Note: [a] The position of C in Figure 2.3 is illustrative: it could be above or below E and indeed above A* in the case where the baseline grows (D is above A) – it is unlikely that C would be above D.

Source: ten Brink (2002).

As an alternative to, or part of, the policy counterfactual, consideration could also be given to the extent of the technical innovation deemed to have been

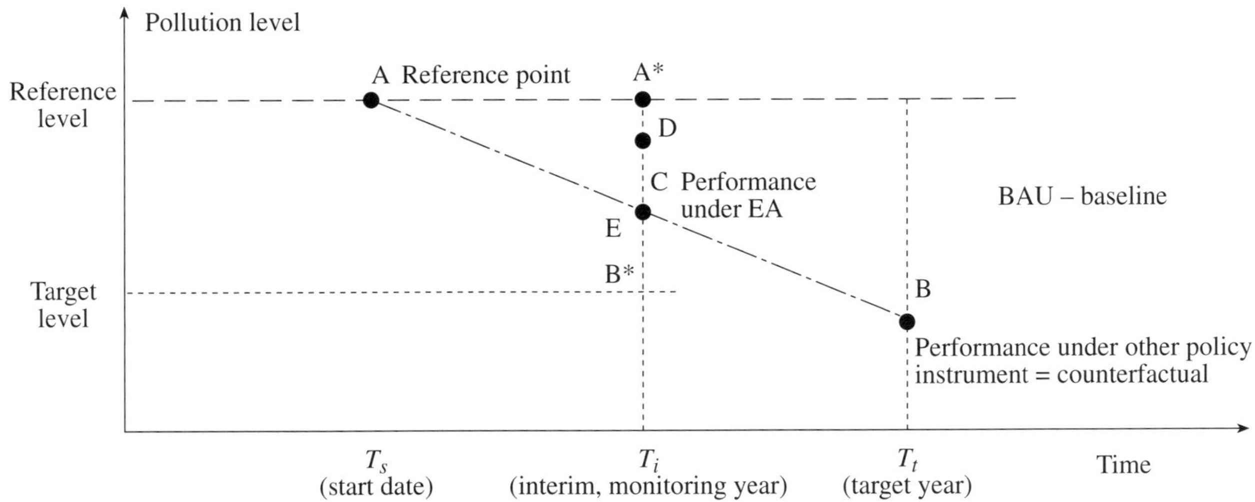

A = Reference point; A* is reference level
B = Target for year T_t; B* is target level
C = Actual level of pollution in year T_i – at interim point/time of assessment
D = BAU level of pollution
E = Interim pollution level expected under alternative instrument

Interim assessment:
A*–C = environmental improvement over reference situation
D–C = environmental effectiveness against business as usual
E–C = environmental effectiveness against alternative policy instrument
C–B* = progress to be made to reach targets

Source: ten Brink (2002).

Figure 2.3 Quantitative assessment framework for EA performance

induced by the EA, as well as the extent of learning and institutional capacity building, as partial measures of the effectiveness of the instrument in stimulating beneficial changes amongst participants.

An assessment of the environmental effectiveness of EAs, especially one undertaken at the present time when the instrument is still in its formative stages, is difficult and requires consideration of quantitative analysis (where monitoring and baseline data are available) and a qualitative review (based on common-sense rules of good practice). The European Commission's guidelines on good practice – as given in the communication on the use of EAs as an instrument for the implementation of environmental policy in the Community (CEC, 1996) – is summarized in Box 2.6 and a more detailed checklist is given in Box 2.7.

BOX 2.6 EUROPEAN COMMISSION GUIDELINES FOR THE DEVELOPMENT OF EAS

The Commission has produced guidelines for EAs at the European level. According to the Communication, voluntary agreements should meet the following requirements:

- prior consultation with concerned sectors;
- legal basis (performance of which is covered by civil or public law);
- where appropriate, stipulations for effective sanctions (fines, penalties, withdrawal of authorizations);
- a gradual approach and quantified objectives (with corresponding deadlines); monitoring of the results;
- periodic reports to the competent authorities;
- provision of appropriate information to the public;
- where appropriate, establishment of an independent committee or body to collect, evaluate or verify the results.

Source: Adapted from CEC (1996).

BOX 2.7 EUROPEAN COMMISSION CHECKLIST FOR EAS

This checklist is valid for environmental agreements concluded at whatever level. Elements such as the legal nature and the competent jurisdiction are only relevant in case of agreements in the form of a contract.

Reasons for the choice of the instrument:

1. Advantages compared to legislative and economic measures (environmental and cost-effectiveness, feasibility)
2. Sector coverage, strength of business associations
3. Public awareness of the issue
4. Previous involvement of legislator in setting objectives

Content

1. Parties to the agreement (associations and/or individual firms)
2. Subject
3. Definition of terms
4. Quantified objectives
5. Staged approach
6. Specification of obligations
7. Monitoring of results
8. Periodic reporting
9. Access to information
10. Arrangements for collection/evaluation/verification of results
11. Sanctions
12. Accession of third parties
13. Duration
14. Revision
15. Termination
16. Legal nature of the agreement
17. Jurisdiction

Compliance with EC Treaty

1. Notification to the Commission required?
2. Free movement of goods affected?
3. Competition affected (competitors excluded, prices fixed, etc.)?
4. State aid rules applicable and respected?
5. Distortion of competition justified on environmental grounds?
6. Distortion of proportionate means to reach the objective?

Source: Adapted from CEC (1996).

In many cases the ability to assess quantitatively the environmental effectiveness of EAs is compromised by: (a) the fact that many of the EAs are ongoing, and sufficient results are not available for *ex-post* analysis; (b) the lack of appropriate monitoring data, given insufficiently strictly stipulated indicators, lack of environmental accounting practices, and concerns over confidentiality; and (c) the inherent difficulties in defining business as usual and counterfactuals under different policy instruments. In the former case it is clear that there is a danger that past improvements may reflect past initiatives within environmentally or energy-conscious companies, such that many of the low-cost or no-regrets options have already been realized – with the consequence that historic annual efficiency gains are not likely to be representative of future business-as-usual energy efficiency improvements. On the other hand, it is clear that it is in some industries' interest to argue this case, as it would allow less stringent targets to be set.

Given the difficulty of the above arguments, and the need to assess whether the EA is the appropriate instrument *ex ante*, rather than *ex post*, a qualitative assessment is also required. This needs to address the following questions:

- In the negotiation of the EA is there sufficient participation by relevant stakeholders?
- Is the design of the EA appropriate for the environmental objective being pursued? This links to the nature of the objective, the type of targets, the type of initiatives embraced by the EA, and the type of success indicators adopted.
- Is there any monitoring and, if so, is it detailed, credible and transparent, and does it ensure true awareness of progress?
- Are the targets explicit, and able to be updated in light of progress, to ensure that the EA targets really represent more than just business-as-usual approaches?
- Are there sufficient incentives (carrot and stick) in place to ensure that the targets will be complied with – that is, legal stature, public pressure, threat of sanctions, alternative instruments?
- Is it likely that the EA will have a comprehensive coverage of the sector and that free-riders outside and within the EA are discouraged? This depends on the structure of the sector concerned, and the number and size of industry.
- Is there sufficient dissemination of the EA internally to ensure that the principles and initiatives of the EA are taken up at all levels, and externally, to help ensure stakeholder pressure or support, depending on the development of the EA?

If these questions, and the related questions in the Commission checklist, can be answered positively for the EA, then an *ex-ante* appraisal would suggest that the EA could have a good chance of being effective. A separate, though connected, question remains as to the complete package of instruments to be presented with the EA – that is, to what extent the EA can usefully substitute other instruments (to avoid excessive regulation, or taxes for sensitive industries), and to what extent the EA can be usefully complemented by other instruments (taxes in place as threat against non-compliance and so on).

Good Practice Guides

Several important initiatives have been developed for the preparation of environmental agreements – notably on the development of good practice guides. Existing good practice guides include :

- European Commission guidelines (see CEC, 1996)
- The Canadian guidelines (see Industry Canada's Voluntary Research Form web-site)
- NGO drafted guidelines from the UK (for example by the Green Alliance[15] in the UK)
- National guidelines, such as the Dutch Guidelines on Covenants[16]
- As noted elsewhere in this chapter, the European Climate Change Programme (ECCP) multi-stakeholder sub working group on voluntary agreements also agreed some guiding principles for climate change voluntary agreements.[17]

In addition, the European Commission is considering developing EU framework guidelines – best practice guidelines for VAs at national level. This is likely to build on and expand the broad ECCP forum-inspired stakeholder agreed guiding principles for: the form of the agreement; the role of reference year and baselines; the role of target setting (as in Box 2.1); and the role of monitoring (see Box 2.8). Finally, see ten Brink (2002) which contains several chapters offering guideline methodologies that complement the formal guidelines note above.

BOX 2.8 ECCP SUB-GROUP ON VAS' CONCLUSIONS ON KEY ELEMENTS FOR NEAS

A Form of Agreement

- The agreement should be signed by associations/federations and/or companies according to the kind of agreement, and their likely effectiveness. Some stakeholders argued that agreements should be signed at the company/installation level, and others argued that this was not necessary. Particularly in sectors with many SMEs it might not be feasible to have all individual companies sign the agreement.
- Where the agreement is negotiated, the relevant public authorities should also sign it.
- The issue of free-riders should be taken into account and an appropriate system to deal with this issue should be put in place.
- Whether an agreement is legally binding or not depends on the country-specific contexts and on the nature and ambition of the agreement. Some stakeholders noted that for VAs to be explicitly linked to ET, and with absolute targets, they would need to be legally binding.

B Reference year and baseline

- The Kyoto requirements should be considered when setting base and target year, but 1990 (1995) should not be required to be base years for CO_2 (other gases), given data practicalities. Comparability of targets to the Kyoto reference years would, however, be valuable.
- Common guidelines on transparent emissions inventories (notably for reference year and target years) are desirable so as to facilitate comparison of VAs between member states.
- It was generally felt that baseline projections could be a useful tool in the negotiation of targets and with respect to monitoring. They should be seen as a working tool, rather than as a 'binding commitment to the future'. Some stakeholders felt that a baseline report could be a useful tool, available to third parties in the interests of transparency and credibility.

C Target Setting

- Concrete absolute targets should be encouraged where possible, but the issue of absolute vs. relative targets should be judged on a case-by-case basis.
- Most stakeholders argued that targets should be reviewed and strengthened according to technological development and the need for stronger climate action. It was stressed that any renegotiation of targets should involve all the key stakeholders.
- Most stakeholders agreed that for long-term VAs, interim targets or milestones would be valuable (for example to verify progress), but would not be practical for shorter VAs.
- Binding targets might be required if there is a possibility of linkage to emissions trading (ET) schemes.
- Given the future potential linkage of VAs to emissions trading and the Kyoto mechanisms (JI and CDM), agreements could usefully consider the role of these in meeting targets and the influence on the targets themselves.

D Monitoring

- Monitoring is a key element of VAs. An appropriate system of monitoring needs to be established for all VAs, be it at EU or national level, and be in place before the agreement is signed.
- The monitoring system should include an independent aspect in order to ensure the credibility of the VA.
- While the core role of monitoring is to assess the progress of the agreement and it therefore focuses on the specific objectives and targets of the agreement, stakeholders agreed that where possible monitoring reports should include calculations of total emissions of GHGs to allow the contribution to Kyoto targets to be assessed.
- Reporting of monitoring results should be regular (for example once yearly), transparent and open to the public, whilst taking appropriate account of concerns of commercial confidence. Transparency requires note of measurement methodologies, including, where relevant, the use of correction factors, how electricity is treated, and how feedstock is addressed.
- Independent and transparent verification would be essential to improve the liability of the VAs and to counteract criticism by the public.

- VAs need a review clause which can allow them to be reviewed and potentially negotiated in light of the findings from the monitoring of the agreement.
- Many stakeholders felt it would be helpful to have common principles on monitoring.

Note: see the European Climate Change Programme (ECCP) multi-stakeholder sub working group on voluntary agreements guiding principles for climate change VAs within the ECCP Report. Web page: http://www.europa.eu.int/comm/environment/climat/eccp_longreport–0106.pdf

7. CONCLUSIONS

Use of NEAs

Within the EU, NEAs are increasingly seen as offering a pragmatic and potentially valuable instrument to help address climate change. Indeed in several countries NEAs have been chosen by governments to be an appropriate instrument for meeting the Kyoto targets (for example France, the Netherlands, Italy, and Germany – at least under the last government). It is clear from the results to date that significant further efforts will be required to ensure that the NEA tool can fulfil this promise and responsibility. Without further efforts, it is likely that countries that are relying on the use of this instrument to help meet their reductions commitments will run into trouble.

Patterns emerging from practice, supported by the case studies analysed here, include:

1. An increasing number of countries are looking to NEAs as an important policy tool in implementing domestic climate change strategies. While it is true that in some countries they are promoted as a key tool, in all countries they are but one instrument of many being used to implement climate change strategies and achieve GHG emissions reductions targets.
2. Many countries use other instruments (for example tax, regulation through permits) as the main instrument for achieving climate change targets, and use of NEAs as a means of allowing ‘exceptions’ in order to preserve the competitiveness of those firms hard hit by such instruments (notably taxes). In the case of the Dutch LTAs, the primary policy instrument is permits, and LTAs offer a cost-effective, flexible alternative for firms. In the case of Danish energy efficiency agreements, there is a blanket policy of taxes, and agreements allow a partial avoidance of the tax burden for

firms. In the UK there is an overall climate change levy and NEAs are a way of allowing a flexible approach and avoiding tax liabilities.

3. NEAs are used as the central policy instrument in many countries: the German Self-Declaration was promoted as a key instrument to meet their emissions reduction targets. However, the initial (weak) threat of regulation/taxes resulted in an unambitious target of the NEAs, until an increasingly credible threat allowed revision of the NEAs, resulting in somewhat more ambitious targets. It is unclear whether the instrument will live up to its role or whether further improvements will be required. In Italy the Pact for Energy and the Environment was launched as a key instrument for Italian climate change strategy. However, the threat of alternative instruments was not credible, and there was resistance from sectors which did not want to be first movers as they could not see the advantages. The performance of *process* in terms of number of agreements concluded is poor. There is a possible impact mainly in terms of resource development and capacity building. Unless further developments occur, it is unlikely that the Italian Pact for Energy and the Environment will deliver the results it promised.
4. NEAs can be used in a policy mix with emissions trading and Kyoto mechanisms. There is increasing interest in the linkage to these other instruments and some examples of practice. The UK allows access to emissions trading for signatories to the climate change agreements (though with some gateway restrictions). The approach in the Dutch benchmarking agreement is to consider the Kyoto mechanisms as a 'last resort' for firms, suggesting that the government would prefer firms to work without immediate resort to them. For the Italian ENEL agreements, the use of domestic trading is foreseen, as is the use of JI and CDM.

There is also an increasing demand for more mature NEAs which are credible and offer a real potential to be successful and have measures in place to encourage success. It is now less possible to conclude a poor agreement on the understanding that it will be improved in due course. Now NEAs tend to need to demonstrate that they will be successful, and allow further improvements in the future. During the ECCP, there was broad agreement on the base elements for successful agreements:

> for a VA to be successful, it should: be carefully designed and monitored; be negotiated in the presence of credible alternative instruments; be implemented with recourse to appropriate sanctions and incentives; adopt transparent and open reporting that does not impinge on concerns of commercial confidence; include independent verification; and be launched where the government has appropriate technical capacity to negotiate and monitor the EA. Furthermore, the quality of target setting is a key success feature, and for longer agreements there should be a

mechanism for reviewing and renegotiating the targets. (ECCP Report: see note to Box 2.8.)

Regulatory Capture and Free-riding

The first generation of NEAs are perceived as entailing some degree of regulatory capture. This has led to some scepticism about the use of the instrument. However, efforts are under way (regulatory authorities, encouraged by NGOs and parliament) to reduce the level of capture, and ensure that targets are indeed realistic and demanding, and that the instrument is also transparent and appropriately monitored and verified. NEAs can usefully be seen as a tool that follows a 'process' of development rather than a one-off tool. One could expect that, with continued pressure, regulatory capture could be reduced further, but the danger will always remain.

There have to date been no known and important cases of free-riding, but this is probably due to the weak targets set to date, the low 'stakes' of the NEAs, and the fact that systems are not in place to detect free-riding. Again, the NEAs are developing over time, and with continued improvements of monitoring, reporting and verification systems, complemented by credible sanctions, detection of free-riding should become easier. This is important, especially when climate change NEAs start to entail real costs, and the incentives for free-riding become real. At the moment only the proposed UK negotiated agreements look to offer real incentives for free-riding.

Impacts

Emissions reductions from NEAs to date have been modest, and it is difficult to split the allocation of the reductions across NEAs and business-as-usual. One expert (Rietbergen et al., 1998) assessed the share due to NEAs in Dutch LTAs as accounting for around one-quarter of the actual emissions reductions.

It is too early to be conclusive as to the quantitative contribution of NEAs to the national GHG reduction targets. Most experts do not expect NEAs, in their present configuration, to contribute significantly to reaching these targets, despite some governments building a strategy around their use. This does not exclude the possibility that, with increased monitoring, increased pressure, and repeated renegotiation, NEAs could prove a useful tool to help address climate change.

Conclusion on Linkage of Instruments

Some NEAs are linked to permits (Netherlands), some to tax exemption (Denmark and the UK); few are explicitly linked to the emissions trading schemes and the Kyoto mechanisms – JI, CDM (UK offers a link to ET). Subsidies and regulation (for example standards, requirement for audits) are also usually linked to NEAs.

It is clear that NEAs cannot realistically – without linkage to other instruments, and at least incentives and sanctions – offer any significant benefits for emissions reductions and meeting GHG emissions reduction targets.

The linkage to ET and eventually the Kyoto mechanism will undoubtedly play an important role, though the nature and extent of this role will only become fully clear in the future – both for reasons of the uncertainty of the future of the Kyoto Protocol and remaining uncertainties in design and policy positions *vis-à-vis* combining the instruments. However, it is clear that where NEA signatories (or a representative body) can have access to trading schemes, one could expect that NEAs will be able to have stricter targets (and total emission targets), and that they can help contribute to the liquidity of the emissions trading market through the sale of permits from over-compliance with NEA targets, and purchase permits where non-compliant with the NEA targets. It is also important that for any linkage or foreseen linkage of instrument the design implications of such a linkage are taken into account. This will be important for the design of the monitoring system, for target setting and for compliance systems, as well as for future review of an agreement to allow substantiation of any linkage.

Future Potential

It is far from clear at this stage whether NEAs are just a transition measure to address climate change or whether they will become a standard instrument, integrated into the portfolio of measures. While NEAs have the potential to become a long-term instrument, this depends on their achievements, on the culture and country-specific contexts, and on the institutional and instruments context that develops around the climate change issue. It also depends on the actual motivation (and success) of stakeholders in the continued process of revision and improvement of NEAs. Finally, the uptake depends on the level of international commitment and collaboration on combating climate change. At one level, with reduced international commitment, there will be less call for action (at least in some countries and some sectors of all countries), but where there is call for action, the NEA may be perceived as a less risky instrument within a trade and competition-led globalization.

It is clear that the climate change NEAs do have a real learning and regulatory capacity development role, but without clear and quantitative early successes identified and communicated through more credible monitoring schemes, the tool is unlikely to live up to its declared potential and instead might live up to its threat of being a tool for exemptions rather than a tool for action.

A discussion of the future potential cannot ignore the huge uncertainty surrounding the future of the Kyoto Protocol and international efforts to combat climate change. Following President Bush's rejection of the Kyoto Protocol, it is unclear whether the Protocol will be ratified, and also unclear what actual level of international support there will be for addressing climate change. The EU position for the moment is clear – the EU will continue in its strategy to put in place policies and measures to combat climate change and, importantly, continue to move to implement ET at EU level. The effect on the future use of NEAs is unclear, as there are two opposing influences: (a) with no Kyoto Protocol and with continued commitment to address climate change, more governments might prefer to use the NEAs as these are less burdensome on industry than taxes – this would lead to a growth in the use of this instrument. Furthermore, linkage to emissions trading schemes might be strengthened given competitiveness concerns; (b) with no Kyoto Protocol and significantly less interest in climate change, the use of NEAs will drop, or if they do not drop, then they might become weaker instruments and not live up to their potential to contribute to combating climate change. In the second scenario, it is of course entirely possible that there will be more NEAs, but that these will generally be paper agreements.

Recommendations for Use of Agreements

Ultimately it is impossible, without looking explicitly at specific sectors, categorically to recommend or oppose the use of NEAs. It is clear that they can be useful, constructive and effective instruments – but only in some cases and only where negotiated, designed, implemented, monitored and reviewed appropriately. It is also clear that NEAs do not guarantee results and that there are many risks in adopting the measure.

If treated appropriately, NEAs can be a serious and valuable instrument, but not a sufficient instrument to meet GHG emissions reduction targets.

The choice of whether to use NEAs, how to integrate them with existing instruments and policies, and how to launch them combined with other instruments need to be assessed carefully on a case-by-case basis.

In practical terms, therefore, it is important within a national climate change strategy setting to:

- clarify the policy objective that NEAs and/or other instruments are to fulfil;
- carry out an assessment of the structures of the sectors for which an NEA could in principle be considered;

- carry out an *ex-ante* assessment of the possible emissions reductions from the sectors;
- explore what share of these reductions the NEA could have in these sectors, with various assumptions of targets, measures, and in various combinations of other instruments, and for various instruments on their own;
- carry out a cost–benefit and cost-effectiveness assessment of the various instruments, as well as a political feasibility analysis of the various options;
- draw conclusions as to whether the NEA is a realistic and appropriate instrument and for which sectors – and whether the NEA should be the leading instrument or a supporting instrument (for example for offering exemptions to sensitive industries);
- then, for each individual sector where there is promise of a successful NEA, explore and subsequently negotiate the details of the NEA given the particular economic, institutional, technological and cultural circumstances of the sector – and ensure that good practice guidelines are followed for the design and implementation of the agreements, and that the linkages to other instruments ensure a practical and effective solution to the environmental challenge.

If these practical steps are followed, then the risk of agreeing impractical NEAs will be reduced, and the likelihood of agreeing credible, promising NEAs with key sectors should be increased. This will help ensure that the use of NEAs is targeted effectively. This, together with the concerted and coordinated implementation of other policy instruments, will significantly support the likely success of national climate change strategies and the meeting of national emissions reduction targets. It is therefore up to the particular national context whether NEAs will be only a marginal instrument or whether they can become a significant instrument in achieving national climate change objectives.

NOTES

1. The COP7 in Marrakesh (29 October to 9 November 2001) obtained final agreement on the rules governing the implementation of the Protocol, which includes the use of flexible mechanisms and carbon sinks. While this agreement was weaker than the earlier proposal in the Hague (COP6) and the agreement reached in Bonn (COP6bis) earlier in 2001, it provides an agreed basis for ratification in the near future. It can therefore be expected that NEAs have a potential role to play in the countries that ratify the Protocol. Ratification requires that 55 countries ratify the Kyoto Protocol, covering 55 per cent of developed country (1990) emissions. At the time of writing 94 countries, including the EU and Japan, have ratified the Protocol. While the first threshold has been attained, full ratification

still requires additional developed-country ratification to meet the 55 per cent of emissions target threshold, although when the recent Russian and Canadian commitments to ratify are realised, the 55 per cent emissions target will be reached. Full ratification is expected for 2002.

2. http://www.bmu.de/english/topics/climateprotection/agreement.htm
3. http://www.defra.gov.uk/environment/ccl/index.htm
4. http://www.keidanren.or.jp/english/policy/pol058/index.html
5. See http://www.mfe.govt.nz/new/ccrelease.htm
6. The agreements were checked, as standard practice, to see whether any conflict under EU State Aid rules, and were accepted in 2001. The Memoranda of Understanding (MoUs) were actually agreed earlier, with the first ten in 1999 and others in 2000.
7. See http://www.defra.gov.uk/environment/ccl/index.htm and http://www.defra.gov.uk/environment/ccl/agreements.htm
8. See http://info.minez.nl/ezenglish/energy.htm
9. Indeed NGOs are actively campaigning for voluntary environmental agreements (VEAs) in some European countries.
10. This is why Greenpeace in Italy refused to sign the Climate Change Pact, claiming that the objectives were gifts to industry and that no quantified targets had been agreed on.
11. To put it more simply, welfare gains can be had:

    (a) if the NEA leads to a reduced environmental burden (hence social welfare increase), and if companies implementing the NEA make real costs savings from the measures. This is a clear win–win situation;
    (b) where the reduction of the environmental burden leads to welfare improvements that are larger than potential welfare losses to industry (for example benefits are larger than additional costs). This is no longer win–win in the strict sense (though arguably win–win if the NEA offers lower costs to industry than an alternative instrument). Welfare losses accrue where: the benefits of reduced pollution burdens are smaller than the new regulatory costs to industry. Finally, net welfare losses might also occur where there are net benefits from the NEA, but where these are smaller than the benefits from an alternative instrument (often known as 'the counterfactual') to the NEA – these can also be called 'sub-optimal welfare gains'. This links to regulatory capture, where the welfare gains under the NEA are less than the welfare gains to be had under a different policy instrument, as the NEA has led to 'sub-optimal welfare gains' through its lobbying influence.

12. Though in practice this might not be the case, notably where the benefits to the signatories are significant, and there is no loss of competitive position *vis-à-vis* the non-signatories; then there is no economic argument to suggest that the NEA should not be signed.
13. In addition to these 'external' incentives, there is also an 'internal' incentive: cost-effective measures can be encouraged by the NEA, which would otherwise not have been realized, with the NEA helping to overcome an internal information barrier or management failure.
14. NEAPOL (Negotiated Environmental Agreements: Policy Lessons to be Learned from a Comparative Case Study) project. See NEAPOL web page: http://fetew.rug.ac.be/neapol/Index.htm
15. See Green Alliance (2001a and 2001b) and Green Alliance web page: http://www.green-alliance.org.uk/publications.htm
16. The Dutch guidelines are a legal document and noted in Regulation no. 95MOO9543 of 18 December 1995 as issued by the Prime Minister and signed by the PM and the Minister for General Affairs in the Hague.
17. For the ECCP see http://www.europa.eu.int/comm/environment/climat/eccp_longreport_0106.pdf

REFERENCES

Bailey, P. (2002), 'The public enquiry procedure as a model for public participation in environmental agreements', in P. ten Brink (ed.), *Voluntary Environmental Agreements: Process, Practice and Future Use*, Sheffield, UK: Greenleaf Publishing.

Barth, R. and B. Dette (2001), 'Cava Policy brief No 2: The integration of voluntary agreements into existing legal systems', FEEM, Milan, forthcoming.

Bastmeijer, Kees (1996), 'The Covenant as an Instrument of Environmental Policy in the Netherlands'. A Case Study for the OECD/PUMA. April, unpublished script.

BDI (1995), *Declaration by German Industry and Trade on Global Warming Prevention*, Cologne: BDI, March.

BDI(1996a), *Updated and Extended Declaration by German Industry and Trade on Global Warming Prevention*, Cologne: BDI, March.

BDI (1996b), *CO_2 monitoring. Concept for Regular Reporting to provide Transparent and Intelligible Verification of the Declaration by German Industry and Trade on Global Warming Prevention*, Cologne: BDI, February.

BDI (1999), *Voluntary Agreement and Self-commitments by Industry in the Field of Environmental Protection. Situation Analysis*, Cologne: BDI.

Blackman, A. and J. Mazurek (1999), *The Cost of Developing Site-specific Environmental Regulations: Evidence from EPA's Project X2*, Discussion paper 99-35, Washington, DC: Resources for the Future.

Brau, R. and C. Carraro (2001), 'CAVA Policy brief No 3. Are VAs a threat to competition?', FEEM, Milan, forthcoming.

Brau, R., C. Carraro and G. Golfetto (2000), 'Endogenous VAs, coalition formation and market structures', paper presented at the CAVA Workshop on Voluntary Approaches, Competition and Competitiveness, FEEM, Milan, 25–26 May.

Cabuguiera, M. (2002), 'Co-regulation performance factors: Lessons from the theory and practice', in P. ten Brink (ed.), *Voluntary Environmental Agreements: Process, Practice and Future Use*, Sheffield, UK: Greenleaf Publishing.

CAVA (1998), Concerted Action on Voluntary Approaches web site, http://www.ensmp.fr/Fr/CERNA/CERNA/Progeuropeens/CAVA/index.html

CEC (1996), *Communication from the Commission to the Council and the European Parliament on Environmental Agreements* (CEC: COM (96) 561 final, Brussels, 27 November).

CEC (1999), *1999/125/EC: Commission Recommendation of 5 February 1999 on the reduction of CO_2 emissions from passenger cars, Official Journal L 040*, 13/02/1999 pp. 0049–0050.

Chidiak, M. (1998), 'Same Name, but Different Policy Instruments. Voluntary Agreements for Energy Efficiency in five EU Countries', paper presented at CAVA Workshop 'The Worldwide Use of Voluntary Approaches: State of the Art and National Patterns', Ghent, Belgium, 26–27 November.

Chidiak, M. (1999), 'The choice between a tax and a "menu voluntary agreement" under asymmetric information', paper presented at CAVA Workshop 'The efficiency of Voluntary Approaches in environmental policy – what can be derived from theory?', Copenhagen, Denmark, 25–26 May.

CNEL (1999), *Rapporto sullo Stato di attuazione del Patto per l'energia e l'ambiente 1999*, Rome: CNEL.

Convery, F. and F. Lévêque (2001), 'CAVA Policy Brief No 5: Applying voluntary Approaches – some insights from research', FEEM, Milan, forthcoming.

De Clerq, M., B. Ameels and C. Suck (2001), 'The influence of the institutional–

economic context on the performance of negotiated agreements: results from the European NEAPOL project', paper presented at the IEA Workshop on Government–Industry Cooperation to Improve Energy Efficiency and the Environment Through Voluntary Action, Washington, DC, 22 February.

Dutch Ministry of Economic Affairs (1998), *Long Term Agreement on Energy Efficiency: Progress in 1996*, The Netherlands.

Dvorak, A., A. Lisa and P. Sauer (2002), 'Negotiated voluntary agreements: cases in the Czech Republic', in P. ten Brink (ed.), *Voluntary Environmental Agreements: Process, Practice and Future Use*, Sheffield, UK: Greenleaf Publishing.

ECOTEC (1999), *The Role of Voluntary Initiatives to Address Climate Change* (Report for ICC, WBCSD, USCIB, Keidanren, WWF and UNEP).

Eichhammer, W. and E. Jochem (1998), *Voluntary agreements for the reduction of CO_2 greenhouse gas emissions in Germany and their first recent evaluation* (Frauenhofer Institute for Systems and Innovation Research, Karlsruhe, Germany, CAVA Workshop Ghent, Belgium, 26–27 November.

ELNI (Environmental Law Network International) (eds) (1998), *The Role and Effect of Environmental Agreements in Environmental Policies*, London: Cameron May.

Energistylrelsen (1999), *The Danish Agreements on Energy Efficiency*, Copenhagen.

European Environment Agency (1997), *Environmental Agreements: Environmental Effectiveness*, Copenhagen.

Finansministeriet (1999), *Evaluation of Green Taxes and Trade and Industry*, Copenhagen.

Glasbergen, P. (ed.) (1998), *Co-operative Environmental Governance: Public Private Agreements as a Policy Strategy*, Dordrecht: Kluwer Academic Press.

Glasbergen, P. et al. (1997), 'Evaluatie Meerjarenafspraken over energy-efficiency', Department of Environmental Studies and Department of Science, Technology and Society, Utrecht University.

Green Alliance (2001a), *Negotiated Agreements – best practice checklist*, London: Green Alliance, April.

Green Alliance (2001b), *Signed, Sealed and Delivered?: The role of negotiated agreements in the UK*, London: Green Alliance, April.

Jeder, P. (2000), 'Position Paper of the Association of the German Chemical Industry (VCI): Self-commitments as an Instrument of Environmental Policy' (CAVA paper no. 2000/2/20, CERNA, Paris, in Convery and Lévêque (2001).

JEP (Joint Implementation Quarterly) (1998), *New Interactive Approaches in the EU and Selected Member States*, Wageningen, The Netherlands: Wageningen Agricultural University.

Jochem, E. and W. Eichhammer (1996), 'Voluntary agreements as an instrument to substitute regulating and economic instruments? Lessons from the German voluntary agreement on CO_2 reduction', paper presented at conference on 'Economics and Law of Voluntary Approaches in Environmental Policy', organized by CERNA and FEEM, Venice.

Johannsen, K. (2001), 'Policy learning and energy efficiency agreement design in Denmark: 1993–2000', presentation for the IEA Workshop on Government–Industry Cooperation to Improve Energy Efficiency and the Environment Through Voluntary Action, Washington, DC, 22 February.

Krarup, S. (2001), 'CAVA Policy brief No 4: Can voluntary approaches be environmentally effective and ecnomically efficient?', FEEM, Milan, forthcoming.

Krarup, S. and S. Ramesohl (2000), *Voluntary Agreements in Energy Efficiency –*

Implementation and Efficiency, Copenhagen: AKF (Institute of Local Government Studies).

Krarup, S. and S. Ramesohl (2002), 'Voluntary agreements: a key to higher energy efficiency in industry?', in P. ten Brink (ed.), *Voluntary Environmental Agreements: Process, Practice and Future Use*, Sheffield, UK: Greenleaf Publishing.

Legerwood, G. (1997) *Greening the Boardroom*, Sheffield, UK: Greenleaf Publishing.

Lévêque, F. (ed.) (1996), *Environmental Policy in Europe*, Cheltenham, UK and Lyme, USA: Edward Elgar.

Lévêque, F. (1999), *The Economic Analysis of Voluntary Approaches, Voluntary Approaches for Environmental Policy: an assessment*, Paris: OECD.

Maxwell, J.W. and T.P. Lyon (2001), 'Voluntary Approaches to Environmental Regulation', Kelley School of Business, Indiana University, USA, forthcoming.

Maxwell, J.W., T.P. Lyon and C. Hackett (1998), 'Self regulation and Social Welfare: the Political Economy of Corporate Environmentalism', Nota di Lavoro FEEM, Fondazione Eni Enrico Mattei, 55.98.

Millock, K. (1999), 'The Combined Use of Carbon Taxation and Voluntary Agreements for Energy Policy – a Model Based on Danish Policy', paper presented at CAVA Workshop 'The efficiency of Voluntary Approaches in environmental policy – what can be derived from theory?', Copenhagen, 25–26 May.

Ministry of Economic Affairs, Netherlands (1999), *Long Term Agreement on Energy Efficiency: Progress in 1998*, The Hague, Netherlands: Dutch Ministry of Economic Affairs.

Moffet, J. and F. Bregha (1999), *The Implications of Voluntary Codes of Conduct for Competition and Competitiveness*, CAVA Working Paper, CERNA, Paris.

Montedison (1998), *Protocollo d'intesa tra Ministero dell'Ambiente e Montedison*, Rome, 1 December.

Mortensen, B.O.G. (2002), 'Voluntary environmental agreements between private and public law', in P. ten Brink (ed.), *Voluntary Environmental Agreements: Process, Practice and Future Use*, Sheffield, UK: Greenleaf Publishing.

OECD (1998), *The Use of Voluntary Initiatives in the European Union: An Initial Survey* (OECD, ENV/EPOC/GEEI (98)29/Final).

OECD (1999) *Voluntary Approaches for Environmental Policy: an assessment* (Working Party on Economic and Environmental Policy Integration, Environmental Policy Committee) ENV/EPOC/GEEI[98]30/rev.1), Paris: OECD.

Ost, François (1994), 'A Game without Rules? The ecological self-organisation of firms', in G. Teubner, L. Farmer and D. Murphy (eds), *Toeksessa: Environmental Law and Ecological Responsibility*, London: John Wiley, pp. 337–61.

Ramesohl, S. and K. Kristof (2002), 'Voluntary agreements: an effective tool for enhancing organisational learning and improving policy-making?' in P. ten Brink (ed.), *Voluntary Environmental Agreements: Process, Practice and Future Use*, Sheffield, UK: Greenleaf Publishing.

Rietbergen, M., J. Farla and K. Blok (2001), 'Do Agreements Enhance Energy Efficiency Improvement? Analysing the Actual Outcome of Long-Term Agreements on Industrial Energy Efficiency Improvement in the Netherlands', paper prepared for the *Journal of Cleaner Production*, presented at IEA Workshop on Government–Industry Cooperation to Improve Energy Efficiency and the Environment Through Voluntary Action, Washington, DC, 22 February.

Rietbergen, M., J. Farla and K. Blok (1998), 'Quantitative Evaluation of Voluntary Agreements on Energy Efficiency', paper presented at CAVA Workshop 'The

Worldwide Use of Voluntary Approaches: State of the Art and National Patterns', Ghent, Belgium, 26–27 November.
Rigsrevisionen (1998), *Beretning om indførelse, administration og kontrol af den grønne afgiftspakke*, Copenhagen: Rigsrevisionen.
RWI (Rheinisch-Westfälisches Institut für Wirtschaftsforschung) (1999), *Second Monitoring Report: CO_2-Emissions in German Industry 1996–1997*, RWI-papiere, Nr. 57. Essen, July.
Sairinen, R. (2000), *Regulatory Reform of Finnish Environmental Policy*, Helsinki University of Technology.
Salmons, R. (2000), 'Case Studies on Negotiated Environmental Agreements: The United Kingdom: Agreement on Energy Efficiency Improvements', CEC (2000) NEAPOL Project conference paper, Ghent, 30 November 30–1 December.
Segerson, K. and T.L. Miceli (1999), 'Voluntary approaches to environmental protection: the role of legislative threats', in C. Carraro and F. Lévêque (eds), *Voluntary Approaches in Environmental Policy*, Dordrecht: Kluwer Academic Publishers.
Sexton, K., B.S. Murdock and A.A. Marcus (2002), 'Co-operative environmental solutions: Acquiring competence for multi stakeholder partnerships', in P. ten Brink (ed.), *Voluntary Environmental Agreements: Process, Practice and Future Use*, Sheffield, UK: Greenleaf Publishing.
Storey, M. (1997), 'Demand Side Efficiency: Voluntary Agreements with Industry', Policies and Measures for Common Action working papers, No. 8, Paris: OECD.
Strachen, P. and B. Clark (1997), *Managing Green Teams: Cases Issues and Concepts*, Sheffield, UK: Greenleaf Publishing.
ten Brink, P. (2002), *Voluntary Environmental Agreements: Process, Practice and Future Use*, Sheffield, UK: Greenleaf Publishing.
ten Brink, P. and M. Morère (2000), 'Voluntary Initiatives to Address Climate Change in the EU', *European Environment*, **10**, 178–92.
ten Brink, P. and M. Morère (2002), 'Monitoring mechanisms for efficient environmental agreements', in P. ten Brink (ed.), *Voluntary Environmental Agreements: Process, Practice and Future Use*, Sheffield, UK: Greenleaf Publishing.
ten Brink, P., V. Pollard, J. Medhurst, M. Morère and P. Repinski (1999), 'Voluntary Agreements: Lessons for Future Use', *Greener Management International*, Issue 26.
Torvanger, A. and T. Skodvin (2002), 'Environmental Agreements in Climate Politics', in P. ten Brink (ed.), *Voluntary Environmental Agreements: Process, Practice and Future Use*, Sheffield, UK: Greenleaf Publishing.
UNEP (United Nations Environment Programme) (1994), *Company Environmental Reporting: A Measure of Progress of Business and Industry towards Sustainable Development*, UNEP IE Technical Report No. 24, UNEP Industry and Environment Programme.
UNEP (United Nations Environment Programme) (1997), *Reducing Greenhouse Gas Emissions: The Role of Voluntary Programmes*, Paris: UNEP Industry and Environment Programme Activity Centre.
Van Luyt, P. (2001), 'LTAs and the recent covenant benchmarking energy efficiency agreements in the Netherlands', presentation and fact sheet prepared for the IEA Workshop on Government–Industry Cooperation to Improve Energy Efficiency and the Environment Through Voluntary Action, Washington, DC, 22 February.

3. Kyoto flexible mechanisms: opportunities and barriers for industry and financial institutions

Josef Janssen

1. INTRODUCTION

While the previous chapter analysed the incentives to negotiate and adopt agreements to reduce GHG emissions at the industry level, this chapter analyses the incentives to reduce emissions provided by the so-called 'flexible mechanisms'.

The Kyoto Protocol to the United Nations Framework Convention on Climate Change (UNFCCC), adopted in December 1997 by more than 150 countries, established global markets for GHG emissions reductions.[1] By means of the market-based Kyoto mechanisms of joint implementation (JI), clean development mechanism (CDM) and international emissions trading (IET), it will be possible to trade emission permits globally. Demand will be driven by the emissions reduction commitments of industrialized and transitional countries (the group of Annex B countries).[2] JI and IET enable international transactions of emission permits among Annex B countries. The CDM allows developing countries, that is, non-Annex B countries, to deliver emission permits to Annex B countries. This emerging global market, the net value of which is estimated to amount to probably several US$10 billion annually,[3] will offer challenging opportunities to a wide variety of different players in industry and business.

Will the reluctance of the US administration to ratify the Kyoto Protocol threaten the emergence of such GHG markets? Yes, but only to some extent.

First, the Intergovernmental Panel on Climate Change (IPCC), in its recent third assessment report, reinforces the need for climate change mitigation. Hence the issue will not go away. Moreover, the findings of the IPCC have been reconfirmed by a recent study commissioned by the Bush administration from the renowned US Academy of Science.

Second, at the national level industrialized countries, especially in Europe, are continuing to design policies aimed at meeting the Kyoto obligations. In most cases, emissions trading is envisaged to play an important role. Indeed, in October 2001 the European Commission released a proposal for a directive that envisages the establishment of a mandatory CO_2 emissions trading system by 2005 for the following sectors and economic activities: combustion installations with a rated thermal input exceeding 20 MW (excepting hazardous or municipal waste installations); mineral oil refineries; coke ovens; production and processing of ferrous metals including iron and steel; production of cement and lime; manufacture of glass and ceramic products; pulp and paper. The emissions trading system of the UK started in April 2002.

Third, any institutional framework that may emerge as an alternative to the Kyoto Protocol (which is not, however, very probable) will rely heavily on market-based instruments such as (international) emissions trading. Most importantly for the present chapter, international emissions trading under any alternative institutional framework may be expected to resemble the Kyoto mechanisms as regards many functional and technical elements. To give an example, any form of project-based emissions trading will involve the issue of how to determine baselines, which is a major point in the present debate about the Kyoto mechanisms. Moreover, any international market for GHG emission permits, be it under the Kyoto Protocol or any alternative framework, may be expected to encompass some form of project-based emissions trading, as represented by JI and the CDM (see Chapter 1 and Section 2 of this chapter). For this reason, the present analysis of barriers and opportunities associated with the Kyoto mechanisms is not specifically tied to the flexible mechanisms of the Kyoto Protocol but has general validity, even if explicit reference is made to the provisions of the respective articles of the Kyoto Protocol. Indeed, such explicit reference does not imply, by any means, that this chapter becomes irrelevant or completely outdated if in the future the parties to the UNFCCC were to agree to abandon the Kyoto Protocol and to search for alternative avenues for consensus.

Fourth, in anticipation of the implementation of the Kyoto Protocol, or any other institutional framework that might emerge as an alternative to the Kyoto Protocol, major companies around the globe from a whole range of industries are taking actions to prepare themselves for emerging national and international markets for GHG emission permits. This group of proactive companies includes BP, Shell, Transalta, Ontario Power Generation, DuPont, Lafarge, Swiss Re, to name some of them.

None the less, the reluctance of the new US administration to ratify the Kyoto Protocol may be a major threat to emerging global markets for GHG emission permits, if other major GHG emitting countries abandon their opposition to Bush's position. In this case, the Kyoto Protocol may fail to enter into force. It will enter into force if the following conditions are fulfilled:

- 55 parties have ratified the Kyoto Protocol.
- These 55 parties must include Annex B countries that in 1990 accounted in total for more than 55 per cent of total CO_2 emissions of the group of Annex B countries.

As of 1 June 2002, 74 parties have ratified or acceded to the Kyoto Protocol, including the following Annex 1 countries: Austria, Belgium, Czech Republic, Denmark, Finland, France, Germany, Greece, Iceland, Ireland, Italy, Japan, Luxembourg, the Netherlands, Norway, Portugal, Romania, Slovakia, Spain, Sweden and the UK. They account for 35.8 per cent of the CO_2 emissions of the group of Annex 1 countries.

Table 3.1 shows that the Kyoto Protocol could enter into force without US ratification. Ratification by the EU, the countries with economies in transition (EIT) and Japan would be insufficient for entry into force: it is also necessary that Russia ratifies. Both Japan and Russia have declared that the Kyoto Protocol has become ratifiable on the basis of the Marrakesh Accords which were agreed upon at the seventh Conference of the Parties (COP7) in November 2001.

Table 3.1 Share of CO_2 emissions of Annex B countries

Country	Share of CO_2 emissions in 1990 in percentage of total Annex 1 emissions
USA	36.1
EU	24.2
Russia	17.4
Japan	8.5
EIT	7.4
Other	6.4

This chapter aims to discuss main opportunities for industry, including the financial sector, associated with the flexible mechanisms. Furthermore, it identifies major barriers companies might face in realizing such opportunities. To this end, the remainder of this chapter is organized as follows. The next section (Section 2) extends the discussion of the nature of the flexible mechanisms in Chapter 1, adding some aspects which are relevant for industry as their prospective user.

Since it is the aim of the chapter to explore business opportunities associated with the global GHG market, it is important to have an idea of its size. Therefore, Section 3 gives an overview of the results of different economic models that estimate the potential size of the global GHG market. Two basic international trading regimes are distinguished: an international trading regime with unrestricted use of the CDM (global trading), and an international trading regime that encompasses only trading among Annex B countries. The latter may be a realistic scenario if the rules on the CDM are very restrictive, which does not seem to be the case after Marrakesh. Since prices are the key parameters in business decisions, the section also presents estimates of permit prices under both regimes.

Section 4 explores opportunities and barriers for GHG-emitting industries that are interested in taking advantage of the flexible mechanisms. As regards opportunities, one has to distinguish between potential buyers and sellers of emission permits. The former would benefit from reduced cost of compliance with domestic climate policy regulations. The latter could take advantage of the Kyoto mechanisms in order to generate additional profits. The main barriers discussed include uncertain rules, guidelines, procedures and modalities that will govern the Kyoto mechanisms. They are still uncertain since the Marrakesh Accords have yet to be implemented at the national level. Other barriers include the lack of incentives at the national level to use the flexible mechanisms, transaction costs and uncertain prices.

Section 5 explores some functions that financial institutions could perform in emerging GHG markets, and Section 6 concludes.

2. SOME ADDITIONAL KEY FEATURES OF THE KYOTO MECHANISMS

The nature of the Kyoto mechanisms has already been discussed to some extent in Chapter 1. In this section, I want to explore some additional points regarding the basic nature of the Kyoto mechanisms that are relevant for industry as prospective users.[4]

To call the Kyoto mechanisms to the reader's mind, they encompass the following instruments:[5]

- Joint implementation – JI (Article 6, Kyoto Protocol): international transactions of emission permits between Annex B countries.[6]
- Clean development mechanism – CDM (Article 12, Kyoto Protocol): international transactions of emission permits from non-Annex B countries to Annex B countries.[7]
- International emissions trading – IET (Article 17, Kyoto Protocol): international transactions of emission permits between Annex B countries.[8]

What strikes one at first glance is that the Kyoto Protocol provides for two forms of international mitigation activities among Annex B countries, that is, JI and IET. Against this observation, the question arises how JI and IET might be distinguished in essence, and how they relate to each other.

2.1 Distinguishing JI and IET

In the relevant literature, three main criteria for distinction have been advanced:

- private sector versus government participation;
- baseline-and-credit versus cap-and-trade or cap-and-allowance trading;
- international investments versus international trade.

In the following, I will briefly discuss these criteria.[9]

Sometimes it has been argued that IET only allows international emissions trading among governments. If private entities, that is GHG-emitting companies, wish to trade GHG reductions internationally, they would need to use JI. According to this view, the main criterion for distinguishing IET and JI refers to participants: regarding IET, only governments are allowed to participate, whereas JI is also available for private entities.

This perception originates from the language of the respective Articles of the Kyoto Protocol: Article 6.3 on JI explicitly states that 'a Party included in Annex I may authorize *legal entities* to participate … in actions leading to the generation, transfer or acquisition under this Article of emission reduction units'.[10] Article 17, which refers to IET, does not mention explicitly any private sector involvement. Instead, it only states that 'Parties included in Annex B may participate in emissions trading …'.

Several arguments might be advanced against the perception that Article 17 only enables government-to-government trading: first of all, it is rather unusual that public international law, as is the Kyoto Protocol, explicitly refers to private sector entities. Hence the language of Article 6, referring to legal entities, is rather unusual or abnormal. The non-reference to private entities of Article 17 is instead the normal case in public international law. And second, the Marrakesh Accords explicitly state that legal entities may be authorized to participate in IET. Concluding, it does not appear to be appropriate to distinguish between JI and IET by a criterion of allowed participation.

Another criterion for distinguishing JI and IET is the type of emissions trading system to be governed by JI and IET. As already discussed in Chapter 1, in the literature on emissions trading systems, a most important distinction is drawn between cap-and-allowance (or cap-and-trade) systems and baseline-and-credit systems. In both schemes, the generic term for the unit of trade is an emissions permit. As already indicated by the names, permits in the former system are called allowances, whereas credits are earned in the latter. The most prominent example of a successful cap-and-trade system is the US SO_2 trading system.[11]

In a baseline-and-credit system, emission credits are earned if actual project emissions are below a reference value, or baseline, against which reductions are measured. The emissions baseline in a credit scheme can be identical to the emissions cap in an allowance scheme. However, the two schemes have different implications regarding the timing and extent of regulatory involvement: cap-and-trade schemes, which are comprehensive by their nature, require an extensive regulatory involvement and hence effort at the beginning to set them up. In contrast, credit schemes require less initial design and inception effort, but baselines need to be determined on an individual basis and individual trades must be certified and/or verified by the regulator. A credit system depends on a project-by-project analysis, whereas an allowance system depends on an inventory analysis of the regulated entities.

Both approaches have their advantages and disadvantages, and there is considerable dispute about which system is more efficient and hence more desirable. At present, the parallel implementation of both systems, cap-and-trade and baseline-and-credit, is being considered in some countries, for example in the UK (DETR, 2000; UK Emissions Trading Group, 2000, and Chapters 1 and 2 of this volume). What is more important in the present context is how JI and IET might relate to these two systems. Indeed, it may be argued that JI can be regarded as international baseline-and-credit trading, while IET constitutes a form of international cap-and-trade system (Janssen, 2001 or 2000a).

Which trading system is more appropriate? It is conceivable that for certain countries JI is the more efficient trading institution, whereas for other countries transactions under IET are more efficient. Indeed, it is sometimes argued that the economies in transition do not have the (financial) resources required to set up a comprehensive cap-and-trade system. Hence it could be more efficient for them to participate in international emissions trading under JI. Other countries which value differently the costs associated with setting up a comprehensive cap-and-trade system could find it more desirable to opt for this trading institution. Moreover, one has to distinguish between different GHGs and different sources. It could be true that IET is more efficient in the context of CO_2 emissions from the power sector. Regarding other GHGs and sources, like methane emissions from agriculture or CO_2 emissions from industrial processes, transactions under JI might be associated with lower regulatory set-up or transaction costs.

Coming to the third criterion, it is frequently argued that JI involves international (equity) investments: Grubb et al., for example, state that 'JI enables emission savings … arising from *cross-border investments* between Annex B parties to be transferred between them' (1999, p. 131). And 'the two project-level mechanisms JI and CDM are intended to use *international* corporate *investment* as an engine for the generation and transfer of emission credits' (ibid., p. 195). From this perspective, another distinction that might be drawn between JI and IET is that the former represents international production of emission permits involving international investments, whereas the latter constitutes international trade in emissions permits. And it is important to note that production and trade are two fundamentally different economic activities.

However, in the Kyoto Protocol there is no provision requiring that JI *must* involve international investments. Hence it seems to be more convincing to distinguish between JI and IET according to the criterion baseline-and-credit versus cap-and-allowances trading. International investments are *not* to be regarded as the distinguishing quality of JI projects, but the international transaction of resulting emissions permits. JI projects *may* involve international investments, but *do not require* them from a regulatory or legal point of view. This is illustrated by Figure 3.1.

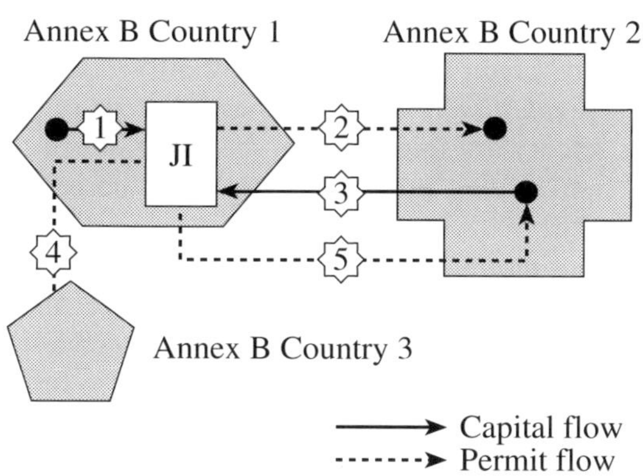

Figure 3.1 International permit flows as necessary attribute of JI projects

JI projects may involve combinations of the following investment and permit flows: (1;2), (1;4), (3;2), (3;4), (3;5). On the other hand, international (and national) investments in climate protection projects the emission credits of which are sold and used for compliance in the host country would not qualify as JI projects. Thus, the combinations of transactions (1;2) or (3;2) in Figure 3.2 would not be governed by JI.

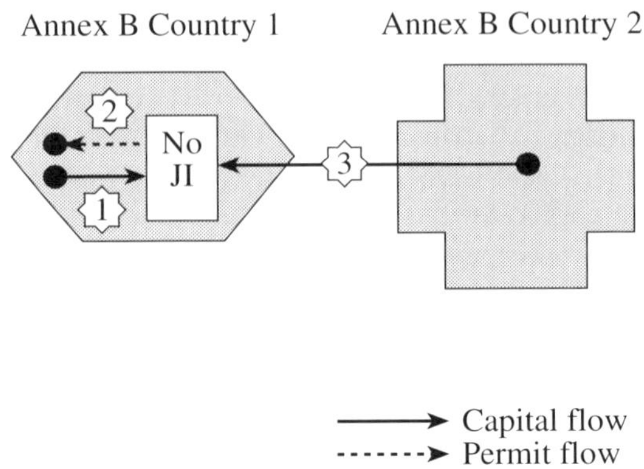

Figure 3.2 International capital flows as insufficient attribute of JI projects

2.2 CDM as Baseline-and-credit Trading and the Issue of a Unilateral Model

What does the previous discussion imply for the CDM? Regarding the first criterion, the Kyoto Protocol clearly allows private sector participation. With respect to the second criterion, the wording of Article 12 on CDM suggests that international GHG transactions under the CDM would constitute some form of baseline-and-credit trading: Article 12.5(c) requires that 'reductions in emissions … are *additional to any that would occur in the absence* of the certified project activity'. In this wording, the emissions that would 'occur in the absence of the certified project activity' may be regarded as baseline emissions.

Concerning the criterion of international production versus international trade, it is frequently said that CDM projects involve international investments, thus constituting international production of emissions permits. On the other hand, it is also argued that CDM projects formally would not necessarily require any foreign investments. Instead, 'a host country could already finance projects on its own and sell credits earned. Article 12 would not prevent this' (Tietenberg et al., 1999, p. 49). Such a financing (and trading) model has been referred to as a unilateral model.

In Figure 3.3 the unilateral model is characterized by the combination of investment and permit flows (1;2). Under the unilateral model 'the developing country would … itself be acting as the main project *investor* and would attain the benefits as well as absorb the associated project risks' (Stewart et al., 1999, p. 28).[12] In such a unilateral model there would be no foreign equity capital involvement, as opposed to the bilateral model which is characterized by international investments by entities from Annex B countries (see

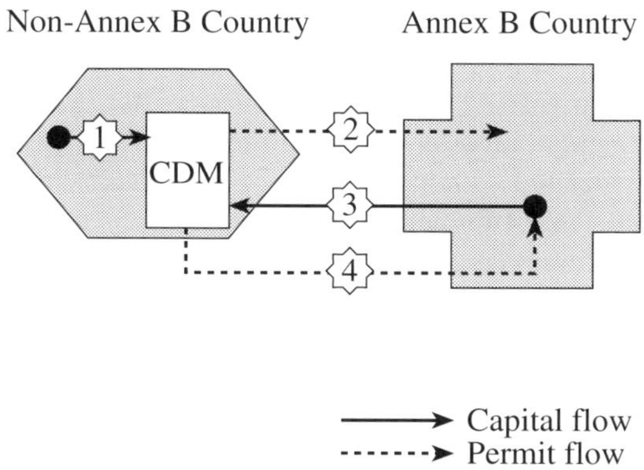

Figure 3.3 Unilateral and bilateral CDM model

combination (3;4)). On the international political agenda, the issue of the unilateral model had attracted some attention. It seems that the Marrakesh Accords do not prevent developing countries from applying the unilateral model. Indeed, it may be expected that the majority of CDM projects will be implemented unilaterally, since capital formation in most developing countries is financed predominantly by domestic sources.

2.3 Which Type of Kyoto Mechanism for International Transactions in GHG Emission Permits?

It seems that JI and CDM would represent some form of international baseline-and-credit system. In addition, they could involve international investments.

Since international transactions of GHG permits in a baseline-and-credit system between Annex B countries seem to be governed by JI, it follows that Article 17 on IET would only apply to international transactions out of a cap-and-trade system. Otherwise, either Article 6 on JI or Article 17 on IET would be redundant due to institutional arbitrage by market participants.

Emission permits of all three Kyoto mechanisms are fungible, at least as regards international guidelines. As a consequence, all three Kyoto mechanisms will compete with each other: prospective buyers want to achieve compliance at lowest cost. This means that they will buy those emission permits that sell at minimum price. Thus, if emission permits generated from a CDM project are less expensive than emission permits available under JI or IET, prospective buyers will use CDM credits for compliance.

Regarding such competition between the different Kyoto mechanisms, it is frequently maintained that CDM projects might involve lower abatement costs as compared to projects in industrialized countries. On the other hand, it is becoming apparent that CDM projects will be burdened by more administrative regulations and procedures such as validation, verification and certification, thus involving higher transaction costs. In addition, from the perspective of potential investors, CDM projects, being located in developing countries, are associated with higher political and country risks.

Regarding the level of implementation of emissions trading schemes, one can distinguish different categories:

- National/domestic, regional (for example EU-wide), and international trading schemes.
- Intra-firm, inter-firm, firm-to-government, and government-to-government emissions trading.

All cross-country transactions will be governed by the Kyoto mechanisms or similar forms of project-based or inventory-based emissions trading.

Concluding, international transactions related to abatement activities could be classified on the basis of the following different criteria:

- Transferring entity is from (1) Annex B country or (2) non-Annex B country.
- Acquiring entity is from (1) Annex B country or (2) non-Annex B country.
- Transferring entity has (1) a baseline or (2) operates under a cap.
- Production of emission permits involves (1) no international investments or (2) international investments.

Table 3.2 shows the 16 different transaction types that emerge when combining these criteria. In the last row an attempt is made to assign the different types of transactions to the three Kyoto mechanisms. Even if policy makers, researchers and practitioners might not agree with this assignment, it would be desirable to be clear about the precise nature of JI, CDM or IET in terms of transaction types when discussing the Kyoto mechanisms. This is true as regards policy making, research and practice: to design efficient rules and guidelines for JI, CDM and IET, policy makers should have a clear idea of the nature of underlying transactions. The same is true for economic research on the Kyoto mechanisms: in order to analyse, for example, the implications of JI, CDM and IET for technology innovation and diffusion, one would obviously need to have a clear understanding about the economic characteristics of these concepts. And firms involved in international GHG transactions need to have a clear understanding about the regulatory framework, in terms of Kyoto mechanisms, that governs these transactions.

Table 3.2 Types of international GHG transactions

	Transaction number															
Transaction features	1	2	3	4	5	6	7	8	9	10	11	12	13	14	15	16
Transferring entity is from:																
Annex B country	X	X	X	X	X	X	X	X								
Non-Annex B country									X	X	X	X	X	X	X	X
Acquiring entity is from:																
Annex B country	X	X	X	X					X	X	X	X				
Non-Annex B country					X	X	X	X					X	X	X	X
Transferring entity has a:																
Baseline	X	X			X	X			X	X			X	X		
Cap			X	X			X	X			X	X			X	X
Production of emissions permits involves:																
No international investments	X		X		X		X		X		X		X		X	
International investments		X		X		X		X		X		X		X		X
Institutional framework	JI	JI	IET	IET	n.a.	n.a.	n.a.	n.a.	CDM u.m.	CDM b.m.	???	???	n.a.	n.a.	n.a.	n.a.

Note: n.a. = not available; u.m. = unilateral model; b.m. = bilateral model.

3. SIZE OF EMERGING GLOBAL MARKETS FOR GHG EMISSIONS PERMITS

The Kyoto mechanisms constitute the instrumental basis of emerging global GHG markets. Examining opportunities for industry that are associated with these markets, it is interesting to have a rough idea about their size.

Several economic studies have tried to quantify the costs of reaching the commitments of the Kyoto Protocol, thereby also focusing on the scope of reducing these costs through the Kyoto mechanisms. In this section, I give an overview of the results of these models.[13]

Different model types as well as diverging projections of emissions growth without climate policy (reference case) are the main reasons for the large differences found between the model results. A common finding of all studies is that trading lowers the costs of achieving the commitments in the Kyoto Protocol. Model results can serve as indicators of the permit prices and the likely size of the emerging market for GHG permits. However, their limitations and implicit assumptions should be noted carefully.[14]

All models (around 20) discussed by Springer (2000) share one common finding: trading lowers costs substantially. If no emissions trading is allowed, each country has to meet its reduction obligation on its own. However, for some countries with high carbon efficiency (such as Japan), this would be very expensive. Therefore, autarkic abatement costs without trading would be very high. Trading allows those countries and industries with lower costs to reduce more and those with higher costs, to abate less. Of course, the overall reduction is the same under trading. But overall costs are substantially lower. This is shown by Table 3.3.

The second, third and fourth column show estimates of marginal abatement cost by different models for the USA, Japan and the EU, respectively, in a situation without international trade. The last column lists estimates of world market prices with global trading. Such prices would be equal to marginal abatement cost if global trading were allowed.

At present, several European countries such as the UK, Germany, Norway, Sweden and Switzerland are considering setting up domestic GHG trading schemes. The European Commission strives for the establishment of an EU-wide trading scheme by 2005. The Danish trading system is already operational. What would such a European market look like as regards market prices?

Table 3.4 reports marginal costs if all EU countries were to meet respective Kyoto targets through domestic action only. In addition, this table shows market prices for different trading scenarios. As can be clearly seen, marginal abatement costs for the EU as a whole could be significantly reduced by opening up internationally an EU-wide trading scheme.

Table 3.3 Marginal abatement costs with domestic action only and with global trading

Model	Marginal abatement costs with domestic action only (1998 US$/tonne CO_2)			World market price (1998 US$/tonne CO_2)
	USA	Japan	EU	
EPPA	64	201	94	8
G-Cubed	26	33	53 (OECD other than USA, Japan, Australia)	7
SGM	51	139	40	8
MS-MRT	91	161	70	10

Sources: Bernstein et al. (1999a); Ellerman et al. (1999); MacCracken et al. (1999); McKibbin et al. (1999).

Coming back to the more aggregate international level, Table 3.5 shows permit prices and net international trade volumes in both physical and value terms estimated by the models for an international trading regime with unrestricted CDM (full global trading scenario).[15]

Average estimated price is US$9.6 per tonne of CO_2, which is quite close to the median (US$8 per tonne of CO_2). The range of estimated prices is broad: the lowest price is US$4; the highest US$22.[16] But this wide price range is not necessarily a problem for decision makers in affected industries, since abatement costs of single projects are not continuous but discrete. Consequently, it may be that known costs of different abatement technologies vary to an even greater extent than the prices indicated above.

Only some models explicitly estimate the number of permits traded. Average annual net international trade volumes in physical terms range from 1540 million tonnes CO_2 (GRAPE) to 3428 million tonnes (EPPA).

Estimates of average net international annual trade volume in value terms are much closer together than estimates of permit prices and trade volumes in physical terms.[17] They range from US$8476 million to US$27 424 million. Under global trading, the average trade volume of all models is US$18 843 million, being very close to the median (US$20 020 million).

Under a scenario in which international trading is restricted to Annex B countries, that is, if the rules on the CDM are extremely restrictive, all prices

Table 3.4 Marginal abatement costs and market price with and without emissions trading within the EU

	Kyoto target	Marginal costs of domestic action EUR 1999/t CO_2	Trading scenario	Market price EUR 1999/t CO_2
Austria	–13	28.4	EU-wide emissions trading among energy suppliers	32.3
Belgium	–7.5	89.3	EU-wide emissions trading among energy suppliers and energy intensive sectors	33.3
Denmark	–21	47.9	EU-wide emissions trading, all sectors	32.6
Finland	0	63.5	Annex-1-wide emissions trading	17.7
France	0	20.6		
Germany	–21	13.5		
Greece	25	39.0		
Ireland	13	53.5		
Italy	–6.5	33.3		
Netherlands	–6	150.7		
Portugal	27	41.1		
Spain	15	27.7		
Sweden	4	39.7		
United Kingdom	–12.5	31.9		

Source: Capros et al. (2000).

Table 3.5 Prices and net international trade volumes under international trading regime with unrestricted CDM

Model	Permit price (1998 US$/ton CO_2)	Net international trade volume as annual average (million tonnes CO_2)	Net international trade volume as annual average (million 1998 US$)
AIM	12	1 833	21 996
ECN	4	2 119	8 476
EPPA	8	3 428	27 424
G-Cubed	7	3 318	23 226
GEM-E3	8	–	–
GRAPE	13	1 540	20 020
GREEN	7	2 427	16 989
MERGE	22	–	–
MS-MRT	10	–	–
POLES	6	2 295	13 770
RICE-98	5	–	–
ZEW	13	–	–
Average	9.6	2 423	18 843
Median	8.0	2 295	20 020

Sources: Springer (2000); based on Bernstein et al. (1999b); Böhringer (2000); Capros (1999); Criqui and Viguier (2000); Ellerman et al. (1999); Kainuma et al. (1999); Kurosawa et al. (1999); Manne and Richels (1999); McKibbin et al. (1999); Mensbrugghe (1998); Nordhaus and Boyer (1999); Sijm et al. (2000).

are higher and trade volumes are lower compared to the global trading scenario (see Table 3.6).

The results again differ significantly: the lowest price estimate is US$5 per tonne of CO_2 and the highest US$70.[18] On average, models predict a price of US$24.6 per tonne of CO_2, which is close to the median (US$18).

As expected, trade volume in physical terms is smaller under an international trading regime with restricted CDM because of the smaller number of potential sellers and a higher permit price. All models estimate that permit trades would decrease significantly. The average trade volume in physical terms is 1613 million tonnes CO_2 under Annex B trading, compared to 2423 million tonnes under global trading.

Interestingly, most models predict a higher trade volume in value terms with restricted than with unrestricted CDM. The average trade volume in

Table 3.6 Prices and trade volumes under international trading regime with restricted CDM

Model	Permit price (1998 US$/ton CO_2)	Trade volume (million tonnes CO_2)	Trade volume (million 1998 US$)
AIM	21	1 467	30 807
ECN	5	–	–
EPPA	44	1 265	55 660
G-Cubed	18	2 017	36 306
GEM-E3	16	–	–
GRAPE	22	1 283	28 226
GREEN	18	1 503	27 054
GTEM	36	–	–
MS-MRT	29	1 852	53 708
Oxford	70	1 074	75 180
POLES	17	1 467	24 939
RICE-98	18	–	–
WorldScan	6	2 592	15 552
Average	24.6	1 613	38 604
Median	18.0	1 467	30 807

Sources: Springer (2000); based on Bernstein et al. (1999b); Bollen et al. (1999); Capros (1999); Cooper et al. (1999); Criqui and Viguier (2000); Ellerman et al. (1999); Kainuma et al. (1999); Kurosawa et al. (1999); McKibbin et al. (1999); Mensbrugghe (1998); Nordhaus and Boyer (1999); Sijm et al. (2000); Tulpulé et al. (1999).

value terms under Annex B trading is US$38 604 million, more than twice as much as under global trading.

Major seller countries under an international trading regime with unrestricted CDM are China, India and the Russian Federation. Important buyer countries are the EU, Japan, and the USA (if the USA participated in the Kyoto Protocol). Within the EU, Greece, Italy and the Netherlands are the largest buyers. In an international trading regime with restricted CDM, major importing countries are about the same. On the supply side, Austria, Eastern Europe, France, Germany and the Russian Federation emerge as exporters of GHG emission permits.

Table 3.7 shows that shares of permit imports in total reductions for three major importing regions, namely the USA (if they participated in the Kyoto Protocol), the EU and Japan. Under an Annex B trading scenario, these regions meet between one- and two-thirds of their reductions requirements

Table 3.7 Share of international permit imports in total reductions (%)

	International trading regime					
	World trading			Annex B trading		
Model	USA	EU	JAPAN	USA	EU	JAPAN
ECN	83	59	86	–	–	–
EPPA	68	76	92	19	35	66
GREEN	50	68	67	29	48	48
POLES	82	78	–	51	52	62

Sources: Springer (2000); based on Burniaux (1998); Criqui and Viguier (2000); Ellerman et al. (1999); Sijm et al. (2000).

through imported permits. Obviously, fewer domestic measures are undertaken under global trading because permit prices are lower. The estimated shares of imported permits in total reductions are from 50 per cent up to 92 per cent.

Several studies have tried to determine the relative size of international imports under the CDM as opposed to international transactions among Annex B countries (see Table 3.8).[19]

Estimates of the value of annual permit trade under JI and IET range from US$1727 million (Zhang) to US$11 920 million (Haites, 1998).[20] The market share of these two Kyoto Mechanisms is expected to lie between a quarter and half of the unrestricted global market for tradable GHG permits in the first commitment period. Emissions reductions generated under the CDM have a larger share of the market: estimates range from 55 per cent to 77 per cent. In value terms, the market volume is between US$3102 and US$21 208 million. The market shares could change due to transaction costs or political restrictions. Political constraints could be placed on the use of Kyoto mechanisms by individual countries, for example by not allowing certain project types under the CDM. Such restrictions could influence the demand for permits of specific project types.

4. OPPORTUNITIES AND BARRIERS FOR GHG-EMITTING INDUSTRIES

Key players on both the demand and the supply side of global GHG markets are GHG-emitting industries. They will benefit from GHG markets by achiev-

Table 3.8 Market share of the CDM

Model/ Author	International emissions trading and joint implementation: Trade volume (million tonnes CO_2)	(million 1998 US$)	Market share (%)	CDM: Trade volume (million tonnes CO_2)	(million 1998 US$)	Market share (%)
ECN	3 227	3 227	42	4 543	4 543	58
EPPA	774	6 189	23	2 651	21 208	77
G-Cubed	1 503	10 523	45	1 815	12 705	55
GREEN	972	6 802	40	1 456	10 190	60
Haites	1 192	11 920	36	2 108	21 083	64
POLES	986	7 891	38	1 606	12 848	62
SGM	1 309	10 472	44	1 665	13 317	56
Zhang	576	1 727	35	1 071	3 212	65

Sources: Springer (2000); based on Criqui and Viguier (2000); Ellerman et al. (1999); Haites (1998); MacCracken et al. (1999); McKibbin et al. (1999); Mensbrugghe (1998); Sijm et al. (2000); Zhang (2000a).

ing compliance at lower cost or by producing and selling GHG emission permits to other companies, thus earning additional income.

Knowing the GHGs controlled by the Kyoto Protocol, it is straightforward to deduce the sectors and industries emitting GHGs.[21] They include (Watson et al., 1996, part III):

- production of natural gas and hydrocarbon oil;[22]
- coal mining;
- transmission and distribution of natural gas;
- refining of hydrocarbon oil and petrochemistry;
- power production based on fossil fuels;[23]
- production of basic metals including iron and steel, aluminium and copper;[24]
- chemical industry;
- pulp and paper production;
- production of construction materials including cement and concrete, brick and glass;[25]
- fossil-fuel-based transportation;

- fossil fuel consumption in residential and commercial buildings;
- waste and wastewater disposal;
- agriculture and forestry.

Depending on the comparative cost advantage of individual firms operating in these industries, they will be potential buyers or sellers of emission permits. In the following, I will explore potential opportunities and barriers existing on the demand and supply side.

4.1 Opportunities and Barriers for GHG-emitting Industries: Demand Side

All sectors and industries in Annex B countries which are emitting GHGs controlled by the Kyoto Protocol will potentially be on the demand side – and only potentially, for several reasons.

First, some governments may decide to regulate only certain GHG-emitting industries by climate policies. Experience in recent energy and environmental taxation initiatives in several European countries indicates that governments tend to exempt energy-intensive and export-oriented industries from taxation (Ekins and Speck, 1999 or Hanisch, 1998). Here is not the place to discuss the efficiency or possible political-economy explanations of such policy approaches. However, they make evident that it cannot necessarily be expected that the intensity of climate policy regulations will coincide with the GHG emission quantities of different industries.

Second, climate policy regulations may be implemented at various stages of the production cycle: the literature distinguishes between upstream and downstream regulation.[26] The former refers to policies that aim at restricting the carbon input to the economy. They address producers and importers of primary energy like natural gas, coal and oil. The latter aim at directly regulating final emitters of GHGs like power plants, and households and firms that use fossil fuels for heating and transportation. Both approaches can be designed effectively. But from an economic perspective, they are not necessarily equivalent, since they may involve different price signals and transaction costs. In any case, it will depend on the concrete policy approaches adopted in Annex B countries whether upstream industries like oil and gas companies or downstream emitters like power producers and processing industries represent the major groups of final buyers of GHG permits.

Third, some regulated industries and companies might be able to reduce emissions internally at marginal abatement costs that are lower than present or future market prices for GHG emission permits. In this case, it is economically rational to reduce GHG emissions at regulated sources instead of buying emission permits on national or international markets.[27]

Fourth, the potential to take advantage of the Kyoto mechanisms will also depend on the size of the firm in question. Most companies outside the USA are not familiar with emissions trading schemes and the Kyoto mechanisms. Thus, many firms would need to build up relevant expertise and know-how. This is especially true in the case of the Kyoto mechanisms, the application of which may be expected to be more complicated than purely national transactions. Now, such investments in required human capital, infrastructure and networks involve additional costs which, to a large extent, are fixed. Such fixed costs may prevent small and medium-sized enterprises at the outset from developing a compliance strategy based on the Kyoto mechanisms. The same considerations apply on the supply side. Some state governments in Germany, for example in Baden-Württemberg, are trying to promote involvement of medium-sized companies in the Kyoto mechanisms by supporting capacity-building initiatives.

4.1.1 Opportunities on the demand side

Opportunities of the Kyoto mechanisms are straightforward for GHG-emitting companies which are potentially on the demand side of global GHG markets: if they are regulated by domestic climate policies, and if they are faced with relatively high internal abatement costs, they could reduce costs of compliance by means of the Kyoto mechanisms. Such a minimization of compliance costs constitutes a most significant opportunity of the Kyoto mechanisms for GHG-emitting firms. Figure 3.4 illustrates this point, assuming that a GHG-emitting firm is regulated by a domestic tradable permit system.

The company in question has an initial allocation of emission permits equal to E_P. Its original emissions are E_0. At the end of the compliance period,

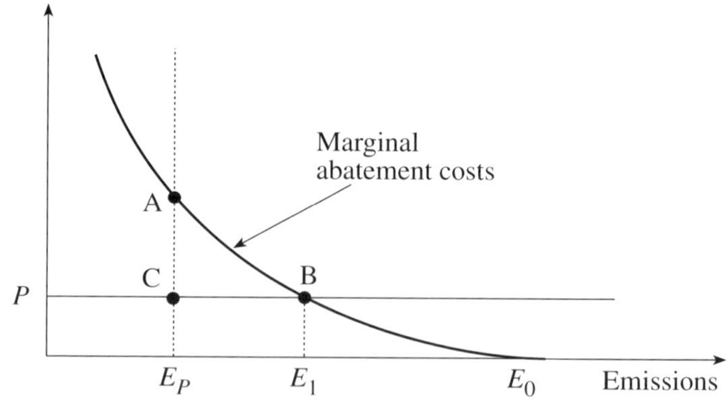

Figure 3.4 Economic benefit of Kyoto mechanisms: demand side

for example at the end of one year, the firm must hold emission permits equal to its actual emissions in order to be in compliance. For the first emission reduction units, it is cheaper to abate internally than to buy emission permits at price *P*. Consequently, the firm reduces emissions internally up to that amount where marginal abatement costs equal the market price of emission permits (E_1). In order to be in compliance, the firm could continue to reduce emissions internally or just buy emission permits on national or international markets, the latter through the Kyoto mechanisms. Since any further internal reduction would be more expensive than purchasing emission permits at price P, a cost-minimizing firm would buy emission permits in a quantity equal to the difference ($E_1 - E_P$). If no trading were allowed, it would have to reduce emissions internally up to E_P. Cost savings achieved through emissions trading amount to the area (ABC).

Cost savings associated with the Kyoto mechanisms are expected to be even higher than those obtained through domestic emissions trading (see above). These compliance cost savings associated with the Kyoto mechanisms represent the main opportunity for GHG-emitting companies that are regulated by climate policies and that face comparatively high abatement costs. And all economic studies on the costs of Kyoto indicated that potential costs savings will be huge (see Section 3).

4.1.2 Barriers on the demand side

Opportunities of the Kyoto mechanisms are straightforward. However, companies may face significant barriers that might prevent them from realizing these opportunities.

For analytical clarity, it is useful to distinguish between barriers associated with the market itself, and barriers induced by regulatory restrictions at the international and national level.

- Market barriers: price uncertainties and liquidity risks.
- Regulatory barriers imposed by international rules: import restrictions, levies on transactions and eligibility requirements for countries.
- Regulatory barriers at the national level: low compatibility of domestic regulations with the Kyoto mechanisms and lack of incentives for legal entities to engage in the Kyoto mechanisms.

4.1.3 Price uncertainties

The discussion of Section 3 has shown that future prices of emission permits are quite uncertain, estimates ranging from US$4 to US$22 per tonne of CO_2 equivalents under an international trading regime with unrestricted CDM, and from US$5 to US$70 under an international trading regime with very restrictive rules on the CDM. On average, models discussed in Section 3 predict a price of approximately US$10 and US$25 per tonne of CO_2 equivalents for both regimes, respectively. Experience with the US SO_2 emissions trading system indicates that future prices might be significantly overstated (Schmalensee et al., 1998).[28] Costs of internal emission reduction programmes may appear to be more certain. Consequently, regulated companies that are faced with the need to develop compliance strategies may regard these price uncertainties as obstacles that prevent them from using the Kyoto mechanisms.

However, it is quite probable that tools for hedging such price risks will emerge, as has been the case in other (emission) markets. Such risk management tools encompass forward and future contracts, call and put options, swaps and others. It may be expected that financial institutions will enter this market and offer such derivative instruments.[29]

4.1.4 Liquidity risks

Liquidity risks are a major obstacle for emissions trading. Low (expected) liquidity of emission markets is generally considered to be one of the most important reasons for their failure: if potential buyers anticipate that it could be very difficult or even impossible to purchase emission permits required to get into compliance, they might adopt a compliance strategy of internal emission abatement from the outset. In addition, low liquidity will increase search costs associated with finding an interested seller, thus raising barriers to trade.[30] Search costs will increase directly or indirectly, depending on the arrangement of trades. If trades are arranged bilaterally between buyer and seller, the buyer will need more time and money to find a counterparty. If they are intermediated through a broker, increases in search costs will be reflected in higher brokerage fees.

Liquidity risk may be mitigated if national markets are integrated at the international level.

4.1.5 Import restrictions

Closely linked to liquidity risks are barriers created by import restrictions. Why import restrictions? The text of the Kyoto Protocol on JI, CDM and IET requires that 'the acquisition of emission reduction units shall be *supplemental* to domestic actions for the purposes of meeting commitments under Article 3'; 'Parties included in Annex B may use the certified emission reductions … to contribute to compliance with *part of* their quantified emission limitation and reduction commitments under Article 3'; and 'any such trading shall be *supplemental* to domestic actions for the purpose of meeting quantified emission limitation and reduction commitments …'.[31]

These provisions are also called supplementarity provisions. The EU, supported by some developing countries, had been strongly advocating definition of supplementarity in quantitative terms. Quantitative restrictions on the use of the Kyoto mechanisms ('ceiling') would constitute import quotas for emission permits, raising the domestic price of emission permits.[32] From the perspective of potential buyers, such import restrictions, if expected to be binding, constitute additional barriers. First, permit buyers will need to pay higher prices as compared to a situation without a ceiling. Second, import quotas involve additional information costs for importers associated with required learning effort about their administration. Third, if rationing is implemented on a first-come, first-served basis, such a ceiling could jeopardize liquidity of the market at an increasing rate, thus involving risks of non-compliance and associated financial penalties for companies counting on the use of the Kyoto mechanisms. Anticipating such risks, potential buyers will be reluctant to rely on the Kyoto mechanisms.

At COP7, the parties agreed that there will be no obligation to impose quantitative restrictions on the use of the Kyoto mechanisms. However, some countries might still wish to do so.

4.1.6 Levies on transactions

Article 12.8 on the CDM states that 'a share of proceeds from certified project activities is used to cover administrative expenses as well as to assist developing country Parties that are particularly vulnerable to the adverse effects of climate change to meet the costs of adaptation'. This provision constitutes the basis for a levy on transactions under the CDM.

At COP7, the parties decided that the share of proceeds for adaptation measures shall be 2 per cent of the CERs (certified emission reductions) issued for a CDM project activity.[33] The level of the share of proceeds to cover administrative expenses of the CDM shall be determined at a later stage. From the perspective of potential buyers, such transaction fees on the CDM increase the barriers to its use.

4.1.7 Eligibility requirements for countries

Some countries may not be eligible to acquire emission permits from other countries. According to the agreement reached at COP7, a country that wants to acquire emission permits for compliance must meet the following criteria:

- It is a party to the Kyoto Protocol.
- It has established its assigned amount.
- It has in place a national system for the estimation of anthropogenic GHG emissions and removals.
- It has in place a national registry for emission permits.
- It has submitted annually the most recent required inventory.

If a country fails to meet these obligations, companies in that country intending to buy emission permits through the Kyoto mechanisms may be disallowed to use such permits for compliance.

4.1.9 Possible incompatibility of domestic regulations with the Kyoto mechanisms

This issue involves two aspects: first, crediting procedures could involve high transaction costs due to technical or institutional reasons. Some domestic climate policy instruments might simply not be compatible with the Kyoto mechanisms in terms of providing incentives for their use (see also Chapters 1 and 2 regarding compatibility of the Kyoto mechanisms with domestic climate policies). Second, domestic climate policy regulations may simply not allow the use of the Kyoto mechanisms for compliance.

Concerning the first point, different domestic climate policy instruments involve different transactions costs in terms of using the Kyoto mechanisms for compliance. In general, governments of Annex B countries can induce their economies to reduce GHG emissions by means of different climate policy instruments.[34] They include (OECD, 1999b):

- regulatory standards, for example energy efficiency standards, automobile fuel efficiency, renewable energy quotas;[35]
- emission and carbon taxes and charges;[36]
- environmental subsidies;
- domestic tradable permits;[37]
- voluntary agreements (see Chapter 2);
- green government purchasing; and
- removal of fossil energy subsidies.

The different policy instruments do not have the same degree of compatibility with the Kyoto mechanisms in terms of technical feasibility and transaction costs.[38] Domestic cap-and-allowances trading systems and carbon taxes are easy to combine with the Kyoto mechanisms. Voluntary agreements may be compatible with the Kyoto mechanisms, if the consequences of agreed actions can be directly expressed in terms of tonnes of GHG emissions. The same consideration applies in the case of regulatory standards (Haites and Aslam, 2000, pp. 27–8). Transaction costs associated with using the Kyoto mechanisms for compliance with domestic climate policies will generally increase if such policies define relative or intensity targets as opposed to absolute targets.

Regarding the second point, governments of Annex B countries are free to decide that regulated firms within their jurisdiction are not allowed to comply with (part of) their commitments by means of the Kyoto mechanisms, for example in order to achieve local environmental objectives.[39]

4.2 Opportunities and Barriers for GHG-emitting Industries: Supply Side

4.2.1 Opportunities on the supply side

All sectors and industries which are on the demand side are potentially also on the supply side, depending on their comparative advantage in reducing GHG emissions. Specifically, those companies which manage to generate surplus GHG emission reductions at marginal costs that are below the governing world market price for GHG permits will potentially be on the supply side.[40] This point is illustrated by Figure 3.5, assuming that a GHG-emitting firm is regulated by a domestic tradable permit system. The company in question has an initial allocation of emission permits equal to E_P. Its original emissions are E_0. At the end of the compliance period, for example at the end of one year, the firm must hold emission permits equal to its actual emissions in order to be in compliance. For the company considered, it is easy to reduce its emissions to an amount which equals its initial allocation of emission permits (E_P). The company is now in compliance. But it could reduce emissions even further at relatively low cost. It is profitable to reduce emissions up to that amount where marginal abatement costs equal the market price of emission permits (E_1). The surplus emission reductions ($E_P - E_1$) can be sold on national or international markets, thereby making a profit which corresponds to the area (ABC).

Sales on international markets would occur under one of the Kyoto mechanisms, depending on the location of the permit-producing entity (Annex B versus non-Annex B country) and the climate policy instrument by which it is regulated.[41] Thus companies in GHG-emitting industries may take advantage of the Kyoto mechanisms to realize additional business opportunities.

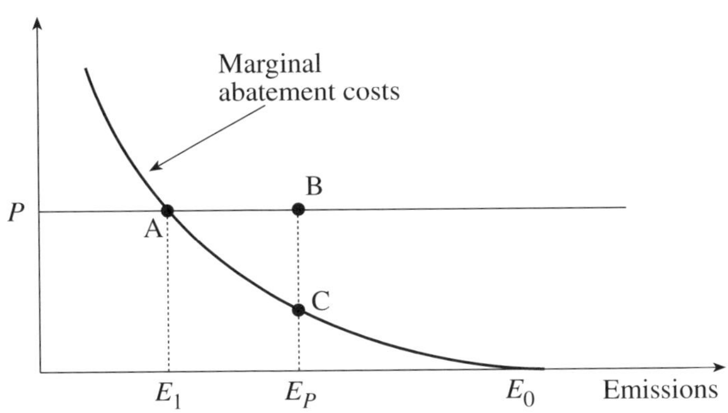

Figure 3.5 Economic benefit of Kyoto mechanisms: supply side

A general question is how to determine surplus emissions. In order to answer this question, it is useful to distinguish between two situations: if a company or a plant is regulated by a cap-and-allowance system, surplus emissions are any positive difference between holdings of emission permits and actual emissions (see above). Under any other climate policy regulation, an emission surplus is any emission reduction below a certain reference scenario or baseline.[42]

In addition to having the potential to generate surplus emission permits at competitive costs, firms need to be allowed by national and international law to trade possible surplus reductions internationally. I will discuss this point in section 4.2.2.

In conclusion, the emerging global GHG market will enable companies in GHG-emitting industries to make additional profits through the production and subsequent sale of surplus emission permits. As the discussion in Section 3 has shown, business opportunities associated with the by-product emission permits are deemed to be quite large.

4.2.2 Barriers on the supply side

Business opportunities on the supply side of the global GHG market are expected to be huge. However, firms that have the potential to deliver emission permits at competitive costs to the market may be prevented from doing so due to the existence of different kinds of barriers. As on the demand side, they may be classified into different categories:

- Market barriers: price uncertainties and liquidity risks.
- Regulatory barriers imposed by international rules: levies on transactions, eligibility requirements for countries, transaction costs associated with baseline determination and so on.
- Regulatory barriers at the national level: transaction costs and complicated procedures associated with exports.

Barriers on the supply side are quite similar in their nature to those pertinent to the demand side (see above). Consequently, I will touch only on some additional or different aspects.

4.2.3 Price uncertainties and liquidity risks

Here, similar considerations apply as those mentioned in the context of the demand side.

4.2.4 Levies on transactions

Levies on transactions, as discussed before in the context of barriers related to the demand side, will also affect the supply side.

4.2.5 Country eligibility

Regarding JI and IET, the same country eligibility criteria apply for transfers as for acquisitions of emission permits (see above). In case a country has no national registry in place, emission reductions to be transferred under JI need to be verified by an independent entity under the supervision of the JI supervisory committee.

Host countries of CDM projects are eligible to transfer CERs if they have ratified the Kyoto Protocol and if it has designated a national authority for the CDM.

4.2.6 Transaction costs involved with baseline determination, verification and certification

Prospective sellers of emission permits to be generated through JI or CDM projects need to determine emission baselines of such projects. Article 6.1(b) on JI states 'any such project provides a reduction in emissions by sources, or an enhancement of removals by sinks, that is additional to any that would otherwise occur'. Article on 12.5(c) on CDM requires 'reductions in emissions that are additional to any that would occur in the absence of the certified project activity'. Those emissions that would otherwise occur or that would occur in the absence of the certified project activity are the baseline emissions.

Several approaches to baseline determination of JI and CDM projects are under discussion.[43] At COP7, an agreement was reached on some guidelines for baseline determination. In any case, baseline determination brings about considerable costs which constitute barriers to the implementation of JI and CDM projects.

CDM projects need to be validated, verified and certified. JI projects require verification, and possibly also validation. In any case, associated fees of accreditated companies that perform these tasks will increase transactions costs. In order to reduce validation, verification and certification fees to a minimum, it is important to establish accreditation procedures that promote intensive competition among these service suppliers.

4.2.7 Domestic climate policies may not allow the export of emission permits

Governments of Annex B countries are free to decide if companies within their jurisdiction are allowed to export emission permits to other countries.

5. OPPORTUNITIES FOR FINANCIAL INSTITUTIONS

Financing and risk management are important functions performed by financial institutions such as banks and insurance companies. The supply of optimized financial products and services is essential for the efficiency of markets. Emerging GHG markets represent challenging new opportunities for financial institutions to supply innovative financial products and services that are tailored to the needs of market participants. In the following, I will highlight some areas that appear to be particularly interesting. They include project finance, carbon investment funds and insurance.

In the past, the role of financial institutions in contributing to implementation of the Kyoto Protocol and the Kyoto mechanisms has rarely been explored. Discussion has centred around the issue of how financial institutions could contribute to sustainable development in general. With regard to climate change and financial institutions, discussion has predominantly been directed towards the question of how climate change might affect financial institutions, and of how financial institutions could adapt to climate change impacts (Berz, 1997; Dlugolecki, 1996; Knoepfel et al., 1999; Leggett 1996; Tol, 1996; Tucker, 1997). An exception is Janssen (2000b and 2001), who discusses some functions financial institutions could perform to facilitate use of the Kyoto mechanisms.[44]

In advance of the academic debate, several major private sector financial institutions like Crédit Lyonnais, Deutsche Bank, Sanpaolo Imi, Société Générale, Swiss Re, among others, are already designing new financial products and services targeted at the needs of corporate clients that wish to take advantage of the Kyoto mechanisms and GHG emissions trading. Against this observation it is interesting to explore how banks and insurance companies could contribute to the use of the Kyoto mechanisms and GHG emissions trading, thus facilitating the implementation of the Kyoto Protocol. This section discusses the role of carbon investment funds set up by financial institutions and explores the possible role of insurance in emerging GHG markets.

5.1 Carbon Investment Funds

5.1.1 Introduction

Emissions trading generally induces investments in abatement activities. In the case of JI and CDM, which represent post-verification or project-based trading, emission permits are generated only if abatement activities reduce project emissions below baseline emissions. In cap-and-allowance trading systems, the transfer of emission permits does not require abatement activities to be undertaken *ex ante*. Still, unless the initial allocation of emission permits exceeds business-as-usual emissions, the transferring party needs to abate emissions within the relevant compliance period in order to be in compliance. As a consequence, the sale of emission permits requires investments in abatement activities at some point in time.

As with all kinds of investments, the willingness of private sector companies to invest in commercial GHG reduction projects (defined as climate protection projects that generate marketable emission permits) will depend on respective risk–return profiles. Commercial climate protection projects represent attractive investment opportunities only if financial risks associated with required investments are not too high. One classical instrument to reduce investment risks is risk diversification through efficient portfolio construction.

This subsection aims to discuss how specific risks associated with JI and CDM investments may be managed by taking advantage of portfolio diversification and carbon investment funds. To this end, I will first identify sources of investment risks associated with JI and CDM projects. The starting-point of risk analysis is the net present value (NPV) of JI and CDM investments. A distinction is made between price and quantity risks. My analysis is limited to project-based emissions trading, that is, JI and CDM, for several reasons. First, it helps to keep the analysis relatively simple and focused. Second, most insights gained may be easily extended to cap-and-allowance trading systems. After having identified specific risks associated with JI and CDM investments, I will discuss from a conceptual angle the potential of risk diversification. The concluding part of this subsection will analyse how risk diversification could be achieved by taking advantage of indirect investment vehicles like carbon funds. In addition, it describes some carbon investment funds that are emerging in practice, including the World Bank's Prototype Carbon Fund.

5.1.2 Identifying investment risks of JI and CDM projects

JI and CDM projects aim at producing emission credits for subsequent sale on international markets or for transfer to the project sponsor. In both cases, the value of such investments is affected by different kinds of risk.

To identify the different risk sources that could affect the value of JI/CDM projects, it is useful to distinguish between different sources of uncertainty:

- **Technological** risks are tied to the process of production. They refer to uncertain output quantities. In the present context, technological risks involve uncertain quantities of emission reductions achieved.
- **Economic** risks depend on uncertainties generated by markets and refer to uncertain input and output prices.
- **Political** risks refer to uncertainty over property rights. Political risks may involve currency control, unexpected changes in tax laws or regulation, or expropriation in the extreme case.

Net present value of JI and CDM investments as starting-point of risk analysis From the investor's perspective the value of a JI/CDM project may be measured by the NPV of the project considered. Investment risks of JI/CDM projects refer to a risky NPV of such projects. Thus, in order to analyse investment risks of JI/CDM projects, one needs to explore the main elements of their NPV.

In a simple form, the NPV of investment projects is defined as the difference between the discounted cash inflow and the discounted cash outflow associated with the project over its economic lifetime. The discount rate is also called the hurdle rate or the opportunity cost of capital. It should be equal to the rate of return offered by comparable investment alternatives. It is called the opportunity cost of capital because it is the return foregone by investing in this project rather than investing in the next-best investment opportunity. In principle there can be a different discount rate for each future period. In the simple discounted cash flow formula, it is assumed that the discount rate is the same regardless of the date of the cash flow.

Concerning the decision to invest in a project, the investor should apply the NPV rule. According to this investment decision rule, investment projects should be accepted if they have a positive NPV. In a situation of uncertainty, it may be appropriate to expand the discounted cash flow in order to take properly into account possible real options embedded in different investment alternatives or project designs (Teisberg, 1995; Trigeorgis, 1995). Real option values reflect the value of flexibility to adapt and revise later decisions in response to unexpected developments. Despite the relevance of real option theory, it is beyond the scope of this chapter.[45]

These basic investment considerations may also be applied in the context of JI/CDM projects. The main challenge is to define costs and revenues appropriately. Generally, the NPV of investments in JI/CDM projects may be stated as

$$NPV = \sum_{t=0}^{T} \frac{R_t - C_t}{(1+r)^t} \qquad (3.1)$$

where

R_t denotes the revenue of the JI/CDM project in period t,
C_t are the costs of the JI/CDM project in period t, composed of abatement and transaction costs,
r is the discount rate and
T denotes the crediting lifetime of the JI/CDM project.

The remainder of the present part of this subsection will discuss the different components and associated risk factors of the NPV of JI/CDM projects. Due to space restrictions I will discuss only some risk sources. For a more comprehensive and detailed analysis the reader is referred to Janssen (2001).

Revenue side The revenue R_t of JI/CDM project investments is composed of

$$R_t = p_t \cdot EP_t \tag{3.2}$$

where

p_t is the unit value or price of emission credits and
EP_t is the amount of emission credits or permits generated by the project in question.

The amount of emission credits generated by JI/CDM investments is defined as the difference between baseline emissions and actual GHG emissions of the project. Various approaches for baseline determination of JI/CDM projects are available.

Several ways to classify baseline approaches have been proposed. One common classification which is based on criteria of aggregation and standardization of baseline setting distinguishes between the following types (Ellis and Bosi, 1999; OECD and IEA, 2000b):

- project-specific baselines,
- multi-project baselines; and
- hybrid baselines.

Project-specific baselines apply to one project only. They are the least aggregate type. These baselines are drawn up by using project-specific assumptions, measurements, comparisons, estimates or simulations for all key parameters.

Multi-project baselines are defined and applied to multiple projects of a similar type. They aim to standardize emission levels and rates or intensities. They are more aggregated than project-specific baselines, but there may be various levels of aggregation such as by technology, sector, country or region. Multi-project baselines correspond to an activity standard or policy target that is aggregated at a certain level. For example, baseline emissions for a new or refurbished power plant could be specified on the basis of the average or marginal emissions per kilowatt-hour of electricity produced in the country or region. At a more disaggregated level, baseline emissions of such a project could be established by determining the average or marginal emissions per kilowatt-hour of the technology to be applied in this project.

Under the hybrid approach, baselines are determined in a hybrid fashion, with some key components or methodologies project-specific, and others standardized.

In the present context it is useful to look at baselines from the angle of uncertainty involved for the investor. To this end it is constructive to distinguish between baselines based on emission levels versus baselines defined in terms of emission intensities or rates, for example as tonnes of CO_2 emissions per kilowatt-hour (see below).

Actual emissions are generally determined by the activity level multiplied by a corresponding emission factor (Houghton et al., 1997). To give an example, process-related CO_2 emissions from cement production are determined by multiplying the output of cement produced (measured in tonnes) by the emission factor 0.4985.

After having discussed the two major determinants of the amount of emission credits earned by JI/CDM projects, it is possible to have a closer look at quantity risks. Where baselines are defined in absolute terms, the quantity of emission credits earned per period is defined as

$$EP_t^{absolute} = E_t^b - E_t^p = e_t^b \cdot x_t^b - e_t^p \cdot x_t^p \tag{3.3}$$

where

$EP_t^{absolute}$	are the emission credits earned in period t in case the baseline is defined in absolute terms, that is, tonnes of CO_2,
E_t^b	are the absolute baseline emissions in period t, measured in tonnes of CO_2,
E_t^p	are the absolute project emissions in period t,
e_t^b	is the baseline emission rate, for example tonnes of CO_2 per kilowatt-hour,
e_t^p	is the actual emission intensity or rate of the project,
x_t^b	is the baseline level of the activity to which the baseline emission rate refers, for example kilowatt-hours, and
x_t^p	is the actual project level of the activity to which the baseline emission rate refers.

This is a reduced formula for calculating emission credits earned by JI/CDM projects that does not make explicit all parameters and technical relations that would be used in an engineering approach.[46] However, for the present purpose such a reduced form is sufficient.

As can be clearly seen, the following factors can lead to uncertainty about the quantities of emission credits actually earned:

- E_t^b : baselines of ongoing projects may need to be adjusted unexpectedly. This is, however, not foreseen by the guidelines agreed upon at COP7.
- e_t^p : the actual emission intensity of the project may deviate from expected values due to technological under-performance (or over-performance) of the mitigation technology employed.

- x_t^p : the actual level of the activity to which the emission rate refers may deviate from the projected level: a power plant may produce less power than projected due to unexpected low demand for power or business interruption.

The last risk factor is particularly interesting, since economic activities associated with GHG emissions, for example power generation or cement production, generally exhibit high fluctuations. Such variability will have a significant impact on the amount of emission credits earned by JI/CDM projects. Will such variances be positively or negatively correlated? This question may be answered by taking the first derivative of the amount of emission credits earned with respect to the activity level. Since in the above formula this derivative is negative, the quantity of emission credits earned will *ceteris paribus* increase with decreasing level of the underlying economic activity. The same affect would apply to projects in a cap-and-allowances trading system which by definition involves allocations of absolute amounts of emission permits.

Where baselines are defined in relative terms, the amount of emission credits earned by a JI/CDM project is defined as

$$EP_t^{relative} = (e_t^b - e_t^p) \cdot x_t^p \tag{3.4}$$

where

$EP_t^{relative}$ are the emission credits earned in period *t* in case the baseline is defined in relative terms, for example tonnes of CO_2 per kilowatt-hour.

The project earns emission credits if actual project emission rates are lower than baseline emission rates. Taking the first derivative with respect to the actual level of the project's economic activity, one finds that the amount of emission credits earned will increase with increasing product output, and vice versa. This is a most interesting result because it is diametrically different from the case where baselines are defined in absolute terms. From an environmental perspective it means that the more product output is produced, the more credits are earned and exported, even if total absolute emissions of the project may be increasing. For a JI host country this could imply that it may get out of compliance if the product output increases beyond a critical value which depends on the difference between the baseline emission rate and the actual project emission rate.[47]

So far, international rules and guidelines on baseline determination of JI/CDM projects provide little guidance regarding the issue of relative emission baselines versus absolute emission baselines. The previous discussion has made evident that both approaches have diametrically different implications regarding investment risks and environmental effectiveness. Thus it is desirable that rules and guidelines address this fundamental issue.

An additional risk factor is the conversion of non-CO_2 GHGs into CO_2 equivalents (CO_2e). If the project in question involves other than CO_2 emissions, then the effect of reductions in non-CO_2 GHGs needs to be compared relative to reductions in CO_2 emissions. This is done by converting the non-CO_2 GHGs into CO_2e by means of the global warming potentials of the different GHGs as provided by the Intergovernmental Panel on Climate Change in its Second Assessment Report, based on the effects of the GHGs over a 100–year time horizon. GHG conversion factors may change during the lifetime of JI/CDM projects, thus affecting the quantity of emission credits earned.

As regards the price component of the revenue side, future prices of emission permits are quite uncertain. None the less, some practical experience has already been gained. Some companies have already entered into option contracts for the first commitment period. The World Bank's Prototype Carbon Fund aims for a portfolio price outcome of US$5.5/t CO_2e at the termination of the fund in 2012 (PCF, 2000a). The public tender of the Dutch government for procurement of emission permits earned from JI projects discovered prices from €5 to 9 /t CO_2e.

On the theoretical side, several economic models have tried to estimate market prices for the first commitment period (see Section 3). On average, the models surveyed predict a price of approximately US$9 (1998)/t CO_2e in a global trading regime. If rules for the CDM are very restrictive this value is much higher, corresponding to US$22 (1998)/t CO_2e. However, in both scenarios estimated prices vary widely.

Cost side Costs of JI/CDM projects are composed of GHG emission abatement or mitigation costs and transaction costs involved. Abatement costs may be calculated by applying the concept of incremental costs (Parkinson et al., 2001). Incremental costs are the difference between the costs of the baseline project and the JI/CDM project. Abatement costs will depend on the concrete technology and the plant location. They involve investment costs, operating and maintenance costs, fuel costs and other variable costs. All abatement cost components are affected by uncertainties.

In addition to abatement costs, one needs to consider transaction costs, which are costs of carrying out transactions. Along the JI/CDM project cycle, a wide variety of different transaction costs may arise which often are uncertain *ex ante*.

Further risk sources of JI and CDM investments Further JI/CDM investment risks comprise the following (Janssen, 2001):

- Uncertain crediting lifetime: the crediting lifetime T in the NPV formula of JI/CDM projects is the length of time emission credits can accrue. This may be uncertain both in the case of JI and CDM.
- Discount rates are uncertain due to financial risks in capital markets.

5.1.3 Risk management through risk diversification

This subsection examines how JI/CDM investment risks may be managed through risk diversification. It first explains the basic economics of risk diversification. Then it discusses how JI/CDM risks as identified in the previous subsection may be diversified.

Basic economics of risk diversification Risk diversification is a fundamental notion in modern financial economics. It is the basis of portfolio theory and modern capital market theory. Risk diversification means that investment risks may be reduced by investing a given investment budget into different assets instead of just one. At its core, risk diversification is based on a simple statistical relation, which implies that the risk of a portfolio of assets is less than the sum of the risks of individual assets, unless they are perfectly positively correlated. This can be seen by the following formula, which describes the risk of a portfolio as measured by the variance of the expected portfolio return:

$$\sigma_P^2 = \sum_{j=1}^{N} \alpha_j^2 \cdot \sigma_j^2 + \sum_{j=1}^{N} \sum_{\substack{k=1 \\ k \neq j}}^{N} \alpha_j \cdot \alpha_k \cdot \sigma_{j,k} \tag{3.5}$$

where

σ_P^2 is the variance of a portfolio consisting of N assets,
σ_j^2 is the variance of the return of asset j,
α_j is the portion of asset j in the portfolio and
$\sigma_{j,k}$ is the covariance of the returns between asset j and asset k.

The first term is the sum of the variances of the single assets, whereas the second term is the sum of terms describing the covariance between any pair of single assets. The covariance is a statistical measure of the relationship between two random variables. That is, it gives information on how two random variables, here uncertain returns, move together. A positive value for covariance indicates that the realization of random variables tends to move in the same direction. For example, a better-than-expected value for the first random variable is likely to occur with a better-than-expected value for the second random variable. A negative covariance indicates a tendency for the realization of two random variables to offset one another. A zero covariance indicates that there is no relationship between the realizations for two random variables.

Investors are interested in both risk and expected return. So, how will diversification affect expected return? Expected return of a portfolio of assets is defined as

$$E(r_P) = \sum_{j=1}^{N} \alpha_j E(r_j) \tag{3.6}$$

where

$E(r_P)$ is the expected return of a portfolio of N assets and
$E(r_j)$ denotes the expected return of asset j.

As can be clearly seen, the expected return of the portfolio is just the sum of the expected returns of single assets in the portfolio, weighted by their portion in the portfolio.

To illustrate the potential of diversification, assume that there are two assets described by the following statistical qualities:

$$\mu_1 = 0.05; \quad \sigma_1 = 0.3; \quad \mu_2 = 0.1; \quad \sigma_2 = 0.4; \quad \sigma_{1;2} = 0.012 \tag{3.7}$$

If both assets are held in the portfolio in equal proportions [$\alpha_1 = \alpha_2 = 0.5$], then the portfolio's expected return and risk are $\mu_P = 0.075$ and $\mu_P = 0.26$, respectively. A rational risk-averse investor would clearly prefer this portfolio over asset 1, because it has a lower risk and a higher expected return (see Figure 3.6). Comparing asset 2 and the portfolio, neither is superior with respect to the other. Thus the choice between asset 2 and the portfolio will depend on the investor's specific preferences.

It is also possible to construct other portfolios characterized by different combinations of asset 1 and asset 2. The dots in Figure 3.6 depict the risk–return profile of portfolios in which the proportion of asset 1 is reduced in steps of 10 per cent (thus the proportion of asset 2 is increased in steps of 10 per cent). Drawing a line between those points, one obtains the set of all feasible portfolios that may be constructed by combining asset 1 and asset 2 in any proportion.

Which combinations are efficient? In the present context, a portfolio is efficient if there is no other portfolio that has a higher return for a given risk, or that involves a lower risk at a given return. It is evident that all portfolios on the line between the points MV and asset 1 are inefficient, because for a given risk a higher return may be achieved by investing in portfolios that lie on the upper part of the curve. Only those portfolios are efficient that lie on the upper half of the hyperbola, here on the curve

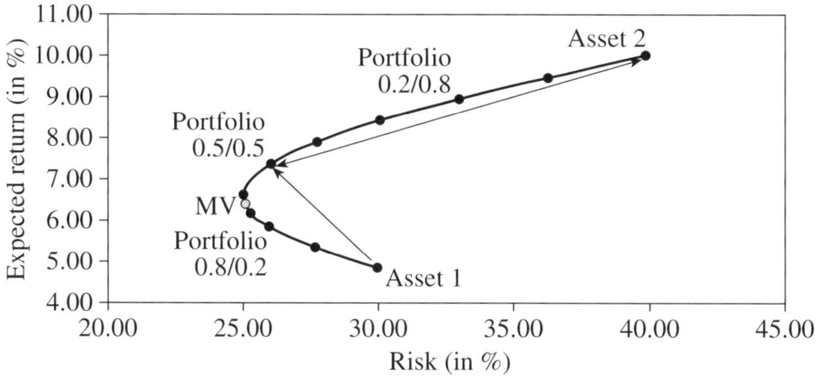

Figure 3.6 Simple portfolio of two assets

between points MV and asset 2. Point MV is the efficient portfolio with minimum variance.

Risk diversification of JI and CDM investments What does the previous discussion on basic portfolio theory imply for risk diversification of investment risks associated with JI/CDM projects?

First, standard statistical information required for portfolio construction, that is, asset prices, is not available for JI/CDM projects since no JI/CDM projects have yet been implemented. One could argue that some experience has already been gained in the JI/CDM pilot phase 'Activities Implemented Jointly', and that quite a lot of projects that potentially could qualify as JI/CDM projects, including renewable energy projects, have already been implemented.[48] However, the relevant asset price data are not available, at least not explicitly, and will probably never be available, even if hundreds or thousands of CDM and JI projects are implemented in the next decade or so.

Why? It is most important to note that the previous discussion has analysed the construction of portfolios of financial assets or securities like stocks or bonds that are publicly traded on (organized) markets or exchanges. This means that there is plenty of information available on asset prices. In contrast, financial assets of JI/CDM projects, or existing projects that may qualify as JI/CDM projects, are not publicly traded. In the real world, only shares of whole companies are traded publicly, but not the shares of single projects that are under legal and economic control of a particular firm. Even in off balance sheet or project finance, where single projects like power plants are legally and economically independent of sponsor companies, shares are not publicly traded. Thus it is difficult to imagine that equity shares in JI/CDM projects will be securitized.

Against these observations, one might raise a question about the value of portfolio theory for JI/CDM investments. To answer this question, it is helpful to consider the following points. First, portfolio theory does not depend *per se* on the existence of markets on which financial assets are traded. Portfolio theory is about selecting between assets of different risk–return profiles. The existence of markets for those assets that are considered for inclusion into the portfolio would facilitate decision making. Second, there exist approaches which allow one to make the necessary statistical estimates by using fundamental data of the company or project in question instead of historical data generated on markets. Third, one could take advantage of experiences gained with another asset class that is not publicly traded, namely real estate. Since single real estates are unsecuritized, as may be expected from assets of JI/CDM projects, experience gained from management of real-estate portfolios in the form of real estate investment trusts (REITs) may provide some guidance in the present context.

What insights might be gained? First, regarding methodology it is helpful to understand how relevant data are being estimated for unsecuritized real-estate assets. Indeed, data on expected returns and respective variances and covariances are usually estimated by means of fundamental variables like average monthly rent per square metre and vacancy.

Second, it has proved useful to identify as many distinct but homogeneous groups of sub-asset classifications as possible and to maximize inter-group heterogeneity. In the context of real estate, the following groups of sub-asset classifications have been proved to result in significant risk diversification effects among different groups (Seiler et al., 1999):

- property type, for example offices, retail, warehouse, apartments;
- city and geographic and economic regions;
- property size; and
- urban versus suburban real estate.

In this context, (empirical) research has been directed towards the question of which classification yields higher diversification effects, in part leading to ambiguous results.

What insights can be gained from real-estate portfolio construction for JI/CDM diversification? First, risk correlation coefficients may be calculated by means of some fundamentals of JI/CDM projects. Second, it could be useful to identify sub-classes of JI/CDM projects with different qualities in order to explore which categories exhibit highest risk diversification potential.

Recalling the results of the previous discussion on risk identification of JI/CDM investments, one remembers that two main risk categories affect the return on JI/CDM investments: price uncertainties and quantity risks. The latter may stem from uncertain baseline and project emissions. Is there any diversification potential associated with these risks?[49]

To examine this question it is useful to define different categories of JI/CDM projects. It is then possible to discuss if risks may be diversified across different groups of a given category. In the following, price risks are not being discussed, but quantity risks only.

Two asset categories will be discussed for illustration, though some additional categories have proved promising for risk diversification (Janssen, 2001):

- Type of Kyoto mechanism: JI or CDM (or commercial climate protection projects under a cap-and-allowance trading system)
- Type of sector:
  - energy:[50] fuel combustion in energy industries; and fugitive emissions from coal mining, and production, processing and transport of oil and natural gas.
  - Industry: energy-related and process-related emissions associated with production of cement, lime, soda ash, glass, ammonia, nitric acid, adipic acid, carbide, iron and steel, ferroalloys, aluminium, pulp and paper.
  - Agriculture: domestic livestock, rice cultivation, agricultural soils.
  - Land-use change and forestry.
  - Waste: solid waste disposal on land, wastewater handling, industrial wastewater and sludge streams, human sewage, waste incineration.

As explored above, quantity risks stem from uncertain baseline emissions and uncertain actual emissions of a JI/CDM project. Is there any scope for diversifying quantity risks across different groups of a given category?

Type of Kyoto mechanism It is quite conceivable that rules for baseline determination for JI and CDM projects will differ. In addition, it is possible that rules for baseline determination of JI and CDM projects may be adjusted at different times and with different consequences. This assertion is based on the observation that baseline determination in the context of JI is less controversial regarding the issue of whether JI projects lead to additional emission reductions due to the fact that total emissions of JI host countries are limited by the Kyoto targets. If rules for baseline determination differ between JI and CDM, baseline risks may be diversified across sub-classes of this category.

What about actual project emissions? They are not affected by the type of Kyoto mechanism. Therefore, risks associated with actual project emissions cannot be diversified across different groups in this category.

Type of sector Baseline rules may be expected to be sector-specific or source-specific. Consequently, changes in baseline rules may affect different sectors and types of emission sources differently; thus they are not perfectly positively correlated.

Variations in actual project emissions will clearly depend on the sector in which the JI/CDM project is realized. As discussed above, project emissions are generally defined as the product of the economic activity level and the corresponding emission factor ($e_t^p \cdot x_t^p$), for example tonnes of cement produced multiplied by tonnes of GHG emissions per tonne of cement. Emission factors are rather fixed in the short and medium term. Therefore actual emissions will strongly depend on the economic activity level x_t^p. Economic activities are subject to economy-wide fluctuations of production, or business cycles. However, different economic activities are not perfectly positively correlated with the general business cycle and with each other. As a consequence, GHG emissions of different industries in a given country may not be expected to be perfectly positively correlated. Thus risks associated with project emission levels may be diversified across different industries.

Investment vehicles for risk diversification and carbon funds in practice
JI/CDM investors can realize diversification potentials in two basic modes. First they can try to build up on their own an efficiently diversified portfolio of JI/CDM projects through direct investments inside or outside the company. The latter seems to be reasonable if the investor operates a sufficiently large number of climate-relevant plants. Second, investors interested in earning emission permits could invest indirectly in a portfolio of viable climate protection projects through dedicated investment vehicles offered by financial institutions.

Indirect investment vehicles may generally be called investment funds even if one needs to remember that in some countries the term 'fund' has a legally well-defined meaning. Such funds may be constituted under the law of contract (as common funds managed by management companies) or trust law (as unit trusts) or under statute (as investment companies).

In the USA, the Investment Company Act classifies investment companies into unit investment trusts and managed investment companies.

Unit investment trusts are registered investment companies that buy and hold a relatively fixed portfolio of stocks, bonds, or other securities. Unit investment trusts have a stated date for termination. To establish a unit investment trust, a sponsor purchases a specific set of securities and deposits them with a trustee. Then a number of units, known as redeemable trust certificates, are sold to the public. These certificates provide their owners with *pro rata* interests in the securities that were deposited with the trustee.[51]

Managed investment companies: the assets of a managed investment company are professionally managed in accordance with the fund's investment objectives and policies. They are divided into closed-end and open-end investment companies:

- **Closed-end investment companies** are normally established with a fixed capital. A closed-end fund is not obliged to redeem issued shares at the request of the shareholder and typically offers its shares for sale for a limited period only. Because a close-end fund is not obliged to redeem issued shares, it may invest in less liquid portfolio assets.
- **Open-end investment companies or mutual funds** are funds with variable capital which can continually issue new units but which must also redeem the units issued upon request at their net asset value.

In the following, I will briefly describe some emerging carbon funds, including the World Bank's Prototype Carbon Fund, on which extensive information material is publicly available. These case studies show how carbon funds could be designed in practice.

The following carbon funds have been launched:

- World Bank's Prototype Carbon Fund
- Dexia-FondElec Energy Efficiency and Emission Reduction Fund
- Clean Energy Fund sponsored by D&B Capital
- Renewable Energy and Energy Efficiency Fund initiated by the International Finance Corporation.

Other financial institutions including Crédit Lyonnais and UBS have announced plans to launch indirect investment vehicles aimed at generating GHG emission permits (Cooper, 2000b and Grice, 1999). Recently, UBS has withdrawn its commitment. There also exist some renewable energy funds including Merrill Lynch's New Energy Technology Fund. So far, these funds have not announced any intentions to capture directly the value of GHG emission permits.

World Bank's Prototype Carbon Fund (PCF): the establishment of the PCF was approved by the World Bank in July 1999. Its operations formally started in April 2000. The PCF is the first indirect investment vehicle whose explicit purpose is to earn GHG emission permits through JI and CDM projects (IBRD, 1999; PCF, 2000a).

The PCF will use the fund resources to support prospective JI and CDM projects. In return, fund investors will receive a *pro rata* share of the emission reductions.[52] The PCF is established as a trust fund of the IBRD, not as a partnership or corporation as other carbon funds (see below). The maximum size of the fund is restricted to US$180 million. After its second closing at the end of 2000, it has attracted investments equal to US$145 million. The PCF is scheduled to terminate in 2012.

In the present context it is interesting to note that the PCF is the sole carbon fund that has established and made publicly available its portfolio selection criteria. They encompass the following:

- Broad balance between JI projects in countries with economies in transition and CDM projects.
- Major emphasis on projects in the area of renewable energy technology such as, but not limited to, geothermal, wind, solar and small-scale hydro energy.
- No less than approximately 2 per cent and not more than 10 per cent of the PCF's assets should be invested in any one project.
- No more than approximately 20 per cent of the fund's asset will be invested in projects in the same host country.
- No more than approximately 10 per cent of the fund's assets should be invested in land-use projects. Unless the parties to the UNFCCC do not decided positively about the eligibility of land use, land-use change and forestry projects under the CDM, no such project shall be located in a non-Annex I country.
- No more than approximately 25 per cent of the fund's assets should be invested in projects using the same technology.

Originally, the PCF was conceived as an investment vehicle. At early implementation stage, the PCF's structure was changed. Under the new structure, its function is one of purchasing emission reductions and credits, as opposed to conventional project finance with investments in specific project components or activities (PCF, 2000a). Since purchase agreements such as power purchase agreements do not involve investment risks but rather contract default risks, one may wonder whether there is still any scope for or benefit from portfolio construction.[53] To answer this question it is instructive to have a closer look at the emission reduction purchase agreement (ERPA) related to the PCF's first project (PCF, 2000b).

This project aims at generating GHG emission reductions through maximum collection and utilization of landfill gas in Liepaja City and Liepaja Region in Latvia. The ERPA is entered into by the IBRD and the Republic of Latvia. It stipulates that the host country delivers a minimum amount of emission reductions to the PCF. The total minimum amount is equal to 105 800 tonnes of carbon equivalents. Annual minimum deliveries are also specified. In addition to the minimum emission reductions, the PCF receives

50 per cent of all additional emission reductions earned by the project, provided that the average market price for emission reductions applicable to the period for which any such additional emissions reductions are being delivered is not above US$6.8/t CO_2e. In case the market price exceeds this value, the percentage is reduced as a function of the market price. At a minimum, the PCF receives 25 per cent of the additional emission reductions, namely at a market price of over US$25.9/t CO_2e.

The PCF pays to the host country a total of US$2.477 million. Disbursements occur as an advanced payment, in five intermediate payments upon the achievement by the project of certain specified milestones, and a final payment.

Coming back to the point of departure of this case study, it is evident that such a purchase agreement clearly shows qualities of a risky investment, even if the time profile of cash flows deviates from that typically observed in common project investments.

First, the ERPA foresees that any additional emission reductions above a well-defined minimum level are to be shared. By its nature, *ex ante* the amount of additional or residual emission reductions is uncertain to both parties. It may be presumed that the PCF had expectations about the amount of additional emission reductions when negotiating the financial terms of the purchase agreement. However, the amount of additional emission reductions that will actually be generated is highly uncertain. The risk associated with the residuum 'additional emission reductions' definitely goes beyond the delivery risk associated with the minimum amount of emission reductions, and resembles the risk associated with equity capital investments.

Second, the purchase agreement may be regarded as a form of investment beyond the agreed advance payment. This is because part of the PCF's assets are already being allocated to a particular future use, and are tied to that use. If future project developments prove unfavourable, thus diminishing the prospects for additional emission reductions, the PCF will face difficulties in withdrawing its commitment.

In conclusion, to minimize the risks associated with the additional emission reductions, it makes perfect sense to take advantage of risk diversification techniques as indicated by portfolio theory and as discussed above.

Compared to the PCF, there is much less information available on the other indirect carbon investment vehicles. Since they are different in certain aspects with respect to the PCF, it is worthwhile mentioning some of their features.

Dexia-FondElec Energy Efficiency and Emission Reduction Fund: this closed-end equity fund was originated and sponsored by the EBRD and Dexia Group, a Franco-Belgian banking group. It will make investments that improve the energy efficiency of existing plant and equipment, including investments in generating plant retrofits and fuel conversions, heat recovery systems, electric transmission grids, gas and district heating system improvements, illumination, and industrial energy efficiency enhancements.

The fund aims at raising €100 to 150 million and will invest primarily in the countries of Central and Eastern Europe. The EBRD and Dexia have each committed €20 million. In addition, the fund has already attracted several other investors, including Marubeni Corporation, Kansai Electric Power Co., Electric Power Development Corporation and Mitsui & Co. Europe Limited. Investors will earn equity return and emission permits. It is intended to establish a number of energy service companies (ESCOs) to assist the fund in the acquisition and disposition of projects. The fund manager is FondElec Group, a private equity fund management firm. The fund is a closed-end equity investment fund and is structured as a ten-year limited liability partnership with domicile in the Cayman Islands.

Clean Energy Fund: this private equity fund is sponsored by D&B Capital, and will invest in sustainable energy projects worldwide. It aims to refinance and restructure existing fossil fuel power plants in order to make them more environmentally friendly. In addition, the fund will finance new power generation projects using renewable energy including hydro, tidal, wind, solar, geothermal and biomass. Completed projects will be sold to third parties. The fund is structured as a ten-year limited liability partnership with domicile in Bermuda. The fund aims to raise US$100 million and, besides common equity return, the fund also expects to generate GHG emission permits from some of its investments (Cooper, 2000a).

Renewable Energy and Energy Efficiency Fund: this private debt and equity fund was initiated by the International Finance Corporation, the private sector arm of the World Bank. It aims at raising US$100 million and will also be able to draw on a US$100 million debt facility. Equity investors in the fund are, among others, private sector utilities and financial institutions. Besides normal returns, the fund will also try to capture the financial value of emission permits that might be generated by the fund's investment projects (Cooper, 2000c).

5.2 Insurance

Very recently, some insurance companies like Swiss Re, Gerling, Jardine Lloyd Thompson or Aon have been showing growing interest in providing risk management tools tailored to the needs of prospective users of the Kyoto mechanisms and GHG emissions trading. Indeed, it might be expected that insurance companies will have a role to play in managing risks associated with the Kyoto mechanisms and GHG emissions trading. The aim of this subsection is to discuss the relevance of insurance for managing risks associated with JI/CDM projects.

To this end, it is organized as follows: first, it discusses insurability of risks. It is a frequent misperception that any kind of risk may be managed through insurance. To be able to apply insurance mechanisms, the risks to be insured need to exhibit certain features. This analysis is then applied to explore if specific risks associated with JI/CDM investments may be managed by insurance.

Like stock markets, insurance is also a way for society to share risk. Insurance mechanisms aim at reducing uncertainty for a group as a whole through the pooling of risks that exhibit certain probabilistic features. The risk that an insurance company faces is not merely a summation of the risks transferred to it by individuals or firms. Instead, insurance companies may achieve risk reduction due to the law of large numbers.

The strong law of large numbers states that the arithmetic average of the realizations of independent and identically distributed random variables tends towards the expected value or mean as the number of random variables becomes very large. This means that the standard deviation of the expected arithmetic average tends towards zero. In other words, the average value is much less uncertain than the individual realization of a random variable.

In the context of insurance the law of large numbers implies that the average loss in an insurance pool will become more certain as more independent and identically distributed exposure units are added to the pool. Figure 3.7 shows that the probability that the average loss will equal the expected loss becomes greater as the number of exposure units (N) increases. Indeed, the risk per exposure unit in an insurance pool, that is, the standard deviation of the average loss, will tend towards zero as the number of exposure units becomes very large. This is shown by Figure 3.8.

This observation on the basic mechanics of insurance leads to the issue of insurability of risks. In contrast to frequent perception, not all risks are insurable. In the insurance literature and practice, the issue of insurability of risks has attracted major interest (Faure, 1995; Holsboer, 1995; Karten, 1997; Vermaat, 1995). A standard reference on insurability criteria is Berliner (1982), a Swiss Re executive. He proposed checking insurability of risks according to the following criteria:

1. Losses occur with a high degree of randomness
2. The maximum possible loss is very limited
3. The average loss amount upon loss occurrence is small

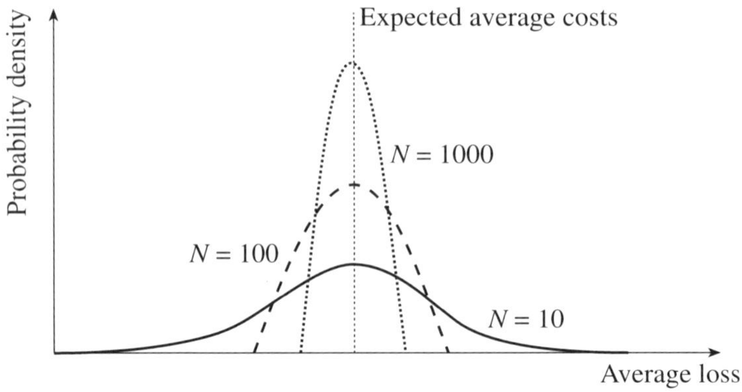

Figure 3.7 Average loss distribution with different number of exposure units

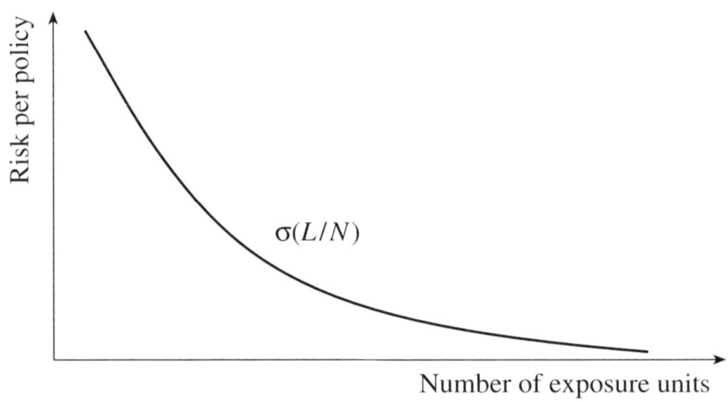

Figure 3.8 Risk per policy as function of number of exposure units

4. The average time interval between loss occurrences is short, losses occur frequently
5. The insurance premium willing to be paid for the coverage is high enough
6. There is a low possibility of moral hazard
7. Coverage of the risk is consistent with public policy
8. The law permits the coverage.

Conditions (1), (2), (3) and (4) concern actuarial issues. Together with condition (5) they basically enable the application of risk-pooling mechanisms by the law of large numbers. Condition (6) is critical from an economic perspective. The last two conditions involve political considerations.

From an economic perspective, the question of insurability is closely linked to the issue of moral hazard and adverse selection. Both problems are caused by asymmetric information between the insurer and the insured.

Given that not all risks are insurable, then which risks associated with JI/CDM investments may be covered by insurance? This question comes at the very beginning of any analysis of JI/CDM insurance.

Based on the general criteria of insurability as mentioned above, it is possible to explore which of the specific JI/CDM risks may be covered by insurance in the future. Recalling the results of the previous sub-section, the following main risk sources will affect the value of JI/CDM investments:

- Uncertain baselines due to uncertain rules and guidelines
- Uncertain crediting lifetime due to uncertain rules
- Uncertain emission factors due to
  - uncertain rules and guidelines and
  - uncertain fuel prices
- Uncertain level of economic activity due to
  - uncertain product demand and
  - uncertain technological performance of production plants or breakdown of machinery
- Uncertain GHG conversion factors
- Uncertain permit prices
- Uncertain abatement and transaction costs.

Obviously, JI/CDM projects also involve other risks which are common to any conventional project. For some of those risks insurance exists. But since they are not specifically related to JI/CDM projects they are not dealt with here.

As discussed above, a fundamental requirement for insurance is that the insurance company can pool a large number of independent and identically distributed events that give rise to a loss. Thus a necessary condition for insurability of JI/CDM investment risks is that respective risk sources approximately exhibit this statistical quality.

Many of the aforementioned risks are probably not covered by insurance (Janssen, 2001). However, some risks appear to be quite promising for insurance.

To give an example, actual project emission levels as one component of the quantity of emission credits earned are determined by both emission factors and the level of economic activity. As explained above, activity level risks are either due to uncertain product demand or to uncertain technological performance or breakdown of production plants and machinery. Uncertain market demand is a typical market risk for which no insurance is available since it affects all agents operating in that market.

The latter risk source may be covered by business interruption insurance.[54] Business interruption insurance indemnifies companies for loss of income during the period required to restore property damaged by an insured peril. Most business interruption insurance policies provide compensation for loss of operating profits and for any fixed costs insured that cannot be reduced if a business interruption occurs. Business interruption insurance is a form of consequential or indirect loss coverage.

What does this imply in the present context? First, net revenue from production and sale of emission credits is part of the operating profit of JI/CDM projects. Second, if the baseline of a JI/CDM project is defined as emission rate relative to the activity level, then business interruption implies a drop or loss of production of both the company's primary product and emission credits. Loss in operating profits due to a loss of production of the primary product is covered by conventional business interruption insurance. It would be straightforward to extend such coverage to additional operating profits earned by the sale of emission credits that are generated by JI/CDM projects.

Statistical data on the frequency of loss occurrence are readily available from traditional business interruption insurance. Additional operating profits earned from the JI/CDM project component may be estimated with the help of data on current and projected permit quantities and prices. As in the case of conventional business interruption insurance, the JI/CDM coverage must specify the insured peril. In this regard, two basic options exist: named perils and all risks cover. With the former the property is only insured against explicitly named and listed events. With all risks cover, all losses are covered that arise from direct and unforeseeable causes as long as they have not been explicitly excluded from the policy.

6. CONCLUSIONS

The Marrakesh Accords agreed upon at the seventh Conference of the Parties (COP7) in November 2001 pave the way for ratification of the Kyoto Protocol. The EU strives for the Protocol's entry into force by the end of 2002. This will be possible despite the persisting reluctance of the USA to ratify the Kyoto Protocol, if the two pivotal countries Japan and Russia ratify the Kyoto Protocol as promised.

Both domestic and international emissions trading will play a key role in national climate policies designed to meet a country's Kyoto target. In Europe, Denmark has already established a system of tradable carbon dioxide permits for its power sector. The UK will launch its comprehensive emissions trading system at the beginning of 2002. And in October 2001, the European Commission released a proposal for an EU-wide trading system to start by 2005. This system would cover around 40 per cent of its projected GHG emissions in the first commitment period and would involve a large number of different sectors.

International transactions of emission permits will be governed by one of the three Kyoto mechanisms. Joint implementation (JI) and international emissions trading (IET), as respectively established by the Articles 6 and 17 of the Kyoto Protocol, enable international transactions of emission permits among Annex B countries. The clean development mechanism (CDM), as defined by Article 12 of the Kyoto Protocol, allows developing countries, i.e. non-Annex-B countries, to earn emission permits and to sell them to Annex B countries.

The difference between JI and IET is that the former involves project-based or baseline-and-credit trading, whereas the latter encompasses inventory-based or cap-and-allowance trading. If governments wish to implement a mix of different climate policies, thus not relying exclusively on cap-and-allowance trading, the two trading systems will coexist. In this case, inventory-based emissions trading may be complemented by project-based emissions trading for those emission sources and GHGs that are not regulated by cap-and-allowance trading, but by other policy instruments like energy efficiency standards or negotiated environmental agreements (see Chapter 2 on the latter).

In contrast to a frequent misperception, from a regulatory perspective JI and CDM projects do not necessarily require international investments. Host countries may be expected to implement the largest number of JI and CDM projects unilaterally, as is the case with common economic activities. The Kyoto Protocol, as put in concrete terms through the Marrakesh Accords, would not prevent them from doing so.

When exploring opportunities and barriers associated with the Kyoto mechanisms, it is useful to distinguish between the demand side and the supply side. Opportunities on the demand side of GHG markets mainly involve significantly lower costs of compliance for industries that are regulated by climate policies. On the supply side, companies with comparatively low abatement costs may benefit from emissions trading and the flexible mechanisms by marketing emission permits, thus making additional profits.

The main barriers that prevent companies from taking advantage of GHG emissions trading and the flexible mechanisms include market risks and regulatory barriers at the international and national level.

Important market risks are permit price uncertainties and liquidity risks. Regulatory barriers induced by international rules include levies on transactions, eligibility requirements for countries and transaction costs caused by the modalities and procedures as stipulated by the Marrakesh Accords. At the national level, low compatibility of domestic climate policy regulations with the flexible mechanisms and lack of incentives for legal entities to engage in the flexible mechanisms may represent major barriers.

Price risks may be managed through derivatives. In contrast, regulatory barriers need to be removed through appropriate design of climate policies and regulations. In particular, governments need to provide incentives for companies to use the flexible mechanisms for compliance. To this end, companies that apply the flexible mechanisms need to get relief from domestic climate policy regulations by a corresponding amount. Transparent and stable regulations governing the Kyoto mechanisms are essential for their use by the private sector.

Financing and risk management are important functions performed by financial institutions such as banks and insurance companies. The supply of optimized financial products and services is essential for the efficiency of markets. Emerging GHG markets represent challenging new opportunities for financial institutions to supply innovative financial products and services that are tailored to the needs of market participants. Areas that appear to be particularly interesting include project finance, derivatives related to GHG markets, carbon funds and insurance.

In the past, the role of financial institutions in contributing to implementation of the Kyoto Protocol and the Kyoto mechanisms has rarely been explored. Discussion has predominantly been directed towards the question of how climate change might affect financial institutions, and of how financial institutions could adapt to climate change impacts.

In advance of the academic debate, several major financial institutions like Deutsche Bank and Swiss Re, among others, are already designing new financial products and services targeted at the needs of corporate clients that wish to take advantage of the Kyoto mechanisms and GHG emissions trading. This chapter has explored some aspects of carbon investment funds and insurance related to JI and CDM projects.

Not all risks are insurable. Indeed, the role of commercial insurance in managing specific risks of JI and CDM projects is limited since many risks are regulatory or market risks which are difficult to cover by classical insurance. A promising area may be business interruption insurance to cover the loss of income associated with the generation of emission permits.

Carbon investment funds have the potential to diversify away some risks associated with investments in abatement activities that aim to result in marketable emission permits. Such funds may be set up and managed by investment banks and other financial institutions. Indeed, some carbon funds such as the World Bank's Prototype Carbon Fund are already emerging in reality.

Concluding, the Kyoto Protocol has established the basis for nascent national and international markets for GHG emission reductions and permits. These markets, whose size is estimated to amount to several ten billion US dollars annually, will offer challenging opportunities to a wide variety of different players in industry and business. Financial institutions may benefit from providing new and optimized financial products and services that increase the efficiency of GHG markets.

NOTES

1. A general overview of the Kyoto Protocol is Grubb et al. (1999) or OECD (1999a). Barrett (1998) explores the political economy of the Kyoto Protocol. Oberthür and Ott (1999) is a detailed discussion of political and legal aspects. Kopp and Thatcher (2000) is a collection of economic papers on the Kyoto Protocol. Yamin (1998) considers the origins of the Kyoto Protocol and analyses its key provisions.
2. Throughout this chapter, to avoid confusion, we refer to Annex B countries (of the Kyoto Protocol), even if the Kyoto Protocol in some provisions refers to Annex 1 countries (of the UNFCCC). For clarification: the Kyoto Protocol prescribes reduction and limitation commitments for Annex 1 countries as quantified in Annex B. Both Annexes are almost identical, except for Belarus and Turkey, which are included in Annex I but not in Annex B. The Articles on JI and CDM refer to Annex I countries, whereas the Article on IET refers to Annex B countries.
3. Springer (2000) gives a comprehensive overview of market and price estimates. See also section 3 of this chapter. The aforementioned figure does not cover transactions in domestic markets or within the EU.
4. A more comprehensive discussion of the points raised in this section is provided in Janssen (2001) or (2000a).
5. Haites and Aslam (2000); Missfeldt (1998); OECD and IEA (2000a); Petsonk et al. (1998) and Skea (1999) provide an overview of the Kyoto mechanisms. Jepma and Gaast (1999) is a collection of various papers on the Kyoto mechanisms.
6. On JI see, for example, Begg et al. (2001) or Jackson et al. (2001).
7. The literature on CDM is enormous. An overview of institutional issues is Haites and Yamin (2000). An extensive treatment of technical, financial and institutional issues is Stewart et al. (2000). Werksman (1998) explores the conceptual roots of different aspects of the CDM. A useful collection of papers is Goldemberg (1998). See also Michaelowa and Dutschke (2000).
8. On IET see, for example, IEA (2001), Kerr (2000); OECD (1999d); or Tietenberg et al. (1999).
9. A comprehensive discussion with references is provided by Janssen (2001) or (2000a).
10. Emphasis added.
11. On the US SO_2 trading system, see for example Burtraw (1999); Ellerman et al. (2000); Joskow et al. (1998); Klaassen and Nentjes (1997); Montero (1999a); Schmalensee et al. (1998) or Stavins (1998).
12. International investments as opposed to international trade are strongly burdened with the risk of contract breach due to weak international compliance regimes, even if there are instruments for stabilizing international CDM contracts (Janssen, 1999a) and managing associated investment risks (Janssen, 2001 and 2000b). However, the possibility of applying the unilateral model could significantly reduce such political risks.
13. This section is based on the excellent survey by Springer (2000a). Another similar market survey is Edmonds et al. (1999), covering however only some of the models included by Springer (2000). The range of models reported in Springer (2000) covers integrated

assessment models, computable general equilibrium models, neo-Keynesian macroeconomic models, energy system models, and hybrid models. The market survey by Marenzi and Varilek (2001) also reports data on real transactions. The results reported here are based on the assumption that the USA would have ratified the Kyoto Protocol. Recently, some research has started to explore the implications of the withdrawal of the USA. Hagem and Holtsmark (2001), for example, predict that permit prices will be reduced to one-third of what they would have been with the USA participating.

14. For a discussion of some modelling issues and main differences between model types see Springer (2000).
15. Full global trading scenario means that developing countries, or non-Annex B countries, can sell any permits that result from emission reductions below their business-as-usual scenario. The results in Table 3.5 are from models that consider only carbon dioxide (CO_2) emissions. Springer (2000a) also considers models that incorporate emissions of other GHGs. The main results are, however, not affected. None of the models includes terrestrial sinks. Prices reported in dollar values of a different year are inflated. Trade volumes in physical terms are stated in tonnes of CO_2. If necessary, they were transformed from carbon to CO_2 by multiplication with the factor 44/12.
16. The low estimate from ECN is due to the fact that a technology-oriented bottom-up approach was used which included a considerable potential of options with negative marginal costs.
17. Since low prices lead to more permit trade and vice versa, the products of these two factors exhibit less variance.
18. The high permit price estimated by the Oxford model can be attributed to the model type: econometric models are based on estimated values of how imperfect markets react to changes in their environment. These frictions and imperfections are not included in bottom-up models like ECN, which report much lower prices.
19. Model results are for scenarios with no restrictions on any of the Kyoto mechanisms and competitive supply. JI/IET includes 'hot air' from Eastern Europe and the Former Soviet Union.
20. Because the models represent emissions trading among regions, they cannot distinguish between the Kyoto mechanisms JI and IET.
21. The Kyoto Protocol controls the following GHGs: carbon dioxide (CO_2), methane (CH_4), nitrous oxide (N_2O), hydrofluorocarbons (HFCs), perfluorocarbons (PFCs), sulphur hexafluoride (SF_6). Milich (1999) discusses in detail economic activities resulting in emissions of methane. OECD (2000) provides a survey of economic activities resulting in emissions of non-CO_2 GHGs.
22. See Janssen (1999b) for a discussion of applying the Kyoto mechanisms at the production stage of oil and gas.
23. On applying the Kyoto mechanisms in the power sector see OECD and IEA (2000b), chapter 3.
24. On applying the Kyoto mechanisms in the iron and steel industry see OECD and IEA (2000b), chapter 5.
25. Regarding the potential to reduce GHG emissions in the cement industry, see, for example, Soares and Tolmasquim (2000) or Worrell et al. (2000). On applying the Kyoto mechanisms in the cement industry see Jansen and Vleuten (1995) and OECD and IEA (2000b), chapter 2.
26. For a discussion of the different systems in the context of emissions trading see, for example, Australian Greenhouse Office (1999a); CCAP (1998a and 1998b); Heinz Center (1998) or TPWG (2000).
27. It might indeed by expected that to some extent companies and industries might find it cheaper to reduce emissions at regulated sources. The survey by Springer (2000a) reveals that Annex B countries will meet their Kyoto targets to a significant extent through domestic action. Some energy experts also argue that there is a wide range of no-regret or no-cost options to reduce GHG emissions. Regarding the debate on no-regret options see, for example, Sutherland (2000).
28. For a detailed analysis of possible explanations of this empirical fact see Montero (1999b).

29. On the use of derivative instruments in emission markets see for example Bailey (1998); Byrd and Zwirlein (1994); ETEI (1999) and Tucker (2001).
30. Liski (2001) provides a theoretical analysis of GHG market liquidity and transaction costs.
31. Article 6.1(d), Article 12.3(b), and Article 17, respectively. Emphasis added.
32. Bernstein et al. (1999b); Bollen et al. (1999); Ellerman and Wing (2000); Gusbin et al. (1999); Ybema et al. (1999) and Zhang (2000b) provide an economic analysis.
33. CDM projects in least developed countries will be exempt from this share of proceeds.
34. Article 2 of the Kyoto Protocol gives Annex B countries high flexibility in the choice of domestic climate policies. Gummer and Moreland (2000) provide an overview on national climate policy programmes in the European Union.
35. Helfand (1991) gives an overview of different modes of defining standards and explores their economic properties. On emerging renewable energy quotas see Berry and Jaccard (2001); Drillisch (2001); EURELECTRIC (2000); Morthorst (2000) or Schaeffer et al. (1999).
36. On carbon taxes see, for example, Baranzini et al. (2000). On the potential for using taxes to reduce non-CO_2 GHG see OECD (2000).
37. On the design of domestic emissions trading markets see, for example, Australian Greenhouse Office (1999a, 1999b, 1999c and 1999d); Hinchy et al. (1998); Ministry of Environment of New Zealand (1998) and TPWG (2000).
38. See for example Hahn and Stavins (1999); Haites and Aslam (2000); Michaelowa (1996) for a general discussion. On the compatibility of tradable renewable energy quotas and GHG emissions trading see Morthorst (2001). On the interaction between the European directive on integrated pollution prevention and control (IPPC), which prescribes the use of best available technologies, and emissions trading see Smith and Sorrell (2001).
39. Obviously, such a policy approach would only be second or third best. A first-best policy would try to tackle a given policy objective with a more targeted instrument.
40. Contrary to frequent perceptions, surplus reductions are not necessarily tied to specific technologies. They are always case-specific.
41. The latter determines whether the company participates in emissions trading under a cap-and-trade or baseline-and-credit system; see above.
42. On baselines see Section 2.
43. Ellis (1999); Ellis and Bosi (1999); OECD (1999c); OECD and IEA (2000b); Parkinson et al. (2001); Sugiyama and Michaelowa (2001); or Willems (2000). Janssen (2001 and 2000b) argues that from the perspective of risk management it makes an important difference if baselines are defined in terms of (relative) emission intensities or rates, or in terms of absolute emission levels.
44. Whitmore (2000) is another but rather curious exception. He argues that compulsory liability insurance for damages caused by emitting GHGs could be a feasible and efficient instrument for dealing with climate change. However, there seem to be some flaws in his argument.
45. Edleson and Reinhardt (1995) apply real option theory in the context of the US SO_2 trading scheme.
46. For a more detailed formula which takes into account all relevant technical parameters and relations see Parkinson et al. (2001).
47. This is why the UK trading scheme envisages a restriction on net exports from those sectors which only have relative emission targets.
48. Springer (2002) explores the potential of risk diversification using data from the JI/CDM pilot phase 'Activities Implemented Jointly'. His findings are indeed positive.
49. The following discussion will only deal with some issues for illustration. For a more comprehensive and thorough analysis see Janssen (2001).
50. This category should not be interpreted as referring exclusively to the energy sector since it also comprises energy-related emissions from any other sector.
51. In some countries, such as the UK, the term 'unit trust' refers to an open-end investment company.
52. Investors or participants in the PCF are the governments of Canada, Finland, the Nether-

lands, Norway and Sweden, Japan Bank for International Cooperation, and the following private sector companies: Electrabel (Belgium), Fortum (Finland), Gaz de France (France), Deutsche Bank and RWE (Germany), Chubu Electric Power Co., Chugoku Electric Power Co., Kyushu Electric Power Co., Mitsubishi Corp., Mitsui, Shikoku Power Co., Tohoku Electric Power Co., Tokyo Electric Power Co. (all from Japan), RaboBank (the Netherlands), Norsk Hydro and Statoil (Norway), and BP (UK).

53. Emission reduction purchase agreements may be quite similar to power purchase agreements (PPAs).
54. On business interruption insurance see, for example, Meier et al. (1998) and Vaughan and Vaughan (1999).

REFERENCES

Australian Greenhouse Office (1999a), *National Emissions Trading: Establishing the Boundaries*, Discussion Paper 1, Australian Greenhouse Office: Canberra, www.greenhouse.gov.au

Australian Greenhouse Office (1999b), *National Emissions Trading: Issuing the Permits*, Discussion Paper 2, Australian Greenhouse Office: Canberra, www.greenhouse.gov.au

Australian Greenhouse Office (1999c), *National Emissions Trading: Crediting the Carbon*, Discussion Paper 3, Australian Greenhouse Office: Canberra, www.greenhouse.gov.au

Australian Greenhouse Office (1999d), *National Emissions Trading: Designing the Market*, Discussion Paper 4, Australian Greenhouse Office: Canberra, www.greenhouse.gov.au

Bailey, E.M. (1998), 'Intertemporal Pricing of Sulfur Dioxide Allowances', *CEEPR Working Paper*, WP-98–006, Center for Energy and Environmental Policy Research (CEEPR), Cambridge, MA: MIT.

Baranzini, A., J. Goldemberg and S. Speck (2000), 'A Future for Carbon Taxes', *Ecological Economics*, **32**, 395–412.

Barrett, S. (1998), 'Political Economy of the Kyoto Protocol', *Oxford Review of Economic Policy*, **14**(4), 20–39.

Begg, K.G., T. Jackson and S. Parkinson (2001), 'Beyond Joint Implementation – Designing Flexibility into Global Climate Policy', *Energy Policy*, **29**(1), 17–27.

Berliner, B. (1982), *Limits of Insurability of Risks*, Englewood Cliffs, NJ: Prentice-Hall.

Bernstein, P., D. Montgomery and T. Rutherford (1999a), 'Global Impacts of the Kyoto Agreement: Results from the MS-MRT Model', *Resource and Energy Economics*, **21**, 375–413.

Bernstein, P.M., W.D. Montgomery, T.F. Rutherford and G.-F. Yang (1999b), 'Effects of Restrictions on International Permit Trading: The MS-MRT Model', *Energy Journal,* Special Issue on 'The Costs of the Kyoto Protocol: A Multi-Model Evaluation', 221–56.

Berry, T. and M. Jaccard (2001), 'The Renewable Portfolio Standard: Design Considerations and an Implementation Survey', *Energy Policy*, **29**(4), 263–77.

Berz, G.A. (1997), 'Natural Disasters and the Greenhouse Effect: Impact on the Insurance Industry and Possible Countermeasures', *Geneva Papers on Risk and Insurance – Issues and Practice*, **22**(85), 501–14.

Böhringer, C. (2000), 'Cooling down Hot Air: A Global CGE Analysis of post-Kyoto Carbon Abatement Strategies', *Energy Policy*, **28**, 779–89.

Bollen, J., A. Gielen and H. Timmer (1999), 'Clubs, Ceilings and CDM: Macroeconomics of Compliance with the Kyoto Protocol', *Energy Journal*, Special Issue on 'The Costs of the Kyoto Protocol: A Multi-Model Evaluation', 177–206.

Burniaux, J.-M. (1998), 'How Important Is Market Power in Achieving Kyoto? An Assessment Based on the GREEN Model', in OECD (1998), pp. 235–60.

Burtraw, D. (1999), 'Cost Savings, Market Performance and Economic Benefits of the US Acid Rain Program', in Sorrell and Skea (1999), pp. 43–62.

Byrd, J. and T. Zwirlein (1994), 'Environmental Protection and Forward Contracts: Sulfur Dioxide Emission Allowances', *Journal of Applied Corporate Finance*, **6**(4), 109–10.

Capros, P. (1999), *European Emission Mitigation Policy and Technological Evolution: Economic Evaluation with the GEM-E3 Model. Final Report*, National Technical University of Athens, Institute of Communication and Computer Systems, E3M-Lab.

Capros, P., L. Mantzos, M. Vainio and P. Zapfel (2000), *Economic Efficiency of Cross Sectoral Emission Trading in CO_2 in the European Union*, National Technical University of Athens, E3M-Lab.

CCAP (1998a), *US Carbon Emissions Trading: Description of an Upstream Approach*, Washington, DC: Center for Clean Air Policy (CCAP).

CCAP (1998b), *US Carbon Emissions Trading: Some Options that Include Downstream Sources*, Washington, DC: Center for Clean Air Policy (CCAP).

Cooper, A., S. Livermore, V. Rossi, A. Wilson and J. Walker (1999), 'The Economic Implications of Reducing Carbon Emissions: A Cross-Country Quantitative Investigation Using the Oxford Global Macroeconomic and Energy Model', *Energy Journal*, Special Issue on 'The Costs of the Kyoto Protocol: A Multi-Model Evaluation', 335–66.

Cooper, G. (2000a), 'New Funds Eye Carbon Credits', *Environmental Finance*, **1**(5), 6.

Cooper, G. (2000b), 'Crédit Lyonnais Eyes Carbon Fund', *Environmental Finance*, **1**(6), 5.

Cooper, G. (2000c), 'Climate Change Funds Exceed Targets', *Environmental Finance*, **1**(8), 7.

Criqui, P. and L. Viguier (2000), 'Kyoto and Technology at World Level: Costs of CO_2 Reduction under Flexibility Mechanisms and Technical Progress', *International Journal of Global Energy Issues*, **14**, 155–68.

DETR (2000), *A Greenhouse Gas Emissions Trading Scheme for the United Kingdom: Consultation Document*, London: Department of the Environment, Transport and the Regions (DETR), http://www.environment.detr.gov.uk/climatechange/index.htm

Dlugolecki, Andrew (1996), 'Financial Services', in Watson et al. (1996), pp. 539–60.

Drillisch, J. (2001), 'Renewable Portfolio Standard and Certificates Trading on the Dutch Electricity Market', *International Journal of Global Energy Issues*, **15**(1–2), 25–48.

Edleson, M.E. and F.L. Reinhardt (1995), 'Investment in Pollution Compliance Options: The Case of Georgia Power', in Trigeorgis (1995), pp. 243–63.

Edmonds, J., M.J. Scott, J.M. Roop and C.N. MacCracken (1999), *International*

Emissions Trading and Global Climate Change: Impacts on the Costs of Greenhouse Gas Mitigation, Arlington, VA: Pew Center on Global Climate Change.
Ekins, P. and S. Speck (1999), 'Competitiveness and Exemptions from Environmental Taxes in Europe', *Environmental and Resource Economics*, **13**, 369–96.
Eliasson, B., P.W.F. Riemer and A. Wokaun (eds) (1999), *Greenhouse Gas Control Technologies*, Amsterdam: Pergamon.
Ellerman, A.D. and I.S. Wing (2000), 'Supplementarity: An Invitation to Monopsony?', *Energy Journal*, **21**, 29–95.
Ellerman, A.D., H.D. Jacoby and A. Decaux (1999), 'The Effects on Developing Countries of the Kyoto Protocol and CO_2 Emissions Trading', *Policy Research Working Paper*, 2019, Washington, DC: World Bank.
Ellerman, A.D., P.L. Joskow, R. Schmalensee, J.-P. Montero and E.M. Bailey (2000), *Markets for Clean Air: The U.S. Acid Rain Program*, Cambridge: Cambridge University Press.
Ellis, J. (1999), *Experience with Emission Baselines under the AIJ Pilot Phase*, OECD Information Paper, Paris: OECD.
Ellis, J. and M. Bosi (1999), *Options for Project Emission Baselines*, OECD and IEA Information Paper, Paris: OECD and IEA.
ETEI (1999), *Emissions Trading Handbook*, Milwaukee: Emissions Trading Education Initiative (ETEI), www.etei.org
EURELECTRIC (2000), *Market Mechanisms for Supporting Renewable Energies: Tradable RES Certificates*, Brussels: Union of the Electricity Industry (EURELECTRIC).
Faure, M.G. (1995), 'The Limits to Insurability from a Law and Economics Perspective', *Geneva Papers on Risk and Insurance Issues and Practice*, **20**(77), 454–62.
Goldemberg, J. (ed.) (1998), *Issues and Options: The Clean Development Mechanism*, NewYork: United Nations Development Programme (UNDP).
Grice, M. (1999), 'UBS Plans Carbon Fund', *Environmental Finance*, **1**(2), 4.
Grubb, M., C. Vrolijk and D. Brack (1999), *The Kyoto Protocol: A Guide and Assessment*, London: Royal Institute of International Affairs.
Gummer, J. and R. Moreland (2000), *The European Union and Global Climate Change: A Review of Five National Programmes*, Arlington, VA: Pew Center on Global Climate Change.
Gusbin, D. et al. (1999), 'Costs of a Ceiling on Kyoto Flexibility', *Energy Policy*, **27**(14), 833–44.
Hagem, C. and B. Holtsmark (2001), 'From Small to Insignificant: Climate Impact of the Kyoto Protocol with and without the US', *CICERO Policy Note*, 2001:1, Oslo: Center for International Climate and Environmental Research (CICERO).
Hahn, R. and R. Stavins (1999), 'What Has Kyoto Wrought? The Real Architecture of International Tradable Permit Markets', *RFF Discussion Paper*, 99–30, Washington, DC: Resources for the Future (RFF).
Haites, E. (1998), 'Estimate of the Potential Market for Cooperative Mechanisms 2010', unpublished manuscript, Toronto: Margaree Consultants Inc.
Haites, E. and M.A. Aslam (2000), *The Kyoto Mechanisms and Global Climate Change: Coordination Issues and Domestic Policies*, Arlington, VA: Pew Center on Global Climate Change.
Haites, E. and F. Yamin (2000), 'The Clean Development Mechanism: Proposals for its Operation and Governance', *Global Environmental Change*, **10**, 27–45.
Hanisch, C. (1998), 'Taking Stock of Green Tax Reform Initiatives', *Environmental Science and Technology*, **32**(23), 540A–544A.

Heinz Center (1998), *Designs for Domestic Carbon Emissions Trading*, Washington, DC: Heinz Center for Science, Economics and the Environment.

Helfand, G.E. (1991), 'Standards versus Standards: The Effects of Different Pollution Restrictions', *American Economic Review*, **81**(3), 622–34.

Hinchy, M., B.S. Fisher and B. Graham (1998), 'Emissions Trading in Australia: Developing a Framework', *ABARE Research Report*, 98.1, Canberra: Australian Bureau of Agricultural and Resource Economics (ABARE).

Holsboer, J.H. (1995), 'Insurability and Uninsurability: An Introduction', *Geneva Papers on Risk and Insurance Issues and Practice*, **20**(77), 407–13.

Houghton, J.T., L.G. Meira Filho, B. Lim, K. Tréanton, I. Mamaty, Y. Bonduki, D.J. Griggs and B.A. Callander (eds) (1997), *Revised 1996 IPCC Guidelines for National Greenhouse Gas Inventories, Volume 3: Greenhouse Gas Inventory Reference Manual*, Bracknell, UK: Intergovernmental Panel on Climate Change (IPCC), Organization for Economic Cooperation and Development (OECD) and International Energy Agency (IEA).

IBRD (1999), *Instrument Establishing the Prototype Carbon Fund*, Washington, DC: International Bank for Reconstruction and Development (IBRD).

IEA (2001), *International Emission Trading: From Concept to Reality*, Paris: International Energy Agency (IEA).

Jackson, T., K. Begg and S. Parkinson (eds) (2001), *Flexibility in Climate Policy: Making the Kyoto Mechanisms Work*, London: Earthscan.

Jansen, J. and F. van der Vleuten (1995), 'Joint Implementation in the Cement Industry', in Jepma (1995), pp. 299–314.

Janssen, J. (1999a), '(Self-) Enforcement of Joint Implementation and Clean Development Mechanism Contracts', *FEEM Working Paper Nota di Lavoro*, 14.99, Milan: Fondazione Eni Enrico Mattei (FEEM), http://www.feem.it/web/activ/wp/abs99/99.html

Janssen, J. (1999b), 'Joint Implementation and Clean Development Mechanism in the Natural Gas Sector', in Eliasson et al. (1999), pp. 483–9.

Janssen, J. (2000a), 'Will Joint Implementation Survive International Emissions Trading? Distinguishing the Kyoto Mechanisms', *FEEM Working Paper Nota di Lavoro*, 60.00, Milan: Fondazione Eni Enrico Mattei (FEEM), http://www.feem.it/web/activ/wp/abs00/00.html

Janssen, J. (2000b), 'Implementing the Kyoto Mechanisms: Potential Contributions by Banks and Insurance Companies', *Geneva Papers on Risk and Insurance – Issues and Practice*, **25**(4), 602–18.

Janssen, J. (2001), *Risk Management of Investments in Joint Implementation and Clean Development Mechanism Projects*, PhD thesis, University of St Gallen, Switzerland.

Janssen, J. and U. Springer (2001), 'Emissions Trading Switzerland: Half a Leap Forward', *Environmental Finance*, **3** (1), Supplement, XVI–XVII.

Jepma, C.J. (ed.) (1995), *The Feasibility of Joint Implementation*, Dordrecht: Kluwer.

Jepma, C.J. and W.P. van der Gaast (eds) (1999), *On the Compatibility of Flexible Instruments*, Dordrecht: Kluwer.

Joskow, P., R. Schmalensee and E. Bailey (1998), 'The Market for Sulfur Dioxide Emissions', *American Economic Review*, **88**(4), 669–85.

Kainuma, M., Y. Matsuoka and T. Morita (1999), 'Analysis of Post-Kyoto Scenarios: the Asian–Pacific Integrated Model', *Energy Journal*, Special Issue on 'The Costs of the Kyoto Protocol: A Multi-Model Evaluation', 207–20.

Karten, W. (1997), 'How to Expand the Limits of Insurability', *Geneva Papers on Risk and Insurance Issues and Practice*, **22**(85), 515–22.

Kerr, S. (ed.) (2000), *Global Emissions Trading: Key Issues for Industrialized Countries*, Cheltenham, UK and Northampton, USA: Edward Elgar.

Klaassen, G. and A. Nentjes (1997), 'Sulfur Trading under the 1990 CAAA in the US: An Assessment of First Experiences', *Journal of Institutional and Theoretical Economics*, **153**(2), 384–410.

Knoepfel, I. et al. (1999), *The Kyoto Protocol and Beyond: Potential Implications for the Insurance Industry*, Geneva: UNEP Insurance Industry Initiative for the Environment.

Kopp, R.J. and J.B. Thatcher (eds) (2000), *The Weathervane Guide to Climate Policy. An RFF Reader*, Washington, DC: Resources for the Future (RFF).

Kurosawa, A., H. Yagita, W. Zhou, K. Tokimatsu and Y. Yanagisawa (1999), 'Analysis of Carbon Emission Stabilization Targets and Adaptation by Integrated Assessment Model', *Energy Journal*, Special Issue on 'The Costs of the Kyoto Protocol: A Multi-Model Evaluation', 157–75.

Leggett, J. (ed.) (1996), *Climate Change and the Financial Sector: The Emerging Threat – the Solar Solution*, Munich: Gerling Akademie Verlag.

Liski, M. (2001), 'Thin versus Thick CO_2 Market', *Journal of Environmental Economics and Management*, **41**, 295–311.

MacCracken, C.N., J.A. Edmonds, S.H. Kim and R.D. Sands (1999), 'The Economics of the Kyoto Protocol', *Energy Journal*, Special Issue on 'The Costs of the Kyoto Protocol: A Multi-Model Evaluation', 25–71.

Manne, A.S. and R.G. Richels (1999), 'The Kyoto Protocol: A Cost-Effective Strategy for Meeting Environmental Objectives?', *Energy Journal*, Special Issue on 'The Costs of the Kyoto Protocol: A Multi-Model Evaluation', 1–23.

Marenzi, N. and M. Varilek (2001), *Greenhouse Gas Price Scenarios for 2000–2012: Impact of Different Policy Regimes*, IWOe Discussion Paper, Institute for Economy and the Environment (IWOe), University of St. Gallen. www.iwoe.unisg.ch

McKibbin, W.J., M.T. Ross, R. Shackleton and P.J. Wilcoxen (1999), 'Emissions Trading, Capital Flows and the Kyoto Protocol', *Energy Journal*, Special Issue on 'The Costs of the Kyoto Protocol: A Multi-Model Evaluation', 287–333.

Meier, W., M. Kuhn, A. Simone and E. Sormani (1998), *Business Interruption Insurance*, Zurich: Swiss Re.

Mensbrugghe, D. van der (1998), 'A (Preliminary) Analysis of the Kyoto Protocol: Using the OECD GREEN Model', in OECD (1998), pp. 173–204.

Michaelowa, A. (1996), 'Incentive Aspects of Joint Implementation of Greenhouse Gas Reduction', *Mitigation and Adaptation Strategies for Global Change*, **1**, 95–108.

Michaelowa, A. and M. Dutschke (eds) (2000), *Climate Policy and Development: Flexible Instruments and Developing Countries*, Cheltenham, UK and Northampton, USA: Edward Elgar.

Milich, L. (1999), 'The Role of Methane in Global Warming: Where Might Mitigation Strategies be Focused?', *Global Environmental Change*, **9**, 179–201.

Ministry of Environment of New Zealand (1998), *Technical Design Issues for a Domestic Emissions Trading Regime for Greenhouse Gases: A Working Paper*, Wellington: Ministry of Environment of New Zealand.

Missfeldt, F. (1998), 'Flexibility Mechanisms: Which Path to Take after Kyoto?', *Review of European Community and International Environmental Law*, **7**(2), 128–39.

Montero, J.-P. (1999a), 'Voluntary Compliance with Market-Based Environmental Policy: Evidence from the U.S. Acid Rain Program', *Journal of Political Economy*, **107**(5), 998–1033.

Montero, J.-P. (1999b), 'Explaining Low Sulfur Dioxide Allowance Prices: The Effect of Expectation Errors and Irreversibility', *CEEPR Working Paper*, WP-98011, Cambridge, MA: Center for Energy and Environmental Policy Research (CEEPR).

Morthorst, P.E. (2001), 'Interactions of a Tradable Green Certificate Market with a Tradable Permits Market', *Energy Policy*, **29**(5), 345–53.

Morthorst, P.E. (2000), 'The Development of a Green Certificate Market', *Energy Policy*, **28**(15), 1085–94.

Nordhaus, W.D. and J.G. Boyer (1999), 'Requiem for Kyoto: An Economic Analysis of the Kyoto Protocol', *Energy Journal*, Special Issue on 'The Costs of the Kyoto Protocol: A Multi-Model Evaluation', 93–130.

Oberthür, S. and H.E. Ott (1999), *The Kyoto Protocol – International Climate Policy for the 21st Century*, Heidelberg: Springer-Verlag.

OECD (ed.) (1998), *Economic Modelling of Climate Change*, Paris: OECD.

OECD (1999a), *Action against Climate Change: The Kyoto Protocol and Beyond*, Paris: OECD.

OECD (1999b), *National Climate Policies and the Kyoto Protocol*, Paris: OECD.

OECD (1999c), *Status of Research on Project Baselines under the UNFCCC and the Kyoto Protocol*, OECD Information Paper, Paris: OECD.

OECD (1999d), *International Emissions Trading under the Kyoto Protocol*, OECD Information Paper, Paris: OECD.

OECD (2000), *The Potential for Using Tax Instruments to Address non-CO2 Greenhouse Gases: CH4, N2O, HFCs, PFCs and SF6*, Paris: OECD.

OECD and IEA (2000a), *Kyoto Mechanisms, Monitoring and Compliance: From Kyoto to The Hague. A Selection of Recent OECD and IEA Analyses on the Kyoto Protocol*, Paris: OECD and IEA.

OECD and IEA (2000b), *Emission Baselines: Estimating the Unknown*, Paris: OECD and IEA.

Parkinson, S., K. Begg, P. Bailey and T. Jackson (2001), 'Accounting for Flexibility against Uncertain Baselines: Lessons from Case Studies in the Eastern European Energy Sector', *Climate Policy*, **1**, 55–73.

PCF (2000a), 'Learning from the Implementation of the Prototype Carbon Fund', *Occasional Papers Series*, No. 1, Prototype Carbon Fund (PCF), Washington, DC: World Bank.

PCF (2000b), *Prototype Carbon Fund Emission Reductions Purchase Agreement: Liepaja Solid Waste Management Project*, Prototype Carbon Fund (PCF), Washington, DC: World Bank.

Petsonk, A., D.J. Dudek and J. Goffman (1998), *Market Mechanisms and Global Climate Change: An Analysis of Policy Instruments*, Arlington, VA: Pew Center on Global Climate Change.

Schaeffer, G.J., M.G. Boots, J.W. Martens and M.H. Voogt (1999), 'Tradable Green Certificates. A New Market-Based Incentive Scheme for Renewable Energy: Introduction and Analysis', *ECN Policy Studies*, ECN-I-99–004, Petten: Netherlands Energy Research Foundation (ECN).

Schmalensee, R., P. Joskow, D. Ellerman, J.P. Montero and E. Bailey (1998), 'An Interim Evaluation of Sulfur Dioxide Emissions Trading', *Journal of Economic Perspectives*, **12**(3), 53–68.

Seiler, M.J., J.R. Webb and F.C.N. Myer (1999), 'Diversification Issues in Real Estate Investment', *Journal of Real Estate Literature*, **7**(2), 163–79.

Sijm, J.P.M., F.T. Ormel, J.W. Martens, S.N.M. van Rooijen, M.H. Vogt, M.T. van Wees and C. de Zoeten-Dartenset (2000), *Kyoto Mechanisms: The Role of Joint Implementation, the Clean Development Mechanism and Emissions Trading in Reducing Greenhouse Gas Emissions*, Petten: Netherlands Energy Research Foundation (ECN).

Skea, J. (1999), 'Flexibility, Emissions Trading and the Kyoto Protocol', in: Sorrell and Skea (1999), pp. 354–79.

Smith, A. and S. Sorrell (2001), 'Interaction between Environmental Policy Instruments: Carbon Emissions Trading and Integrated Pollution Prevention and Control', *International Journal of Environment and Pollution*, **15**(1), 22–41.

Soares, J.B. and M.T. Tolmasquim (2000), 'Energy Efficiency and Reduction of CO_2 Emissions through 2015: The Brazilian Cement Industry', *Mitigation and Adaptation Strategies for Global Change*, **5**, 297–318.

Sorrell, S. and J. Skea (eds) (1999), *Pollution for Sale: Emissions Trading and Joint Implementation*, Cheltenham, UK and Northampton, USA: Edward Elgar.

Springer, U. (2000), *The Market for GHG Permits under the Kyoto Protocol: A Survey of Model Studies*, IWOe Discussion Paper, Institute for Economy and the Environment (IWOe), University of St Gallen (Switzerland), www.iwoe.unisg.ch

Springer, U. (2002), *Can the Risks of the Kyoto Mechanisms Be Reduced through Portfolio Diversification? Evidence from the Swedish AIJ Program*, IWOe Discussion Paper no. 97, Institute for Economy and the Environment IWOe), University of St Gallen.

Stavins, R.N. (1998), 'What can we Learn from the Grand Policy Experiment? Lessons from SO_2 Allowance Trading', *Journal of Economic Perspectives*, **12**(3), 69–88.

Stewart, R., D. Anderson, M.A. Aslam, C. Eyre, G. Jones, P. Sands, M. Stuart and F. Yamin (2000), *The Clean Development Mechanism: Building International Public-Private Partnership under the Kyoto Protocol. Technical, Financial and Institutional Issues*, New York and Geneva: United Nations.

Sugiyama, T. and A. Michaelowa (2001), 'Reconciling the Design of CDM with Inborn Paradox of Additionality Concept', *Climate Policy*, **1**, 75–83.

Sutherland, R.J. (2000), '"No Cost" Efforts to Reduce Carbon Emissions in the U.S.: An Economic Perspective', *Energy Journal*, **21**(3), 89–112.

Teisberg, E.O. (1995), *Methods for Evaluating Capital Investment Decisions under Uncertainty*, in Trigeorgis (1995), pp. 31–46.

Tietenberg, T., M. Grubb, A. Michaelowa, B. Swift and Z. Zhang (1999), *International Rules for Greenhouse Gas Emissions Trading: Defining the Principles, Modalities, Rules and Guidelines for Verification, Reporting and Accountability*, New York and Geneva: United Nations.

Tol, R.S.J. (1996) 'Climate Change and Insurance: A Critical Appraisal', *Energy Policy*, **26**(3), 257–62.

TPWG (2000), *Using Tradeable Emission Permits to Help Achieve Domestic Greenhouse Gas Objectives: Options Report*, Canada: Tradeable Permits Working Group (TPWG), www.nccp.ca

Trigeorgis, L. (ed.) (1995), *Real Options in Capital Investment: Models, Strategies, and Applications*, Westport, CT: Praeger Publishers.

Tucker, M. (1997), 'Climate Change and the Insurance Industry: The Cost of Increased Risk and the Impetus for Action', *Ecological Economics*, **22**(2), 85–96.

Tucker, M. (2001), 'Trading Carbon Tradable Offsets under Kyoto's Clean Development Mechanism: The Economic Advantages to Buyers and Sellers of Using Call Options', *Ecological Economics*, submitted.

Tulpulé, V., S. Brown, J. Lim, C. Polidano, H. Pant and B.S. Fisher (1999), 'The Kyoto Protocol: an Economic Analysis Using GTEM', *Energy Journal*, Special Issue on 'The Costs of the Kyoto Protocol: A Multi-Model Evaluation', 257–85.

UK Emissions Trading Group (2000), *Outline Proposals for a UK Emissions Trading Scheme*, London: UK Emissions Trading Group, http://www.uketg.com/

Vaughan, E.J. and T.M. Vaughan (1999), *Fundamentals of Risk and Insurance*, 8th edn, John New York: Wiley and Sons.

Vermaat, A.J. (1995), 'Uninsurability: a Growing Problem', *Geneva Papers on Risk and Insurance*, **20**(77), 446–53.

Watson, R.T., M.C. Zinyowera and R.H. Moss (eds) (1996), *Climate Change 1995: Impacts, Adaptations and Mitigation of Climate Change: Scientific–Technical Analyses. Contribution of Working Group II to the Second Assessment Report of the Intergovernmental Panel on Climate Change*, Cambridge: Cambridge University Press.

Werksman, J. (1998), 'The Clean Development Mechanism: Unwrapping the "Kyoto Surprise"', *Review of European Community and International Environmental Law*, **7**(2), 147–58.

Whitmore, A. (2000), 'Compulsory Environmental Liability Insurance as a Mean of Dealing with Climate Change Risk', *Energy Policy*, **28**, 739–41.

Willems, S. (2000), *Framework for Baseline Guidelines*, OECD Information Paper, Paris: OECD.

Worrell, E., N. Martin and L. Price (2000), 'Potentials for Energy Efficiency Improvements in the US Cement Industry', *Energy*, **25**(12), 1189–214.

Yamin, F. (1998), 'The Kyoto Protocol: Origins, Assessment and Future Challenges', *Review of European Community and International Environmental Law*, **7**(2), 113–27.

Ybema, J.R., T. Kram and S.N.M. van Rooijen (1999), 'Consequences of Ceilings on the Use of the Kyoto Mechanisms: A Tentative Analysis of Cost Effects for EU Member States', *ECN Working Paper*, ECN-C-99-003, Petten: Netherlands Energy Research Foundation (ECN).

Zhang, Z.X. (2000a), 'Estimating the Size of the Potential Market for the Kyoto Mechanisms', *Review of World Economics*, **136**, 491–520.

Zhang, Z. (2000b), 'An Assessment of the EU Proposal for Ceilings on the Use of Kyoto Flexibility Mechanisms', *FEEM Working Paper Nota di Lavoro*, 99.00, Milan: Fondazione Eni Enrico Mattei (FEEM), http://www.feem.it/web/activ/wp/abs99/99.html

4. Traditional environmental instruments, Kyoto mechanisms and the role of technical change

Marzio Galeotti and Carlo Carraro

1. INTRODUCTION

The previous chapters explored the role of different policy mixes in providing the right incentives for industries and firms to reduce their GHG emissions. However, most of the analysis focused on short-term actions taken by a single industry or firm. In this chapter, we would like to broaden the analysis of the incentive mechanisms that may lead to effective emission reductions, by stressing the role of climate-related technical change and by analysing which incentive mechanisms can be designed to foster its adoption and diffusion. In particular, we would like to identify which policy mixes are most suitable to induce firms to adopt long-term innovation policies aimed at reducing GHG emissions.

The perspective chosen in this chapter is both a microeconomic and a macroeconomic one, because we would like to take into account also the systemic effects of growth and innovation, including their links with environmental degradation. Hence impacts on growth of different climate policies will be considered jointly with impacts on innovation and emissions.

As already stressed in previous chapters, one important aspect of the Kyoto Protocol, retained in the recent agreement struck in Marrakesh, is the introduction of the so-called flexibility mechanisms, that is, instruments designed to facilitate the achievement of the emission reduction targets agreed upon by the signatory countries. Emissions trading (ET) by permits, joint implementation (JI) and clean development mechanisms (CDM) serve the purpose of reducing the costs of complying with the agreement, thus lessening the feedback of improved environment into slower economic growth and perhaps reduced welfare.

As a consequence, a special effort should be devoted to the study of the role of both traditional instruments and flexibility mechanisms in set-ups which are as realistic as possible. Given the dynamic nature of the problem, it is essential to gain as thorough an understanding as possible of the role of technological change in the process of countries' economic development. In the light of the growth–environment trade-off, it is also crucial to assess the effects of mitigation policies when technological change is explicitly accounted for, both in terms of costs and benefits of these measures and in terms of their appropriate design.

This is the basic goal of the present chapter, which develops along three main lines. First, we consider the process of technological change and its impact on the economy and the environment. Economic activity has potentially adverse effects on the environment, but technological change may play a relevant role in mitigating those consequences. We therefore next look at the effect of environmental policies on the development and spread of new technologies which, in the long run, may be the most important determinant of success or failure in environmental protection. The third step consists of studying the role of different climate policies in models which explicitly incorporate a specification of technological change.

This chapter adds to a mushrooming literature on the role of technological change for the study of climate change. It draws from very recent survey papers, including Carraro (1999), DeCanio et al. (2000), Edmonds et al. (2000), Fischer (2000), Grubb et al. (2000), Jaffe et al. (2000), and Löschel (2001). However, by focusing on the incentive mechanisms and on their quantitative modelling in climate models, this chapter provides a new perspective and some new results. From this standpoint, it complements the work – carried out especially at the International Institute of Applied System Analysis (IIASA) and at the Wuppertal Institute – which looks more closely and deeply at the technological dimension of technical change in the case of IIASA, and at the broad socioeconomic implications of changing current production modes in the case of the Wuppertal Institute.

2. THE IMPORTANCE OF TECHNICAL CHANGE

Technological change is a major force in a country's economic growth. Since before the Industrial Revolution, economies and societies have evolved as a result of technological change. A long sequence of inventions – engines, power generation systems, industrial processes and appliances – has changed people's lives. Society has moved from a reliance on wind, water, animal power and wood to reliance first on coal, and then on natural gas and petroleum. Today, many technologies utilize fossil fuels, which has led to the release of large amounts of carbon to the atmosphere, and the scientific consensus is that these releases will cause the earth's climate to change. Atmospheric concentrations of carbon dioxide (CO_2), a greenhouse gas (GHG), are up by 33 per cent from pre-industrial levels. Stabilizing these concentrations at, for example, twice the pre-industrial levels requires per capita global emissions to peak and then decline to (at least) half their 1990 value by the end of the twenty-first century. This would seem challenging enough with current technologies, but, fortunately, technology does not stand still.

As said, carbon emissions from economic growth depend upon the amounts and types of energy sources and energy technologies that are used, and other consequences of growth such as land clearing. There are three different ways in which technological change can affect carbon emissions:

- technological change can increase the rate of improvement in alternatives to carbon-emitting energy technologies;
- technological change can increase the rate of improvement in the efficiency with which carbon-based fuels are used;
- technological change can favour the development of more effective modes of carbon sequestration.

As for the first mode, improvement in alternatives to carbon-emitting technologies, non-carbon-emitting technologies, such as biomass, wind and hydro, were the dominant energy forms before the age of fossil fuels. New versions of these technologies, plus technologies that would allow the capture and sequestration of carbon from fossil fuels, could provide the technological mechanisms for controlling carbon accumulation in the atmosphere. For example, the IIASA reckons that it takes between 50 and 100 years for an energy form to increase its global market share from 1 per cent to 50 per cent (Grubler, 1998): oil, for example, took 90 years to grow from 1 per cent to 40 per cent of the energy market and has still not reached the 70 per cent share coal reached in 1913; natural gas has captured an increasing share of the market, rising from a 1 per cent to a 25 per cent share of the global energy system over the past 85 years; finally, despite the increased growth in oil and natural gas, other fuels such as hydro and nuclear power have also grown, while coal and wood still account for nearly 25 per cent of the global energy market share. Improvement in the efficiency of energy use, instead, can occur as a result of price changes or technological change, or a combination of both.

The second mode of technological change, the efficiency of energy use, will improve over time as new investments combine with new knowledge to provide the capital that uses energy more efficiently. These improvements may be stimulated by price increases in energy, such as the one the USA experienced in both the early and the late 1970s. Improvements are usually reported as energy intensity declines, that is, as reductions in the amount of energy required to produce a specified level of output are achieved. As an example, US industrial energy intensity declined from about 25 000 British thermal units (Btu)/dollar of output in 1972 to about 17 000 Btu/dollar of output in 1987 (see Edmonds et al., 2000; Jaffe et al., 2000, for additional empirical evidence). Relative to the *Second Assessment Report* (SAR) of the UN Intergovernmental Panel on Climate Change (IPCC) released in 1995, the *Third Assessment Report* (TAR) notes in chapter 3 that several technologies have advanced more rapidly than was foreseen in the earlier analysis. As a result, greater energy efficiency is now available, often at lower cost, for buildings, industry and transportation than was expected during the SAR. For the year 2010 most of the opportunities to save energy are associated with fostering energy efficiency in the end-use sector. More specifically, even though progress will continue to be made in all categories, it is expected that energy efficiency will make the major contribution in the first decade of the next century, whereas physical carbon sequestration from fuels will be the last category to come on line because of cost considerations.

The TAR chapter presents and discusses numerous technologies that are either already commercialized or that show good prospects of being in the commercial market by the year 2025 (a horizon that must be considered since the turnover period, especially in the energy supply sector, is very long). The chapter also considers technologies which will probably contribute to GHG abatement by 2010 in consideration of the relevance of the Kyoto Protocol. A number of technologies and practices have gained significant importance since the preparation of SAR, including:

- hybrid electric vehicles, which have actually been introduced into the market;
- fuel cell vehicles, which seem ready for introduction later in the next decade;
- combined cycle gas turbines, which have become the new choice for standard electric power production because of low cost and pollution;
- fuel cells for distributed power and low-temperature heat applications are demonstrating their technological potential although their cost remains high;
- integrated building design for greater efficiency;
- conversion of cellulosic materials for production of ethanol;
- faster than anticipated growth of wind-based electricity generation;
- carbon sequestration in aquifers;
- increased coal bed methane and landfill gas use;
- advanced sensors and controls for optimizing industrial processes;
- large reductions in process gases such as CF_4 (carbon tetrafluoride), N_2O (nitrous oxide) and HFCs (hydrofluorocarbons) through improved industrial processes;

- biotechnology development for crop improvements (including energy crops), alternative fuels other than biomass, carbon cycle manipulation/sequestration, bioprocessing for fuels and chemicals and biological/biochemical hydrogen production;
- minimum tillage practices in agriculture to reduce energy requirements and soil erosion, and improved management systems that lower N_2O emissions;
- improved containment and recovery of CFC substitutes and the use of low Global Warming Potential alternatives.

Technical progress holds the key not only to pollution abatement in the course of economic growth, but also to reducing the costs of that abatement. As noted by Grubb et al. (2000), if we look at the history of major pollutants, not only have technologies and fuels been developed that are able to reduce pollution per unit of energy use by factors of ten to several thousand or more (such that even with rising energy consumption, dramatic reductions have taken place in emission levels in countries that have introduced policies to encourage pollution abatement), but also the costs have often been far less than originally anticipated, sometimes becoming less than those of the fuels they displaced – the substitution of gas for coal in electricity generation and as a domestic and industrial fuel is a well-known example – or have been offset by efficiency gains elsewhere in the activity in question. Actually, the IPCC and many others have consistently suggested that a low-carbon future can be reconciled with high and rising levels of world energy use in the long term through a shift to such technologies.

Table 4.1 provides evidence on the above points by showing the emissions intensities of technologies and practices for reducing emissions and environmental damage relative to the (polluting) alternatives they displace, together with some indicators of the effects on costs, which are generally small, and sometimes, in retrospect, have turned out to be negative. For each source and individual pollutant, in the second column the percentage of emissions from available low-polluting practices/technologies compared to the current average levels is displayed. The third column shows the percentage of marginal costs of low-polluting options compared to the marginal cost of supply. The fourth column lists the corresponding low-polluting alternatives.

Recalling that emissions are equal to energy consumption times energy efficiency (that is, emissions/energy consumption), it is clear from the table that very high levels of energy consumption can be sustained in practice with greatly reduced emissions once the low-polluting technologies and practices are in place. Experience suggests that a set of new and improved technologies will become available over time, both through incremental improvements and through ‘macro-inventions’. Incremental technological improvement, or ‘learning curves', affects existing technologies, and takes advantage of existing infrastructure and supporting technological systems. Macro-inventions open new markets and stimulate the creation and dissemination of new infrastructure and complementary technologies. The large-scale introduction of low-cost carbon capture and sequestration technologies would constitute a macro-invention. This is the case for instance of electricity, where the costs associated with various technologies have declined in the last decades as the level of installed capacity has increased.

A nice example of learning is provided by a recent study of the International Energy Agency considering the average unit costs (1990 ECU per kilowatt-hour) of a series of electricity-producing technologies in the EU, 1980–95. Figure 4.1 has unit costs on the vertical axis and cumulative electricity production at successive dates on the horizontal axis.

With greater 'experience' (cumulative production), there is a pronounced tendency for a decline in the unit costs of novel technologies such as photovoltaics and wind power, but there is no obvious decline in the unit costs of more conventional methods such as supercritical coal and NGCC (natural gas – combined cycle). Significantly enough, nuclear power is *not* plotted on this diagram. If it were plotted, it would almost surely illustrate an increase in unit costs with additional experience – and with additional concerns about reactor safety.

In Figure 4.1 note that the newer technologies tend to be higher in unit costs than the conventional ones. If investors base all their decisions on immediate costs, there would be little tendency to support the newer technologies that are currently more expensive. Their cumulative experience is too small, and they could be 'locked out' permanently. This is the rationale for public intervention in the market. Leaning-by-doing entails the acceptance of high near-term costs in return for an expected lowering of future costs. It is an investment choice, and it depends critically upon the rate of discount. Associated with each technology, Figure 4.1 shows a 'progress ratio' entered in parentheses. This measures the percentage decline in unit costs that is associated with a doubling of experience. In the case of wind, for example, this parameter is shown as 82 per cent.

With all the ways that technological change can affect the path of growth and resource use, which is most amenable to human intervention? What can be done to stimulate innovation and technological change? While analysts cannot completely solve this problem, it is possible to shed some light on what might be most effective in reducing the costs of GHG emissions mitigation via technological change. It is therefore necessary to turn the attention to environmental policies, recognizing that alternative types of instruments can have significantly different effects on the rate and direction of technological change. At the same time it must also be acknowledged that there are oppos-

Table 4.1 Pollution intensities and costs for selected energy-related activities and pollutants

Source and pollutant	Emissions ratio (%) of low-polluting to polluting practices	Net private MC of low-polluting option as % of MC of supply	Nature of low-polluting alternatives
Electricity generation from coal			
PM emissions	< 0.1	< 0 to ≈ 2	Natural gas; ESP, BHF, FGD, IGCC (for coal); 'low NO_x' combustion and catalytic methods
SO_2	0 to 0.5	5	
NO_x	5 to 10	5	
Gasoline engines			
Pb	0	Combined costs of 4–5% of vehicle and fuel used	Unleaded/reformulated fuels; catalytic converters
CO	5		
NO_x	20		
VOCs	5		
Marine pollution from oil			
Wastes	10	n.a.	Improved ballasting and discharge reduction, improved operating, navigation and safety practices since 1970s
Spills	< 10		
Traditional household fuels (wood and charcoal) in low-income countries			

Smoke (PM, CO and S)	< 0.01	< 0	Gas and kerosene
Soil erosion			
Sediment yield	–5 to 5	≈ –20	Agroforestry; contouring, terracing, vetiver grasses, mulching, and other. Indirectly: substitution of modern energy forms for dung and firewood as cooking fuels
CO_2 emissions from burning fossil fuels			
Electricity (LDCs)	0	≈ 0	Advanced solar energy, wind and other renewable energy technologies for power generation; biomass for liquid fuels and power generation; hydrogen from renewable energy sources and fuel cells for power generation and vehicles; decarbonization of fossil fuels for hydrogen production with carbon sequestration
Electricity (industrial countries)	0	≈ 20	
Liquid fuel substitutes	0	≈ 30–50	

Notes: Aside from CO_2 emissions, all estimates are based on technologies and practices in widespread use. The CO_2 emissions reduction technologies have been well demonstrated; the cost estimates are based on best projections found in engineering studies. Acronyms are for particulate matter (PM), electrostatic precipitators (ESP), bag-house filters (BHF), flue gas desulphurization (FGD), and integrated coal gasification combined-cycle (IGCC) technologies. The negative cost figure arises if gas is available for power generation as a substitute for coal.

Source: See Grubb et al. (2000) and the sources cited therein.

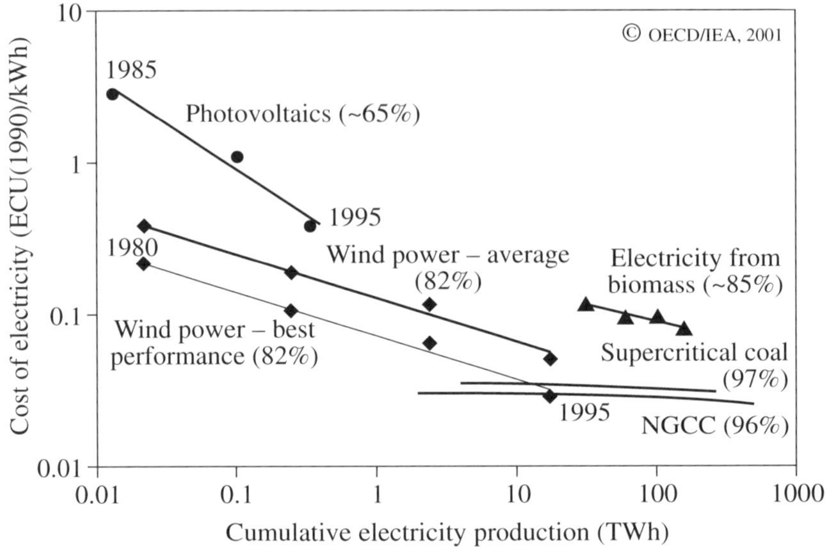

Source: IEA (2000); reproduced in Manne and Barreto (2001).

Figure 4.1 Electric technologies in the EU, 1980–95

ing views about the effects of the adoption and diffusion of low-emitting technological innovations.

3. THE TECHNOLOGIST'S VIEW AND 'WIN–WIN' OPTIONS

While there is little debate over the importance of energy efficiency in limiting GHG emissions, there is intense debate about its cost-effectiveness and about the government policies that should be pursued to enhance energy efficiency. Analysts have pointed out for years that there is an 'energy efficiency gap' between the most energy-efficient technologies available at some point in time and those that are actually in use. On this basis, debate has centred upon the extent to which there are low-cost or no-cost options for reducing fossil energy use through improved energy efficiency. Jaffe et al. (1999) note that this debate opposes 'technologists' and economists, who hold very different views about whether and to what degree it is the result of 'market failures' that might be amenable to policy intervention or simply 'market barriers' that would be surmountable only at relatively high cost.

This debate is illustrated in the IPCC SAR report: one part of it states that energy efficiency improvements of the order of 10 to 30 per cent might be possible at little cost or even with net benefits (ignoring climate benefits), while another part highlights the fact that most economic models indicate a significant cost for stabilizing or cutting OECD emissions below 1990 levels.

Technologists believe that there are plentiful opportunities for low-cost, or even 'negative-cost' improvements in energy efficiency, and that realizing these opportunities will require active intervention in markets for energy-using equipment to help overcome barriers to the use of more efficient technologies. This view implies that with the appropriate technology and market creation policies, significant GHG reduction can be achieved at very low cost. Most economists, on the other hand, acknowledge that there are 'market barriers' to the penetration of various technologies that enhance energy efficiency, but that only some of these barriers represent real 'market failures' that reduce economic efficiency. This view emphasizes that there are trade-offs between economic efficiency and energy efficiency – it is possible to get more of the latter, but typically only at the cost of less of the former. The economic perspective suggests that GHG reduction is more costly than the technologists argue, and it puts relatively more emphasis on market-based GHG control policies like carbon taxes or tradable carbon permit systems to encourage the least costly means of carbon efficiency (not necessarily energy efficiency) enhancement available to individual energy users.

Let us first pursue further the technologist's view and the issue of barriers. The next section will tackle the economic view in more detail.

Chapter 5 of the IPCC TAR, entitled 'Barriers, Opportunities, and Market Potential of Technologies and Practices', explicitly deals with the engineering and economic perspectives and illustrates the concept of barriers to more widespread use of energy-efficient technologies using the notion of technological potential. This issue is illustrated by means of Figure 4.2, also taken from the TAR chapter.

At the top of the figure, there is some theoretical physical limit that can be approached but never reached. The technological potential represents the efficiency that would be achieved if the most efficient technologies that have been demonstrated with prototypes to be technically feasible were universally used, regardless of cost, reliability, availability of materials, or other attributes of economic desirability. The techno-economic potential represents the efficiency that would be achieved if all consumers and firms used the most efficient technologies that appear to be economically viable, overcoming whatever preferential, social, cultural or institutional sources of resistance there may be. The economic potential represents the efficiency that would be achieved if all market failures and government distortions in energy markets could be removed. Finally, the market potential indicates the efficiency level

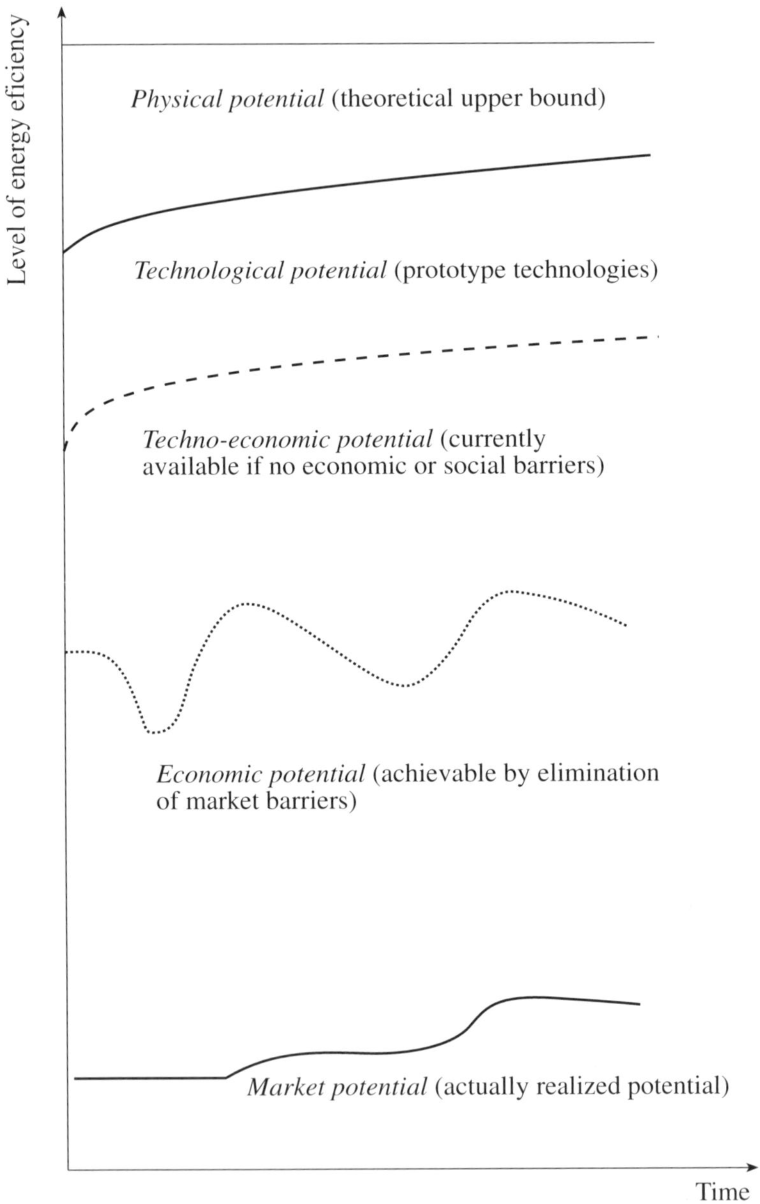

Figure 4.2 *Energy efficiency potentials*

that is achieved by the marketplace with existing market failures and distortions in place.

The technical potential can be illustrated with reference to the fuel cell as a power source for private vehicles. Current fuel cell technology, making use of hydrogen manufactured from natural gas, can offer GHG emission reductions of around 50–60 per cent relative to conventional vehicles. The technical potential is expected to improve over time, as shown in Figure 4.2, through scientific discovery and technological development. However, numerous difficulties would have to be overcome before such vehicles could enter widespread use and offer more substantial emission reductions. Broadly, these challenges or barriers are related to technology performance, cost, and other characteristics that affect safety or that would prevent the technology from fitting into the current system of fuel and transport infrastructure. Additional research and development on fuel cells might address some of these challenges. The economic potential can be similarly illustrated, for example, with reference to energy conservation opportunities in buildings. Engineering-based analysis indicates that measures such as replacing tungsten filament bulbs with compact fluorescent lamps (CFLs), insulating hot water tanks and introducing more energy-efficient refrigerators, could reduce residential electricity by about 40 per cent and deliver a net saving to consumers. This is described as the economic potential. It is smaller than the technical potential, as illustrated in Figure 4.2, because many technologies that are available, such as rooftop solar photovoltaic electricity supplies, would not pay for themselves in energy savings. However, even cost-effective technologies such as CFLs are not used for a variety of additional reasons. For example, many people find them ugly and do not like the tone of the light they generate; they are not as widely available as tungsten filament lightbulbs; they do not fit in some light fittings and cannot be used with dimmer switches; most householders are not aware, or confident, of the financial savings they offer. These are often described as the barriers that policies must overcome to achieve the economic potential.

Identification of the nature of the barriers that separate each of the levels in the figure allows us to formulate the appropriate policy responses to overcome them. Since the technological potential can only be improved upon by solving scientific and technological problems, policies to overcome this category of barriers must be aimed at fostering the research and development (R&D) process, in either the public or private sectors. Similarly, R&D can eventually help to bridge the gap between techno-economic potential and technological potential by reducing the cost and improving the performance of existing feasible but not commercially viable technologies. In addition, because production costs typically fall as experience with a particular technology accumulates, which is one aspect of learning-by-doing (LbD), policies that foster adoption of new technologies can, over time, produce cost reductions and thereby facilitate achievement of the technological potential.

While all perspectives refer back to the path-breaking Schumpeterian theory of technological change as composed of invention, innovation and diffusion activities, the technologist view has its roots in the evolutionary model of Nelson and Winter (1982), which uses Herbert Simon's idea of boundedly rational firms that engage in 'satisficing' rather than optimizing behaviour (as distinctive of the neoclassical model) to build an alternative model of the R&D process. In this 'evolutionary' model, firms use 'rules of thumb' and 'routines' to determine how much to invest in R&D, and how to search for new technologies. Because firms are not optimizing, the evolutionary model implies that it cannot be presumed that the imposition of a new external constraint (for example, a new environmental rule) necessarily reduces profits. There is at least the theoretical possibility that the imposition of such a constraint could be an event that forces a satisficing firm to rethink its strategy, with the possible outcome being the discovery of a new way of operating that is actually more profitable for the firm. This possibility of environmental regulation leading to a 'win–win' outcome in which pollution is reduced and profits increased underlies a stream of studies and of modelling exercises in which the basic trade-off between economic growth and environmental quality is negated.

Replacing optimizing firms with satisficing firms, the evolutionary approach admits greater scope for a variety of consequences when the firm's environment is modified. Satisficing firms may miss opportunities for increased profits simply because they don't look very hard for such opportunities as long as things are going reasonably well. An external shock such as a new environmental constraint can therefore constitute a stimulus to new search, possibly leading to discovery of previously undetected profit opportunities. This observation forms the basis for the normative observation that environmental regulation may not be as costly as we expect, because the imposition of the new constraint may lead to the discovery of new ways of doing things. In the limit, these new ways of doing things might actually be *more* profitable than the old ways, leading to an asserted win–win outcome. As noted by Jaffe et al. (2000), advocates of the win–win view of the consequences of environmental regulation seem unaware of the connection between their argument and the evolutionary school of technological change. Porter and other win–win theorists argue that in this non-optimizing world, regulation may lead to 'innovation offsets' that 'can not only lower the net cost of meeting environmental regulations, but can even lead to absolute advantages over firms in foreign countries not subject to similar regulations' (Porter and van der Linde, 1995, p. 98).

It should be emphasized that win–win theorists do not claim that all environmental regulations generate significant innovation offsets. Indeed, they stress that regulation must be properly designed in order to maximize the chances for encouraging innovation. More generally, in terms of policy prescriptions, while the evolutionary perspective does not provide precise recipes for government policy, it can help to identify areas where policies might be applied. It is suggested that government policies to harness technological and social innovation for climate mitigation should favour:

- development and discovery of new ideas, technology, concepts and behaviours. New ideas and technology may come from conventional R&D but also from outside the R&D process. This highlights the importance of R&D and of knowledge accumulation;
- exchange of ideas, technology, concepts and behaviours among firms, communities, local governments, departments of national government, and so on. New ideas and discoveries often occur in places where they cannot be used. The chance of an idea leading to a successful innovation is greatly increased by a continual exchange of information. This highlights the role of spillovers of knowledge;
- experimentation with new ideas, possibly selecting those that could contribute to GHG mitigation and other policy objectives. This highlights the importance of LbD;
- replication of successful innovations. Innovations often occur in special situations where they are particularly appropriate, but after some replication and development to 'move up the learning curve' and achieve economies of scale, they may be more broadly applicable. This highlights the role of learning from doing and from experience.

The above considerations are consistent with the main conclusion of the evolutionary approach: the effects of government policies are variable, depending on the nature of the sector and on other circumstances, and policies have to be carefully adapted. Hence it is possible for advocates of a variety of policy solutions to find examples to support their cases, as demonstrated in the debate over the 'Porter Hypothesis' that environmental regulation can stimulate innovation.

4. THE ECONOMIST'S VIEW AND THE DISTINCTION BETWEEN ENERGY EFFICIENCY AND ECONOMIC EFFICIENCY

In general, economists have been sceptical of the win–win theory (Oates et al., 1993; Palmer et al., 1995). An implication of that perspective – they argue – is that comparisons of an engineering ideal for a particular energy use with average practice for existing technology are inherently misleading, since the former does not incorporate all the real-world factors influencing energy technology decision making. The overall economic costs of switching to more energy-efficient technology constitute what can be thought of as a market barrier to their use, in that individual consumers and producers will not have incentives to use more costly technologies unless policy measures (such as technology standards or carbon taxes) are employed to compel or induce behavioural changes. Unlike market failures, however, market barriers cannot be lowered in a win–win fashion. Constraining consumers to purchase appliances with a higher level of efficiency based on simplistic analysis will in effect impose extra costs on consumers. It can result in a higher level of energy efficiency, but in reduced economic efficiency since consumers are forced to bear costs that they would otherwise have avoided. Although it is possible that this may be justified by some larger societal goal to address certain environmental externalities associated with energy consumption, the problem should be approached from that broader perspective, rather than from the narrow perspective of constraining energy efficiency decisions. Taking this broader perspective leads one to focus more directly on the real problem – climate change associated with CO_2 emissions – rather than constraining available technology options.

For the purpose of examining the link between environmental policy instruments and technological change, policies can be characterized as either command-and-control or market-based approaches. Market-based instruments (MBIs) are mechanisms that encourage behaviour through market signals rather than through explicit directives regarding pollution control levels or methods. These policy instruments – such as pollution charges, subsidies, tradable permits, and some types of information programmes – have been described as 'harnessing market forces'. This is because if they are well designed and implemented, they encourage firms (and/or individuals) to undertake pollution control efforts that are in their own interests and that collectively meet policy goals.

Conventional approaches to regulating the environment are often referred to as command-and-control (CAC) regulations, since they allow relatively little flexibility in the means of achieving goals. Such regulations tend to force firms to take on similar shares of the pollution control burden, regardless of the cost. CAC regulations do this by setting uniform standards for firms, the most prevalent of which are technology- and performance-based standards. Technology-based standards specify the method, and sometimes the actual equipment, which firms must use to comply with a particular regulation. A performance standard sets a uniform control target for firms, while allowing some latitude in how this target is met.

Holding all firms to the same target can be expensive and, in some circumstances, counterproductive. While standards may effectively limit emissions of pollutants, they typically exact relatively high costs in the process, by forcing some firms to resort to unduly expensive means of controlling pollution. Because the costs of controlling emissions may vary greatly among firms, and even among sources within the same firm, the appropriate technology in one situation may not be appropriate (that is, cost-effective) in another.

While on theoretical grounds MBIs are generally preferred because they are more cost-effective, in practice CAC policies have been predominantly used. In recent years, however, economic instruments have played an ever-increasing role reflecting their perceived superiority. This process has culminated in the official recognition of these policy tools in the Kyoto Protocol. Indeed, much of the debate surrounding the Protocol has centred on the use of MBI mechanisms and on ET in particular precisely because of the large potential cost savings that economic tools offer relative to CAC measures.

All these forms of intervention have the potential to induce or force some amount of technological change, because by their very nature they induce or require firms to do things they would not otherwise do. It is the purpose of this chapter to consider the comparison of the above instruments of environmental policy when the state of the technology evolves in an endogenous manner, looking both at theoretical work and to empirical analyses based on numerical models.

Within the realm of CAC policies, performance and technology standards can be explicitly designed to be 'technology forcing', mandating performance levels that are not currently viewed as technologically feasible or mandating technologies that are not fully developed. One problem with these approaches, however, is that while regulators can typically assume that *some* amount of improvement over existing technology will always be feasible, it is impossible to know how much. Standards must either be made unambitious, or else run the risk of being ultimately unachievable, leading to great political and economic disruption. Technology standards are particularly problematic, since they tend to freeze the development of technologies that might otherwise result in greater levels of control. Under regulations that are targeted at technologies, as opposed to emissions levels, no financial incentive exists for businesses to exceed control targets, and the adoption of new technologies is discouraged.

In contrast with such CAC regulations, market-based instruments can provide powerful incentives for companies to adopt cheaper and better pollution control technologies. With MBIs, it pays for firms to clean up a bit more if a sufficiently low-cost method (technology or process) of doing so can be identified and adopted.

There are two principal ways in which environmental policy instruments can be compared with regard to their effects on technological change. First and foremost, scholars have asked – both with theoretical models and with empirical analyses – the most direct question: what effects do particular instruments have on the rate and direction of relevant technological change? In keeping with the Schumpeterian trichotomy, such investigations can be carried out with reference to the pace of invention, innovation, or diffusion of new technologies.

It is also possible to ask whether environmental policies encourage *efficient* rates (and directions) of technological change, or more broadly, whether such policies result in overall economic efficiency (that is, whether the efficient degree of environmental protection is achieved). This second principal mode for comparison is linked more directly with criteria associated with welfare economics, but such comparisons have been made much less frequently than have direct assessments of technology effects. Within the limits of the existing literature, we consider both sets of criteria.

Most of the work in the economics literature on the dynamic effects of environmental policy instruments on technological change has been theoretical, rather than empirical, and so we consider the theoretical literature first.

The starting-point is to ask why firms would want to develop cleaner technologies themselves. Firms do so not only to directly pursue new product and process innovation but also to develop and maintain their broader capacity to assimilate and exploit externally available innovation. Behind the 'innovative' reason for R&D are two motivating forces: profitable investment and strategic advantage.[1]

Firms realize greater profits if they can produce more and better products for the same cost, or if they can make the same products with less. Thus they are willing to spend resources on R&D if the results will be lower costs and higher productivity. In addition, firms engage in R&D in order to give themselves a strategic advantage over their rivals. A better process or a better product can enhance a firm's market share. If a firm knows that its rivals are engaging in R&D, then it will see its own competitive position as being under threat.

How do GHG abatement policies compare in creating incentives for innovation? Policies designed to abate GHGs can also have significant impacts on incentives for innovation. However, the incentives can differ greatly across the range of possible measures. CAC policies that dictate what specific technologies or processes a firm must use by design give firms little leeway. As a result, the firm has little or no incentive to employ a more cost-effective way to achieve the same emissions reduction, since it is not allowed to deviate from the specified technology. A policy mandating the 'best available technology' might allow the firm to update or retrofit its production process with a more cost-effective technology. However, the firm would not want to develop a better technology that would achieve more abatement if that switch entailed higher total costs – even if from a social standpoint the extra reductions in pollution would justify the added costs. Performance standards, which dictate a certain level of abatement but leave the methods up to the firm, give the firm an incentive to develop a cheaper method of attaining the standard. A problem with performance standards, however, is that firms have no incentive to perform beyond the standard, even if lower abatement costs make more abatement desirable from a social perspective.

Market-based instruments, such as emissions taxes or permits, create a price for emissions. Firms then reduce their emissions as long as the cost of an additional amount of abatement is less than the tax (or permit) cost of additional emissions. For the remainder of their emissions, they pay the tax, buy permits, or use permits that they could otherwise sell. This system ensures that marginal abatement costs are equalized across firms, meaning that total abatement costs cannot be lowered by shifting some abatement from one firm to another (see the seminal work by Kneese and Schultz, 1975). By developing a more cost-effective way to reduce pollution, an innovating firm can then save in two ways: on its costs for achieving its current level of abatement, and on its tax or emissions permit payments by performing additional abatement at the new, lower cost. Thus far, the incentives discussed here are restricted to the firm's own gains from lower pollution abatement costs. Looking at this single-firm context, earlier analyses of environmental policy and innovation showed that emissions taxes and emissions permits generally provide more incentives for technological innovation than CAC policies like technology mandates and performance standards (see Magat, 1978; Downing and White, 1986). However, new technologies or processes may be useful for more than just the one firm performing the R&D. When innovations are more widely applicable, other incentives come into play and further differences arise among environmental policies. Recently environmental innovation has been analysed in the context of oligopolistic markets where strategic incentives and interactions among firms play a relevant role (see Carraro, 1999, for a survey).

A technology developed to reduce one firm's costs of abating emissions may also help other firms do the same. As a result, the social gain from this technology consists of the cost reductions to all firms (and potentially gains from additional abatement once it is less costly to implement). Furthermore, if adopting the new technology helps these other firms' bottom lines, they will be willing to pay to license it. Thus, with a well-functioning patent protection system, for example, the innovator can sell the fruits of its R&D to the potential adopters and reap the full social gains from the invention (Parry, 1995).

However, reaping those gains may be difficult. New knowledge can create 'spillover benefits' for other parties for which the innovator is not compensated. Not every advance may receive a patent: basic research may not have specific, patentable applications. Still, broadening the overall knowledge base can make future valuable applications possible. Even if a new technology or process does receive a patent or copyright, it may be hard to enforce. For example, other firms may also have the opportunity to imitate the innovation despite the patent, such as through reverse engineering. Studies for commercial (or non-environmental) innovations suggest that appropriation rates vary considerably over different types of innovations, with an average rate of around 50 per cent (Griliches, 1980; Nadiri, 1993). The easier it is for other firms to imitate the innovation on their own, the less they will be willing to pay for a licence, and the lower the gains the original innovator will be able to appropriate. As a result, the innovator will have less incentive to conduct R&D than would be best from a societal perspective.

On the other hand, the market for R&D might also encourage too much research. For example, several innovating firms competing for the rewards of a patent can collectively spend too much on redundant research. Thus these 'patent races' can lead to too much R&D. The consensus view, however, is that the private rate of return to R&D is well below the social rate of return (see Mansfield et al., 1977; Griliches, 1984; Bernstein and Nadiri, 1989). Thus positive spillover effects prevail, and markets will tend to provide too little R&D.

How do environmental policies compare when other firms may adopt a new technology? When we take into account interactions among innovating firms and other firms potentially adopting the new technology (at a price), the analysis of how different abatement policies affect innovation becomes more complex and ambiguous. Table 4.2 summarizes the various incentives for innovation and gives an idea of their relative importance.

Command-and-control policies that specify a technology allow other adopters no more leeway than the innovator. Therefore, little or no incentive remains to employ more cost-effective pollution reduction techniques. However, a policy mandating the 'best available technology' offers the possibility that the innovator's new technology will be made the standard. Then the other firms in the industry would have to pay to adopt it, if they could not come up with their own equivalent technology. Therefore, although the innovating firm's own pollution reduction costs could rise due to the new, stricter technology standard, the gains from other firms licensing the technology could be large. Under performance standards, adopting firms receive the same gains from abatement cost reduction as the innovator. They are willing to pay for a cheaper way to abate their emissions up to the standard, but they will not want to push reductions beyond the standard.

Table 4.2 Incentives for innovation created by environmental policies

Policy	Direct gains to innovating firm	Potential rents from adoption
Command-and-control	None	None
Best available technology	– New standard raises overall compliance costs	+++ Tighter standard raises incentives to adopt
Performance standard	++ Limited to existing abatement costs	++ Limited to existing abatement costs
Emission tax	+++ Lowers abatement costs and taxed emissions	+++ Lowers abatement costs and taxed emissions
Auctioned emission permits	+++++ Lowers abatement costs and costs of all permits purchased	+ Buying permits becomes cheaper alternative
Grandfathered emission permits	++ Lowers abatement costs	+ Buying permits becomes cheaper alternative

Source: Fischer (2000).

With market-based mechanisms, the development and industry-wide adoption of an innovation can affect not only individual firms' costs, but also the prices and quantities prevailing in the industry's market. Because different market-based mechanisms allow for different changes in prices and quantities of emissions, they have different implications for innovation (see Parry, 1996; Fischer et al., 1999). Taxes set the price of emissions, allowing total abatement to vary, while permits set the total amount of emissions, allowing the marginal cost of abatement to vary. Therefore, while one can choose a tax and an emissions cap that would generate the same outcome before innovation, product and technology markets will adjust differently after innovation, depending on the policy. Under an emissions tax, as innovation makes abatement cheaper, the total amount of abatement will rise because it will be less expensive to reduce emissions some more than to pay taxes on all current emissions. With emissions permits, on the other hand, firms as a group will not abate more after the innovation, since that total amount is set by the cap. Therefore, the total abatement cost savings are less than under the tax, where more abatement was performed after innovation. Furthermore, as innovation lowers abatement costs, the price of emissions permits will fall. From the point of view of the adopting firms, their individual decision of whether to license the technology does not affect the permit price; however, they will attempt to anticipate the fall in the permit price caused by collective adoption of the innovation. The lower permit price means that they now have a less costly option of foregoing the new technology and buying cheaper permits instead. Thus adopting firms will not be willing to pay as much for the innovation under permits as under the tax. This 'adoption price effect' lowers the maximum royalty the innovating firm can charge to license the technology.

However, the price fall can benefit the innovator directly, if permits are auctioned. Then, a lower permit price means the innovator will not have to pay as much for the rest of its emissions. This 'emissions payment effect' raises the gains to innovation. If permits are freely distributed, on the other hand, the price fall lowers the value of the innovator's allocated permits, which counteracts the emissions payment effect. Thus free permits give less incentive to innovate than auctioned permits and less than taxes due to the smaller abatement cost savings and the adoption price effect.

It is harder to say whether auctioned permits or taxes generate more innovation, since the incentives depend critically on the adopting firms' ability to imitate the innovation. In general, with little imitation, unless the emissions payment effect is very large, taxes provide more innovation incentive. However, if imitation is substantial, preventing the innovator from recouping most gains, if any, from the adopters, auctioned permits provide more incentive.

Of course, it is important to look beyond which environmental policy generates the most R&D and think about overall impacts on societal well-being. R&D spillovers and global warming pollution are two separate problems. To correct for both ideally requires two separate policy tools. Short of that, policy makers concerned about climate change must recognize that incentives for abatement and innovation are inexorably linked. If failures in the market for R&D exist, and available policies cannot correct them, the choice of climate policy will then be affected. If innovation incentives are lacking, the best climate policy response is likely to be looser regulation, reflecting the higher expected abatement costs, not stricter regulation to promote innovation. Similarly, if emissions are underpriced and too little abatement is performed, less investment in R&D may actually be warranted, since the new climate-friendly technologies will not be sufficiently exploited. Since GHGs accumulate over time, incentives for their abatement and innovation are further intertwined, since current and future abatement can depend on each other as well as innovation, which in turn depends on abatement. Thus a technology policy is no substitute for environmental policy. The primary gains to environmental protection always come from reducing environmental damages in the most cost-effective manner, regardless of innovation. Innovation then has the potential to increase those gains, but the benefits of emissions-reducing innovations are by definition limited to the potential environmental gains (Parry et al., 2000). More than other policy instruments, market-based mechanisms offer the incentives both to protect the environment efficiently and to develop better methods of protection. As Fischer (2000, p. 9) puts it, 'rather than look for a free lunch, perhaps we should make a nutritious lunch'.

5. TECHNOLOGY INNOVATION VS TECHNOLOGY DIFFUSION

The discussion so far has mainly focused on the potential for environmental policy to bring forth new technology through innovation. However, there is also a widely held view that significant reductions in environmental impacts could be achieved through more widespread diffusion of *existing* economically attractive technologies, particularly ones that increase energy efficiency and thereby reduce emissions associated with fossil fuel combustion. Here the attention turns to the sluggishness and inertia of diffusion processes. These may be important problems to be faced when designing an innovation policy whose goal is to achieve environmental benefits.

The observation that energy-efficient technologies that are cost-effective at current prices are diffusing only slowly dates back to the 1970s, having been identified as a paradox at least as far back as Shama (1983). There is a large body of economic research on the subject of technology diffusion. The single most important conclusion of this literature, well summarized in Jaffe and Stavins (1994), is that diffusion of new, economically superior technologies is a gradual rather than instantaneous process. Specifically, diffusion is often portrayed as a *sigmoid* curve over time. That is, the rate of adoption begins slowly, speeds up, then eventually slows down again as market saturation approaches.

Many sources of inertia govern the rate at which techno-economic systems change. These include low capital stock turnover rates in some sectors (for example electric power generation); the time needed for innovations to incubate; institutional barriers to diffusion; weak mechanisms incapable of translating political or societal imperatives into effective economic signals; and self-reinforcing loops between particular technical options and consumption patterns, which create technological 'lock-in' and discourage radical innovation.

In particular, *increasing returns* to adoption – in the form of *learning curves* and *positive adoption externalities* – are a significant feature of market penetration processes for many technologies. LbD describes how cumulative production experience with a product leads to reduced production costs, while learning-by-using captures how the value of a good increases for consumers as they gain experience using it. Positive adoption externalities arise when a non-user's probability of adoption is increased when a greater number of potential users have already adopted. This could occur because of fad or herding effects, or because of 'network externalities'.

Network externalities exist if a product is technologically more valuable to an individual user as other users adopt a compatible product (for example, telephone and computer networks). These phenomena can be critical to understanding the existing technological system, forecasting how that system might evolve, and predicting the potential effect of some policy or event.

Furthermore, increasing returns to adopting a particular technology or system have been linked with so-called technology 'lock-in', in which a particular product, technical standard, production process or service is produced by a market, and it is difficult to move to an alternative competing technology. Lock-in implies that, once led down a particular technological path, the barriers to switching may be prohibitive. This situation can be problematic if it would have been in the broader social interest to adopt a fundamentally different pattern of technological capacity. In turn, it raises the question of whether policy interventions – possibly involving central coordination and information assessment, direct technology subsidies, or publicly funded research, development, demonstration and procurement programmes – might avoid undesirable cases of technology lock-in by guiding technological paths in directions superior to those that would be taken by the free market.

Finally, a related characteristic of products or systems subject to increasing returns or 'positive feedbacks' is that history can be critical. While other markets can often be explained by current demand and supply, markets subject to increasing returns may not be fully understandable without knowing the pattern of historical technology adoption. This 'path-dependence' emphasizes the lasting role that chance historical events can play in leading market outcomes down one rather than another possible path.

To the extent that diffusion has been and is limited by market failures, it is less clear that policies that operate by increasing the economic incentive to adopt such technology will be effective. On the other hand, if such market

Table 4.3 Factors influencing technology diffusion

Factor	Likely direction of effect	Potential policy/ institutional instrument
Increased relative price of resource conserved by the technology	(+)	Tax on the resource
Decreased cost and/or increased quality of technology	(+)	Technology subsidy
Inadequate information, uncertainty, agency problems regarding costs and benefits of technology adoption	(-)	Information dissemination, technology demonstration
Learning-by-doing and learning-by-using	(+)	Technology demonstration and deployment, tax/subsidy
Network externalities	(?)	Standards, planning, coordination
Characteristics of potential adopters	Varied	Flexible regulation

Source: Jaffe et al. (2000).

failures are important, then policies focused directly on correction of such market failures provide, at least in principle, opportunities for policy intervention that are social-welfare increasing, even without regard to any environmental benefit.

Table 4.3 summarizes the potential influence on technology diffusion of many of the factors discussed in this section.

In general, remedying structural inertia is difficult. Decision making within complex socio-technical systems takes place at many levels (households, industry, public authorities and so on), all governed by different priorities, and diverting complete systems on to less environmentally harmful technological trajectories is a daunting task. Price signals will undoubtedly have some effect on the technological paths taken by the various actors involved in complex systems, but there are genuine fears that these prices will reflect short-term expediency rather than longer-term societal and environmental needs, and that they will be too low to overcome structural inertia as far as innovation is concerned. There is also the added complication that many parts of complex socio-technical systems are comparatively insensitive to price variations, with decisions on future behaviour driven primarily by other factors, for example cost/speed ratios in the transportation sector; aesthetics in architecture; the price of land in urban planning.

In such situations, the important task is to make sure that future technological options are attractive to even the 'weakest link' sectors, that is, those with high structural inertia and low price sensitivity. In fact, since 'chains' are dependent on their 'weakest links', targeting these sectors (transport, buildings, urban forms) becomes a high priority. There is therefore the need for socioeconomic and policy research into the complementary economic and regulatory policy mechanisms which will be needed to ensure that these new technological options are adopted and diffused.

In addition, apart from research, there is also a great need for public policies supporting experimentation with new technologies, especially radical technologies for which a market is not defined. This would help suppliers to get a better idea of user requirements and the performance of technologies in real-life situations. It would also help users to develop a better understanding of new technologies and how they might benefit from them. It would also help to build a constituency for products and guide private and public actors in their policies. User experiments offer a way of breaking the gridlock of structural inertia by facilitating assessments of the claims for and against new technologies.[2]

6. EMPIRICAL EVIDENCE

Much of the debate surrounding the engineering and economic views of the win–win hypothesis has been ideological or political. Ultimately this is an empirical question and reliable empirical evidence is rather limited (see for an extensive review Jaffe et al., 2000). Recently, however, we are witnessing a tendency for the two views to try to integrate each other and these attempts are taking place at the level where the issues are most relevant and are best studied.

The potential long-run consequences of today's policy choices create a high priority for broadening and deepening our understanding of the effects of environmental policy on innovation and diffusion of new technology. Unfortunately, these issues cannot be resolved at a purely theoretical level, or on the basis of aggregate empirical analyses. For both benefit–cost and cost-effectiveness analysis, modern economics relies on a set of powerful analytical and computational tools that support quantitative modelling of economy- and society-wide policies over the long run. The formidable challenges posed by the complexity of climate policy have made economic modelling an especially attractive means of organizing and applying a range of scientific, economic and social research to analysing how we should respond to the threat of climate change. In practice, such analysis is typically carried out through the construction and application of large-scale computer models that combine scientific and economic theories and data into unified quantitative frameworks. These 'integrated assessment' models have emerged as decision makers' primary tool for quantitative climate policy analysis.

Models that explain how technological change relates to climate change can be divided into two idealized approaches – 'top-down' and 'bottom-up' models (even though many models contain elements drawn from more than one approach), and there are sometimes important distinctions within these groups. Because it is difficult to predict the future of technology, all models addressing this issue rely on some sort of assumption regarding the future course of technological progress.

Bottom-up models are based on engineering, and come to the problem of emissions limitations from the perspective of the cost and performance of emissions-reducing equipment and practices (such as energy-efficient space heating, lighting, or motors). They generally begin by assuming that a set of advanced technologies either does or will exist, with predetermined cost and efficiency characteristics. They then compare the world as it is now to the world that would exist if the assumed technologies were to be commonly used. Thus their depiction of climate change policy will depend strongly on the assumptions they make regarding new technologies. Very importantly, because of the narrow emphasis on the comparative cost and performance of individual technologies, these models usually do not reflect many other aspects of the economy's response to climate change or climate change policy, such as broader price-induced changes in energy demand, or the way households work or save. They are poorly suited, therefore, to estimating the societal economic cost of climate change or related policies. Instead, their strength lies in forecasting the near-term impacts of specific advances and in illuminating the economic value of possible technological improvements.

Top-down models, in contrast, are generally broad economic models – they depict the way the economy and environment interact in the aggregate. They reproduce the history of technological change in its broad outline but do not necessarily provide details about design, costs, or performance of specific technologies. Rather, they start with a set of initial conditions based on the current state of the economy, and then extrapolate from past experience to look at the future implications of major economic and technological forces. As they contain little or no explicit technological detail, many 'top-down' models represent technological change in terms of a single societal rate – the autonomous energy efficiency improvement (AEEI) – at which energy efficiency will continually improve in the future, a rate usually based on observed values in the past. This 'exogenous' depiction (so named because its description of technology comes from outside the model), therefore, requires the modeller to make an assumption regarding the value of this ongoing, 'autonomous', improvement. Other top-down models have attempted to replace this exogenous assumption with a more detailed representation of the very process by which technology is created and adopted by firms in the economy. This approach starts by assuming that the amount of innovative effort in the economy is a direct function of current and anticipated economic conditions, and is called endogenous technological change, because technological change is projected within the model. When this innovative effort impacts directly on energy or carbon efficiency and depends on climate policy measures we speak of induced technological change.

The various types of models represent in different ways the relationship between economic stimuli such as R&D expenditures, energy prices, taxes and subsidies, and the direction and rate of technological change, and the subsequent effect on societal cost. In bottom-up models, the rate of technological change depends on how different the alternatives are to the present technology, as well as how quickly they substitute for one another. For top-down models, the rate of change is determined by assumptions about the autonomous energy efficiency improvement (AEEI) or elasticities (responsiveness to price). In top-down models employing endogenous technological change, technological change is often represented as a function of past production (LbD), the amount of past R&D (stock of knowledge), or the extent of energy price changes. Thus, even these endogenous models operate on the basis of an assumption regarding what determines technological effort and how that effort translates into progress. So, regardless of their different structures, all models of technological progress rely on some fundamental assumption regarding future technology, an assumption that plays a central role in determining model results.

Several alternative criteria have been proposed in the literature for classifying quantitative tools designed for climate change policy analysis. Short of making a long and detailed list, we can follow Grubb et al. (2000), who identify the following three categories:

1. Energy systems/engineering and energy policy models. These are specifically focused on the energy sector and/or climate change policy. These embed the spirit and main assumptions of evolutionary theories and technologists' assessments and are of the bottom-up variety.
2. Macroeconomic models, which are based on modelling the economic

systems, but include specific consideration of energy and emissions. These are top-down models sometimes referred to as E3 models (energy–environment–economy models).

3. The new generation of so-called 'integrated assessment models' (IAMs). These combine highly aggregated models of climate change, economic activity and energy systems and emissions in one model. These models are of the top-down variety but, in their most recent versions, try to incorporate several key ingredients of bottom-up modelling.

In several instances these modelling efforts have concentrated upon the costs of GHG mitigation policies in the presence of penetration and diffusion of new technologies. The definition of costs varies across different models, ranging from direct engineering costs, to financial costs, to sectoral costs, and more generally macroeconomic costs (in terms of GNP lost, decreased welfare, and the like). Rigorous and extensive comparisons of the cost of curbing CO_2 emissions have been undertaken across models, often based on autonomous or exogenous notions of technical change. Here we are instead interested in the same issues but analysed with the help of models that either: (1) incorporate LbD in which technology costs are a function of cumulative investment, recently extended to include learning interactions between different technology clusters; or (2) direct modelling of investment in knowledge/innovation.

The first option has been especially pursued by bottom-up energy systems/engineering and energy policy models. Initial installations of technological innovations are often expensive. Costs decline as individuals, enterprises and industries gain experience with them. The learning or experience curve describes technological progress as a function of accumulating experience with production (LbD for manufacturers) and with use (learning-by-using – LbU – for consumers) of a technology during its diffusion. Technological learning has been observed historically for many different industries, an example being given in Figure 4.1 above, and is a well-established concept. In its basic formulation, technological progress is expressed in terms of decreasing specific costs of a technology as an exponential function of cumulative (installed) capacity. This variable measures the knowledge accumulation that occurs during the production (LbD) or the use (LbU) of the technology. The exponent is the learning elasticity upon which is based the familiar 'progress ratio' that captures the speed of learning. This approach was first applied to energy system scenarios by Anderson and Bird (1992). The authors found that by including LbD, scenarios could readily be developed in which renewable energy technologies became competitive and penetrated the global energy system on a large scale by the middle of the twenty-first century, leading to stabilization of global atmospheric concentrations at around 500 ppm CO_2.

This work was particularly influential in affecting Shell's thinking and development of global energy scenarios. A recent review of experience accumulation and cost reduction for a number of energy technologies is given by McDonald and Schrattenholzer (2001).

Recent developments have considered two-factor learning functions in which there is a separate effect, besides cumulative capacity, of R&D expenditures on the costs of specific energy technologies. Preliminary results do not support this addition, termed 'learning-by-searching': in four of the eight technologies considered cumulative R&D expenditures increased rather than reduced investment costs (Criqui et al., 2000).

A number of bottom-up models have integrated endogenous technological change that assumes LbD. Examples are MESSAGE (Messner and Schrattenholzer, 1998) and MARKAL (Barreto and Kypreos, 1999), dynamic linear programming models of the energy sector that are generally used in tandem with the MACRO macroeconomic model which provides economic data for the energy sector (Manne, 1981; see also Seebregts et al., 1999). They optimize a choice between different technologies using given abatement costs and carbon emission targets. Both models account for the uncertainty associated with the time of arrival and performance of new technologies by employing a stochastic rather than a deterministic optimization technique. Cost and performance information of a variety of advanced renewable and fossil-fuel-based alternative energy technologies are put into the model. MARKAL considers 13, MESSAGE 77 technologies. Both include six learning technologies, which are displayed in Table 4.4.

Recent simulation studies have explored possibilities with alternative values of the learning curve parameter applied to low carbon energy technologies. They suggest that the long-run costs of substituting renewable energy forms (including linked technologies such as hydrogen and fuel cells) for fossil fuels range from being significantly more than those of the latter (though not

Table 4.4 Learning rates in MARKAL and MESSAGE (%)

Energy technology	MARKAL	MESSAGE
Advanced coal	6	7
Natural gas combined cycle	11	15
New nuclear	4	7
Fuel cell	13	–
Wind power	11	15
Solar photovoltaics	19	18
Solar thermal	–	15

unaffordably more) to significantly less – that is, an economic surprise cannot be ruled out.

This is borne out by two distinct sets of studies. Gritsevskyi and Nakicenovic (1999), building on several years' work at IIASA, present numerical results of a model which incorporates a very wide range of energy technologies and resources, coupled with LbD, and which incorporate learning spillover effects in technology clusters. The model itself was run stochastically under widely defined uncertainties in the input scenarios, generating hundreds of thousands of energy system scenarios for the next century. The resulting scenarios span a huge range of CO_2 emissions by the end of the next century. When the 53 least-cost scenarios were selected from this analysis, the range of results scarcely narrows; indeed, it was the central scenarios that tended to disappear. The interest of this result is not in the specific numbers and assumptions, but in the extraordinary range of possible least-cost outcomes and their bimodal distribution. It illustrates numerically that that there were two broad directions for innovation in the global energy systems: one based mostly around coal-based and carbon-intensive resources (including shale conversions, liquefaction, and so on), and the other based around low and zero carbon resources (methane and hydrogen utilized with dispersed technologies including fuel cells, and with major roles for renewable energy and/or nuclear as primary sources alongside methane). The costs were indistinguishable within the range of uncertainties.

Very recent research has again used the notion of technology clusters in energy systems. IIASA's Riahi et al. (2001) purport to identify those groups of clusters that are characteristic of a particular group of scenarios with respect to GHG emissions reductions. In particular, their objective is to identify those principal technology clusters that could accomplish a smooth and efficient transition path from the present energy structure to eventual sustainability. Generally speaking, technology clusters are groups of technologies that have something in common. A technology can be the aggregation of other technologies (for example, a passenger car powered by a diesel engine). Of course, one technology can be used as a constituent in several technologies (the diesel engine is used not only in passenger cars, but also in trucks and for stationary applications, and so on). Therefore, one improvement in one technology can also mean technological progress in all the other technologies that are related to the given technology. Another relation between technologies can be introduced by the common use of infrastructures. For example, gasoline cars and all the technologies related to room heating with light fuel oil belong to one cluster because they both rely on oil refinery products. In long-term energy scenarios, infrastructures are characterized by a particularly long life cycle between build-up and phase-out. As noted in the previous section, this inertia limits the speed at which fundamental changes in the energy system occur, for example a change from the existing gasoline infrastructure for cars to a hydrogen gas infrastructure. In addition to the interdependencies among technologies, the inertia of the energy system to fundamental change is also increased by the fact that the accumulation of knowledge stock leads to increasing returns: for example, the more cars that are built over time, the cheaper their production cost becomes. As previously seen, this is usually called 'lock-in' effect leading to the typical path-dependence of technological change. These two types of relations classify the technologies into groups, and the two classification criteria very naturally lead to the definition of the first two types of technology clusters.

In general, induced technical change has a very large impact indeed in these explicit technology-based models of LbD. Since addressing climate change does involve such large-scale and long-term changes, the results suggest that the long-run economics of climate change policy cannot plausibly be considered without taking LbD fully into account. This is a striking contrast to the aggregate economic models of investment in explicit knowledge sectors considered hereafter.

A very recent model – called DEMETER, which stands halfway between bottom-up and top-down strategies – is due to Van der Zwaan et al. (1999) (see also Gerlagh and Van der Zwaan, 2000). A macroeconomic (top-down) model is expanded with learning curves previously used in energy systems (bottom-up) models. Technological change is represented through a learning curve describing decreasing non-carbon energy prices as a function of cumulative installed capacity. The authors compare several scenarios with taxes on the carbon and subsidies on the non-carbon technology. During the first decades, they find that carbon taxes reduce energy consumption. At a later stage, however, when the GHG policies have enhanced the maturing of the carbon-free technology, energy prices decrease and energy consumption reaches values higher than under business-as-usual. Moreover, overall consumption decreases in the first decades, with respect to business-as-usual, because of transition costs, while the availability of a progressively cheaper non-carbon technology increases total consumption in later periods.

In the model of knowledge accumulation of Goulder and Mathai (2000), a central planner chooses time paths of abatement and R&D efforts in order to minimize the present value of the costs of abating emissions and of R&D expenditures subject to an emission target. The abatement cost function depends both on abatement and on the stock of knowledge that increases over time via R&D investment. By assuming a central planner, this model sidesteps the problem of explicitly modelling innovation incentives and appropriability. A second problem studied by Goulder and Mathai (2000) assumes that the rate of change of the knowledge stock is governed by abatement efforts themselves. This form of technological change is termed

LbD. In this respect, another recent development of the literature is constituted by models based on the concept of learning curves (see Grubler and Messner, 1998) describing future costs and performance improvements of new technologies as a function of accumulated R&D, and learning and experience gained in diffusion of new technologies. Thus technological learning depends on previous, accumulated investments in R&D, demonstration plants and gradually expanding niche markets.

Other top-down models containing a direct modelling of investment in knowledge/innovation are applied general equilibrium (AGE) models and neoclassical growth models.

One example of AGE models is the OECD's GREEN model, which allows for three back-stop options: a carbon-based synthetic fuel, and two carbon-free possibilities. The main hypotheses concern prices and timing of diffusion: the prices are exogenous and the back-stop technologies, once they come on stream, are available in all regions in unlimited quantities at constant marginal costs. The only key variable of this approach is the relative price of the technological substitution options, which is exogenously imposed at current levels; on the contrary, the technological innovation possibilities are assumed to be fixed at the present level of knowledge for the entire simulation path.

As previously said, new technologies are typically developed by the most innovative firms and not immediately available to all. Factors that influence the rate and timing of diffusion are of fundamental importance in assessing the ultimate effectiveness of the innovation. Modelling this factor is obstructed by certain characteristics of empirical environmental models. Usually, top-down models do not provide the degree of sector disaggregation that would be required for analysis at the level of the firm, while bottom-up studies do not consider strategic market behaviour that may delay the diffusion of innovation. There are, however, some attempts to model spillovers and diffusion.

One significant example is the AGE model by Goulder and Schneider (1999). The model quantifies the costs of carbon taxes in the presence of induced technical change and crowding-out. If there are no prior inefficiencies in R&D markets, policies that stimulate environmental R&D crowd out other forms of R&D, which implies that the gross costs (before accounting for environment-related benefits) of carbon taxes are higher than they would be if there were no induced technical change. The result may be reversed in the presence of inefficiencies in R&D markets. More importantly, the environmental benefits are likely to be larger when induced technical change is taken into account (because there is more abatement) and the benefits from the additional abatement more than compensate for the higher gross costs (whenever this is actually the case).

The RICE model proposed by Nordhaus and Yang (1996) is one of the most popular integrated assessment tools for the study of climate change. It is basically a single-sector optimal growth model suitably extended to incorporate the interactions between economic activities and climate. There is one such model for each macro-region into which the world is divided (USA, Japan, Europe, China, Former Soviet Union (FSU), Rest of the World (ROW)). Within each region a central planner chooses the optimal paths of fixed investment and emission abatement that maximize the present value of per capita consumption. Technology levels affect output production (non-environmental technical change) and the ratio between emissions and output (environmental technical change) but grow over time in an exogenous fashion. Nordhaus (1999) starts from this aspect and lays out a model of induced innovation brought about by R&D efforts. In particular, technological change displays its effects through changes in the emissions–output ratio. This aspect is then embedded in the non-regional version of the author's RICE model for climate change policy analysis, called DICE (Nordhaus, 1993).

Buonanno et al. (2000b, 2000c) extend the RICE model by endogenizing both environmental and non-environmental technical change and by allowing for trading of emissions permits. In the model, called ETC-RICE, each country plays a non-cooperative Nash game in a dynamic setting, which yields an open-loop Nash equilibrium. It is assumed that innovation is brought about by R&D spending which contributes to the accumulation of knowledge. The stock of existing knowledge is a factor of production, which therefore enhances the rate of productivity. This is a form of endogenous technical change. Besides this channel, however, knowledge also serves the purpose of reducing, *ceteris paribus*, the level of carbon emissions. This is referred to in the literature as induced technical change.

To begin with, the authors compare the costs of complying with the CO_2 targets agreed upon in Kyoto with and without induced technical change. In Buonanno et al. (2000a) the authors analyse this issue under several policy options, that is, when emissions trade is not allowed, is restricted to (all) Annex B countries only and, finally, is extended to all world countries. This last case is often likened to CDM in the literature. The difference in total mitigation costs between endogenous and endogenous plus induced technical change is computed. They therefore show the benefits – in terms of reduced total costs of complying with the Kyoto Protocol – provided by the presence of induced technical change. Total costs comprise abatement costs, R&D spending and the costs resulting from the algebraic difference between buying and selling pollution permits.

The overall costs of complying with the Kyoto Protocol under different policy options and for different specifications of technical change are pre-

Table 4.5 Total compliance costs under alternative specifications of technical change and for alternative policy options

Specification of technical change	Policy option	USA	Japan	Europe	FSU	China	ROW
ETC	Kyoto	0.480	0.308	0.512	0.080	0.044	0.356
	Et-A1	0.478	0.293	0.500	0.056	0.044	0.356
	Et-All	0.412	0.242	0.387	0.050	0.033	0.335
ETC+ITC	Kyoto	0.550	0.321	0.549	0.066	0.045	0.360
	Et-A1	0.546	0.292	0.513	0.002	0.045	0.360
	Et-All	0.409	0.239	0.380	0.050	0.050	0.379
ETC+ITC with spillovers	Kyoto	0.575	0.296	0.583	0.091	0.058	0.412
	Et-A1	0.591	0.279	0.558	0.047	0.058	0.412
	Et-All	0.446	0.224	0.408	0.067	0.046	0.393

Notes: the figures reported are expressed in 1990 trillion US$ and are averages over the period 2010–2100. ETC refers to the model with only environmental technical change; ETC+ITC also incorporates induced technical change; ETC+ITC with spillovers is self-explanatory. For an explanation of the policy options, see the text.

sented in Table 4.5. For ease of presentation we only display average figures over the simulation period 2010–2100.

The general picture that emerges from the table is consistent across model specifications. First of all, the Kyoto scenario entails the highest overall costs as no emission reduction can be traded away and emission limits cannot be overcome. The possibility of emissions trading leads to a reduction in compliance costs for all Annex 1 countries when trade is restricted as in the Et-A1 scenario, and for all countries under global trade (Et-All case).

There are a couple of exceptions to this rule. The first is the case of the FSU in the two versions of the model with induced technical change. Notice that FSU total costs increase in the global trade scenario with respect to the Et-A1 scenario. This situation can be explained by the fact that, under Et-A1 trading, the FSU sells permits, thus receiving a relevant financial transfer. Under global trading, China and ROW become the main sellers of permits. Moreover, supply increases and the price of permits becomes lower. Hence financial transfers to the FSU are much smaller under global trading and the total costs increase.

A second case concerns China and ROW and depends on the strategic use of R&D. As shown below, sellers of permits have an incentive to invest in

R&D to increase their supply of permits and the related revenue. This incentive, and the non-cooperative framework through which decisions are taken, may lead some countries – notably sellers of permits – to over-invest in R&D. This may induce an excess supply of permits and a too low price in the market for permits. The consequence is that total costs for seller countries under global trading – total costs include the cost of R&D investments – may increase, as shown in Table 4.5 for China and ROW in the ETC+ITC case. Notice that the incentive to over-invest in R&D is lower in the presence of spillovers, because of the well-known free-riding incentive. As a consequence, the effect described above is smaller and global trading induces a reduction of total costs even for China and ROW in the model versions with spillovers.

The results of Table 4.5 can also be analysed from the viewpoint of alternative models of technical change. When induced technical change is allowed for, R&D is also used to lower the emission–output ratio. In this case, spending in domestic abatement can be replaced by investments in R&D activities. The net impact on total compliance costs of this substitution effect of induced technical change cannot be predicted a priori. Costs turn out to be higher for all countries under Kyoto and Et-A1 regimes, the only exception being the FSU. Costs are instead lower under global trading, but only for the three industrialized regions. Considering the role of knowledge spillovers, our a priori assumption is that free-riding induces countries to undertake less R&D relative to the absence of this type of externality. This entails more domestic abatement activities and therefore costs. Again, however, the net effect on compliance costs is not a priori defined and the results show that for all countries, with the single exception of Japan, costs go up when spillovers are part of the picture.

These results suggest that in almost all countries total costs tend to be lower whenever R&D – rather than domestic abatement (through regulation, taxation, and so on) – is used to reduce emissions. However, this also depends on the trading scheme adopted. Hence a clearer understanding of the dynamics of total costs under different specifications of technical change and/or under different scales of trading can be obtained only by looking at the different components of total costs. As a consequence, we now move to consider the individual components of total compliance costs in more detail.

In a subsequent paper, Buonanno et al. (2000c) look again at the efficiency and equity implications of the Kyoto Protocol. In this respect some arguments have been offered in support of the introduction of ceilings on emissions trading, leaving room for domestic abatement through the usual instruments, whether market-based or regulatory ones. These restrictions are often advocated for equity reasons: in particular, developed countries should not be allowed to trade freely in the permit market, in order to be induced to abate their own emissions through domestic policy and measures, rather than by 'exploiting' the lower abatement costs of developing countries. From the perspective of technical change it is for instance argued that the widespread adoption of flexible mechanisms reduces the incentives to carry out environmental R&D, thereby reducing the effectiveness and increasing the costs of abatement options in the long run. Moreover, the incentives to R&D induced by the presence of ceilings on the use of flexible mechanisms may spill over into other sectors, thus speeding up the 'engine of growth', and reducing the impact of climate change control on long-run per capita income and welfare. Ceilings and spillover effects under the Kyoto Protocol are empirically investigated using a further extension of the (ETC) RICE model. The policy options analysed are presented in Table 4.6.

Table 4.6 Policy options

	Flexibility (ceilings)			
Coalition structure	0%	15%	33%	100%
No coalitions	X			
Annex 1 regions	X	X	X	X
All regions		X	X	X

First of all, the authors look at whether all countries increase their R&D effort in the presence of ceilings. This is indeed the case as far as developed countries are concerned, but the behaviour of the other world regions appears to be quite different. In the USA, Japan and the EU the lowest R&D effort, as measured by the percentage ratio between R&D expenditure and GNP, is made in the business-as-usual scenario and when all countries are allowed to trade without any restrictions. The ratio then increases when Annex 1 countries are allowed to trade without ceilings, and further increases in the presence of ceilings. The R&D effort is highest in the Kyoto scenario, which corresponds to a 0 per cent ceiling (all abatement is carried out through domestic measures). Hence these results support the conjecture of those who are in favour of ceilings, namely that these would stimulate R&D expenditure.

The implications for costs and growth are, however, another matter. In the other world regions, the R&D effort depends on the role of these regions in the permit markets. For example, the FSU carries out the largest R&D effort in the case in which Annex 1 countries trade without ceilings. The reason is that this is the most favourable situation for the FSU, because the demand for permits is high and the FSU is the only seller. Hence it increases R&D to reduce its own emission levels, thus increasing the amount of permits that can be sold in the market. When ceilings are introduced, the demand for permits is lower and the FSU's R&D effort is consequently reduced. It becomes even lower when all countries are allowed to trade. In this case, the FSU is no longer the unique seller, nor the one with the lowest abatement costs. Hence, China and the Rest of the World sharply increase their R&D effort to exploit the benefits arising from selling permits. This R&D effort is obviously reduced in the presence of ceilings. The conclusion is that, in general, ceilings are likely to increase R&D expenditures (relative to GNP) in OECD countries, that is, in countries which are going to buy permits, but they reduce them in the FSU, China, and other developing countries – the seller countries – where the greatest stimulus to carry out R&D comes from the possibility to trade emission permits without restrictions.

The evidence just presented notwithstanding, the next question is: when ceilings on emissions trading foster R&D expenditures and increase R&D efforts, do they also reduce abatement costs? What is the overall effect on economic growth? It appears that in all OECD regions the highest cost is achieved when the whole abatement is carried out domestically and there is no trade (the Kyoto scenario), whereas the lowest cost is achieved when all world countries are allowed to trade without restrictions. Ceilings generally increase costs for OECD countries when applied to the case of global trade, while this is not true when the Annex 1 trading is considered. In fact, a 33 per cent ceiling in the case of only Annex B trade may be slightly beneficial at least for the USA and the EU. This may be due to the fact that, since that ceiling is not binding for these countries, its introduction is beneficial because it reduces the equilibrium prices of permits (given the lower demand coming from Japan), without affecting the quantity demanded by both the USA and the EU. Instead, for the developing countries the situation is less clear-cut: it seems to emerge that the lowest costs are achieved when these countries do not participate in the trading (that is, in the Annex B trade case).

In summary, the authors' analysis provides little support to quantitative restrictions on emissions trading. Even if the introduction of ceilings increases the R&D efforts of buyer countries and fosters technological innovation, the overall effect on abatement costs and economic growth is negative. The reason is that the benefits from technological innovation are lower, even in the long run, than the costs of adopting a more costly approach to climate change control. In other words, firms benefit more from a low impact of climate policies on their costs than from the stimulus to innovation that these policies induce.

Even equity is not positively affected by ceilings. Buonanno et al. (2000c) find that flexibility mechanisms in the presence of endogenous technical change increase equity and that the highest equity levels are achieved without ceilings, both in the short and in the long run. The main reason is that developing countries receive important transfers from developed countries through the trading of permits, and this tends to reduce income inequalities. Moreover, the introduction of R&D and technical change also gives developing countries the possibility to use R&D strategically to increase their sale of permits.

Like several others, the model described above has been mainly used to study the effects of the implementation of the Kyoto Protocol, particularly as far as the consequences of the emission targets and of the flexibility mechanisms are concerned. Some authors, however, have been quite sceptical about the viability of that agreement and have done work purporting to show the limits of the Protocol provisions and the superiority of some alternative options. These sceptical notes on the future of the Kyoto Protocol have been recently supported by the outcomes of climate negotiations in Bonn and Marrakesh. The COP meetings in Bonn and Marrakesh and, above all, the US decision to withdraw from the Kyoto Protocol, have shown how fragile the original Kyoto agreement was. In the light of these recent events, a few papers are circulating containing simulations of the consequences of the US defection and of the decisions taken in Bonn and Marrakesh. The main conclusions achieved by this recent literature (see, for example, Hagem and Holtsmark, 2001; Eyckmans et al., 2001; Den Elzen and Manders, 2001; Den Elzen and de Moor, 2001; Kopp 2001; Manne and Richels, 2001) can be summarized as follows:

- all studies conclude that the US withdrawal and the sink provisions of the Bonn agreement highly reduce the environmental effectiveness of the Kyoto/Bonn agreement;
- all studies coincide in finding a decline in the demand for emission permits and therefore in their price;
- however, even if the lower demand implies both a decrease of the permit price and a decline in the compliance costs, the possibility of strategic market behaviours modifies the size of the expected changes in prices and costs. In particular, these changes are much smaller than initially suggested when banking and monopolistic behaviour in the permit market are taken into account.

However, these results are again based on models in which the dynamic of technical change is not adequately taken into account. Does the presence of endogenous and induced technical change modify the above results? Buchner et al. (2001) address this issue by employing the ETC-RICE model described above to assess the consequences of the 'new' Bonn/Marrakesh agreements on permit prices and quantities exchanged, domestic abatement, R&D activity and emissions.

Their results prove again the important influence that technical change exerts on the effects of climate policy and therefore the importance of modelling the interplay between technical change, economic dynamics and climate. Buchner et al. (2001) show that, in the presence of endogenous and induced technical change, the US decision of withdrawing from the Kyoto Protocol, by reducing the demand for permits and their price, lowers the incentives to undertake energy-saving R&D. As a consequence, emissions increase and feed back on the demand and supply of permits. Indeed, in order to meet their targets, countries which stick to the Kyoto commitments must increase their demand for permits or reduce their supply. As a consequence, the fall of the price of emission permits after the US withdrawal from the Kyoto Protocol is much smaller than the one identified in previous studies.

In addition, the US defection, by reducing R&D investment and increasing emissions, also increases damages from climate changes. This leads to increased investments and therefore increased energy demand and GHG emissions. As a consequence, this second-order effect further enhances the previous one by further increasing the demand for emission permits and reducing their supply.

The presence of technological spillovers provides an additional contribution to the feedback effects just described. Indeed, the US defection induces a strong reduction of domestic energy-saving R&D investments. This reduction spills over on to the other countries by reducing the world stock of knowledge, thus increasing the emission–output ratio. Therefore, emissions increase in all countries. In order to meet their Kyoto targets, the remaining Annex B countries must increase their demand for emission permits or reduce their supply. As a consequence, the price of permits increases. This feedback effect also partially offsets the initial fall of the permit price induced by the US defection. Therefore, the final equilibrium price is higher than the one usually estimated in previous studies.

Note that, albeit smaller, a fall of the permit price is also predicted by Buchner et al. (2001). As a consequence, the benefits for Russia from participating in the Kyoto agreement become smaller. In addition, the costs for the EU and Japan of a possible Russian defection becomes very large. Without Russia, abatement costs in the EU and Japan would indeed be much higher. These two effects give Russia a larger bargaining power that can be exploited in two ways: (i) either by obtaining further economic incentives to participate in the Kyoto agreement; (ii) or by reducing the costs of participating in the Kyoto agreement through less stringent targets or increased flexibility (for example increased sinks allowances).

This latter prediction seems to be confirmed by the recent outcome of the Marrakesh Conference of the Parties. Notice, however, that this might not be the end of the story. If the USA realizes that, either for domestic or for international political reasons, a large amount of GHG emission abatement must be undertaken and that these emission reductions are too costly if undertaken through domestic measures only, then they may decide to re-enter the game, either within the Kyoto framework or outside it. In the latter case, they could offer Russia better conditions than the ones offered by the EU and Japan, thus opening the way for an emission market in which the USA and Russia are the main actors. Or, alternatively, the USA could negotiate an agreement with China, in order to trade emission permits at a low price, thus reducing abatement costs.

The fascinating dynamics of climate negotiations, and therefore the analysis of these and other policy scenarios, will be the subject of the next chapter.

7. CONCLUSIONS

In opening this chapter, we suggested that an understanding of the process of technological change is important for the economic analysis of environmental issues for two broad reasons. First, the environmental impact of social and economic activity is greatly affected by the rate and direction of technological change. This linkage occurs because new technologies may either create or mitigate pollution, and because many environmental problems and policy responses are evaluated over timeframes in which the cumulative impact of technological changes is likely to be large. This observation is a natural bridge to the second broad linkage between technology and environment, the effect of environmental policy interventions on the process of technological change. The empirical evidence to date is generally consistent with theoretical findings that market-based instruments for environmental protection are likely to have significantly greater, positive impacts over time than command-and-control approaches on the invention, innovation, and diffusion of desirable, environmentally friendly technologies.

The potential long-run consequences of today's policy choices create a high priority for broadening and deepening our understanding of the effects of environmental policy on innovation and diffusion of new technology. Unfortunately, these issues cannot be resolved at a purely theoretical level, or on the basis of aggregate empirical analyses. For both benefit–cost and cost-effectiveness analysis, modern economics relies on a set of powerful analytical and computational tools that support quantitative modelling of economy- and society-wide policies over the long run. In practice, such analysis is typically carried out through the construction and application of large-scale computer models that combine scientific and economic theories and data into unified quantitative frameworks. These 'integrated assessment' models have emerged as decision makers' primary tool for quantitative climate policy analysis.

Only a handful of such recent models – reviewed in this chapter – have tackled the difficult issue of explicitly modelling the process of technical change. Most of them have then been applied to study the effects of the implementation of the Kyoto Protocol, particularly as far as the consequences of the emission targets and of the flexibility mechanisms are concerned.

In view, however, of the likely collapse of the Kyoto Protocol and of the diffusion among scholars of alternative proposals for early action at the international level designed to mitigate the effects of climate change, we expect that various existing climate models will be used or new ones will be developed in order to study and simulate the effects of these alternative proposals, with the aim of uncovering their relative merits. The message this chapter has tried to deliver is that any new modelling exercise – to reach credible conclusions – cannot avoid the issue of explicitly modelling the process of technical change.

The policy implications that can be drawn from the analysis presented here hinge upon the following basic conclusion: technological innovation and diffusion are the main engine of growth. Growth is correlated with increasing environmental efficiency. Hence innovation accomplishes a twofold task: it stimulates growth and competitiveness and reduces pollution levels. However, several market failures characterize existing innovation systems: spillover effects, inefficiency in the banking sector and financial markets (which have the task of channelling the funds to finance innovation), risk aversion, the nature of knowledge as a public good, and lack of coordination are all elements that reduce private incentives to undertake R&D and innovation and call for public intervention. These market failures are even stronger in the case of international and global environmental problems, for example climate change. Climate change is a global, long-term, uncertain phenomenon. These features crucially reduce firms' incentives to undertake R&D and innovation.

The global dimension of many environmental problems and the increasing mobility of capital and goods call for action in both developed and developing countries. It is therefore crucial to increase:

- knowledge diffusion, by disseminating information about new environment-friendly technologies;
- basic learning and innovation, by stimulating interactions between academic research and applied environmental innovation;
- technology diffusion, by improving the quality of the role of multinationals in the global economy and by implementing technological cooperation agreements.

The long-term dimension calls for action on R&D and innovation that enhances long-run competitiveness in environmentally sensitive sectors, that increases financial resources provided by the banking sector, for example by partly involving banks in the liability process, that guarantees protection for energy-efficient innovation, and supports cooperation and research joint ventures (RJVs) designed to undertake environmental R&D.

In this framework, environmental innovation policies are structurally linked and interact with competition, industrial and trade policies. These should support R&D cooperation, increase the competitiveness of the banking system, protect innovation and introduce lender liability schemes, and support trade rules that favour technology transfers.

NOTES

1. In line with the bulk of the theoretical literature, but not limiting our review to such cases, most innovation to which we refer takes place through deliberate investment by firms in R&D. Innovation deriving from past experience, usually referred to as learning-by-doing, is much less studied, possibly because it lacks the strategic incentive element of R&D investment, being in essence a by-product of a different (abatement or production or investment) decision.
2. One final strand of research related to technological change is the one which looks at the potential savings of the material resource inputs needed in the current production modes. The Wuppertal Institute has been championing this approach. A growing number of enterprises already commit themselves to reducing the material intensity per unit of service by rethinking and reorganizing their production process. Beside soft measures and instruments, which don't always turn out to be efficient, the Wuppertal Institute advocates and proposes (Hinterberger and Luks, 1998) financial incentives that are necessary to support companies on their way to achieve sustainability and dematerialization.

REFERENCES

Anderson, D. and C.D. Bird (1992), 'Carbon Accumulations and Technical Progress: A Simulation Study of Costs', *Oxford Bulletin of Economics and Statistics*, **54**, 1–29.

Barreto, L. and S. Kypreos (1999), 'Technological Learning in Energy Models: Experience and Scenario Analsysis with MARKAL and the ERIS Model Prototype', PSI Report N.99–08.

Bernstein, J. and M.I. Nadiri (1989), 'Rates of Return on Physical Capital and R&D Capital and Structure of Production Process: Cross Section and Time Series Evidence', in B. Raj (ed.), *Advances in Econometrics and Modelling*, Amsterdam: Kluwer.

Buchner, B., C. Carraro and I. Cersosimo (2001), 'On the Consequences of the US Withdrawal from the Kyoto Protocol ', FEEM *Nota di Lavoro*, December, Milan.

Buonanno, P., C. Carraro and M. Galeotti (2000a), 'Endogenous Induced Technical Change and the Costs of Kyoto', FEEM, *Nota di Lavoro*, No. 61.01, Milan.

Buonanno, P., C. Carraro, E. Castelnuovo and M. Galeotti (2000b), 'Efficiency and Equity of Emission Trading with Endogenous Environmental Technical Change', in C. Carraro (ed.), *Efficiency and Equity of Climate Change Policy*, Dordrecht: Kluwer.

Buonanno, P., C. Carraro, E. Castelnuovo and M. Galeotti (2000c), 'Emission Trading Restrictions with Endogenous Technological Change', FEEM, *Nota di Lavoro*, No. 43-00, Milan.
Carraro, C. (1999), 'Environmental Technological Innovation and Diffusion', in H. Folmer, H.L. Gabel, S. Gerking and A. Rose (eds), *Frontiers of Environmental Economics*, Cheltenham, UK and Northampton, USA: Edward Elgar.
Criqui, P., G. Klaassen and L. Schrattenholzer (2000), 'The Efficiency of Energy R&D Expenditures', mimeo, IIASA, Laxenburg.
DeCanio S.J., R.B. Howarth, A.H. Sanstad, S.H. Schneider and S.L. Thompson (2000), 'New Directions in the Economics and Integrated Assessment of Global Climate Change', PEW Center on Global Climate Change Report, October.
Den Elzen, M.G.J. and A.P.G. de Moor (2001), 'Evaluating the Bonn Agreement and some Key Issues', RIVM Report No. 728001016/2001.
Den Elzen, M. and T. Manders (2001), 'The Environmental and Cost Implications of the Kyoto Protocol – An Analysis with the FAIR Model', mimeo, RIVM.
Downing, P.G. and Lawrence J. White (1986), 'Innovation in Pollution Control', *Journal of Environmental Economics and Management*, **13**, 1029.
Edmonds, J., J.M. Roop and M.J. Scott (2000), 'Technology and the Economics of Climate Change', PEW Center on Global Climate Change Report, September.
Eyckmans, J., D. van Regemorter and V. van Steenberghe (2001), 'Is Kyoto Fatally Flawed? An Analysis With MacGEM', mimeo, Katholieke Universiteit Leuven.
Fischer, C. (2000), 'Climate Change Policy Choices and Technical Innovation', Climate Issue Brief No. 20, Washington, DC: Resources for the Future.
Fischer, C., I.W.H. Parry and W.A. Pizer (1999), 'Instrument Choice for Environmental Protection when Technological Innovation is Endogenous', Discussion Paper No. 99-04, Washington, DC: Resources for the Future.
Gerlagh, R. and B.C.C. Van der Zwaan (2000), 'Gross World Product and Consumption in a Global Warming Model with Endogenous Technological Change', mimeo, IVM Institute for Environmental Studies, Amsterdam.
Goulder, L.H. and S. Schneider (1999), 'Induced Technological Change and the Attractiveness of CO_2 Abatement Policies', *Resource and Energy Economics*, **21**, 211–53.
Goulder, L.H. and K. Mathai (2000), 'Optimal CO_2 Abatement in the Presence of Induced Technological Change', *Journal of Environmental Economics and Management*, **39**, 1–38.
Griliches, Z. (1980), 'Returns to Research and Development Expenditures in the Private Sector', in J.W. Kendrick and B. Vaccara (eds), *New Developments in Productivity Measurement*, Chicago: University of Chicago Press.
Griliches, Z. (1984), *R&D, Patents, and Productivity*, Chicago: University of Chicago Press.
Gritsevskyi, A. and N. Nakicenovic (1999), 'Modeling Uncertainty of Induced Technological Change', mimeo, IIASA, Laxenburg.
Grubb, M., J. Kohler and D. Anderson (2000), 'Induced Technical Change: Evidence and Implications for Energy-Environmental Modeling and Policy', Report to the Deputy Secretary-General of OECD, forthcoming in the *Annual Review of Energy and Environment*.
Grubler, A. (1998), *Technological and Global Change*, Cambridge: Cambridge University Press.
Grubler, A. and S. Messner (1998), 'Technological Change and the Timing of Abatement Measures', *Energy Economics*, **20**, 495–512.

Hagem, C. and B. Holtsmark (2001), 'From Small to Insignificant: Climate Impact of the Kyoto Protocol With and Without US', CICERO Policy Note No. 2001:1, Oslo.

Hinterberger, F. and F. Luks (1998), 'Dematerialization, Employment and Competitiveness in a Globalized Economy', paper presented at the Fifth Biennial Conference of the International Society for Ecological Economics, Santiago de Chile, 15–19 November.

International Energy Agency (2000), *Experience Curves for Energy Technology Policy*, Paris: OECD/IEA.

Jaffe, A.B. and R.N. Stavins (1994), 'The Energy Paradox and the Diffusion of Conservation Technology,' *Resource and Energy Economics*, **16**, 91–122.

Jaffe, A.B., R.G. Newell and R.N. Stavins (1999), 'Energy Efficient Technologies and Climate Change Policies: Issues and Evidence', Climate Issue Brief No. 19, Washington, DC: Resources for the Future.

Jaffe, A.B., R.G. Newell and R.N. Stavins (2000), 'Technological Change and the Environment', Paper No. 00-47, Washington, DC: Resources for the Future.

Kneese, A. and C. Schultz (1975), *Pollution, Prices, and Public Policy*, Washington, DC: Brookings Institution.

Kopp, R. (2001), 'A Climate Accord Without the U.S.', Washington, DC: Weathervane/ Resources for the Future.

Löschel, A. (2001), 'Technological Change in Economic Models of Environmental Policy: A Survey', ZEW and University of Mannheim.

Magat, Wesley A. (1978), 'Pollution Control and Technological Advance: A Dynamic Model of the Firm', *Journal of Environmental Economics and Management*, **5**, 1–25.

Manne, A.S. (1981), 'ETA-MACRO: A User's Guide', EPRI Report No. EA-1724.

Manne, A.S. and L. Barreto (2001), 'Learn-by-doing and Carbon Dioxide Abatement', IIASA Interim Report No. IR-01-057, IIASA, Laxenburg.

Manne, A.S. and R.G. Richels (2001), 'U.S. Rejection of the Kyoto Protocol: The Impact on Compliance Costs and CO2 Emissions', AEI–Brookings Joint Center for Regulatory Studies Working Paper No. 01-12.

Mansfield, E., J. Rapoport, A. Romeo, S. Wagner and G. Beardsley (1977), 'Social and Private Rates of Return from Industrial Innovations', *Quarterly Journal of Economics*, **41**, 221–40.

McDonald, A. and L. Schrattenholzer (2001), 'Learning Rates for Energy Technologies', *Energy Policy*, **29**, 255–61.

Messner, S. and L. Schrattenholzer (1998), 'Experience form Including Technological Progress Dynamics into the Global Optimization Model MESSAGE', ECS Contribution to the Mid-term Assessment Report of the TEEM Project, IIASA, Laxenburg.

Nadiri, M.I. (1993), 'Innovations and Technological Spillovers', National Bureau of Economic Research Working Paper No. 4423.

Nelson, R. and S. Winter (1982), *An Evolutionary Theory of Economic Change*, Cambridge, MA: The Belknap Press of Harvard University Press.

Nordhaus, W.D. (1993), 'Rolling the "DICE": An Optimal Transition Path for Controlling Greenhouse Gases', *Resource and Energy Economics*, **15**, 27–50.

Nordhaus, W.D. (1999), 'Modeling Induced Innovation in Climate-Change Policy', paper presented at the IIASA Workshop on Induced Technological Change and the Environment, Laxemburg, 21–22 June.

Nordhaus, W.D. and Z. Yang (1996), 'A Regional Dynamic General-Equilibrium

Model of Alternative Climate-Change Strategies', *American Economic Review*, **4**, 741–65.

Oates, W.E., K. Palmer and P.R. Portney (1993), 'Environmental Regulation and International Competitiveness: Thinking About the Porter Hypothesis', Discussion Paper No. 94-02, Washington, DC: Resources for the Future.

Palmer, K., W. Oates and P. Portney (1995), 'Tightening Environmental Standards: The Benefit–Cost or the No-Cost Paradigm?' *Journal of Economic Perspectives*, **9**, 119–32.

Parry, I.W.H. (1995), 'Optimal Pollution Taxes and Endogenous Technical Progress', *Resource and Energy Economics*, **17**, 69–85.

Parry, I.W.H. (1996), 'The Choice between Emission Taxes and Tradable Permits when Technological Innovation is Endogenous', Discussion Paper No. 96-31, Washington, DC: Resources for the Future.

Parry, I.W.H., W.A. Pizer and C. Fischer (2000), 'How Important is Technological Innovation in Protecting the Environment?' Discussion Paper No. 00-15, Washington, DC: Resources for the Future.

Pizer, W. (1999), 'Choosing Price or Quantity Controls for Greenhouse Gases', Climate Issue Brief No. 17, Washington, DC: Resources for the Future.

Porter, M.E. and C. van der Linde (1995), 'Toward a New Conception of the Environment–Competitiveness Relationship', *Journal of Economic Perspectives*, **9**, 97–118.

Riahi, K., R.A. Roehrl and L. Schrattenholzer (2001), 'Technology Clusters in Sustainable Development Scenarios', Progress Report to the International Forum on Collaboration Projects, ECS Project, mimeo, IIASA, Laxenburg.

Seebregts, A.J., T. Kram, G.J. Schaeffer, A. Stoffer, S. Kypreos, L. Barreto, S. Messner, and L. Schrattenholzer (1999), 'Endogenous Technological Change in Energy System Models, Synthesis of Experience with ERIS, MARKAL and MESSAGE' ECN Report No. ECN-C-99-025.

Shama A. (1983), 'Energy Conservation in U.S. Buildings, Solving the High Potential/Low Adoption Paradox from a Behavioral Perspective', *Energy Policy*, **11**, 148–68.

Van der Zwaan, B.C.C., R. Gerlagh, G. Klaassen and L. Schrattenholzer (1999), 'Endogenous Technological Change in Climate Change Modelling', Report D-99/15, IVM Institute for Environmental Studies, Amsterdam, forthcoming in *Energy Economics*.

5. The future evolution of the Kyoto Protocol: costs, benefits and incentives to ratification and new international regimes

Carlo Carraro and Marzio Galeotti

1. INTRODUCTION

The diverging views of the science of climate change and the need for fair and politically feasible mechanisms for implementing an international agreement to control climate change have been lurking under the surface of the UNFCCC negotiations and have become more explicit in the process of the ratification of the Kyoto Protocol. This process has been and is difficult and controversial. A few years ago, it appeared that the EU countries were waiting for other developed countries – the USA in particular – to take action at the same time (see, for example, the EU Council Conclusions of 1997). But, more recently, at the Bonn and Marrakesh Conferences of the Parties, the EU was strongly committed to implementing the Kyoto Protocol even independently of the decision – taken by the USA at the beginning of 2001 – to withdraw from the Kyoto Protocol. In 1999, the US Senate linked ratification of the Kyoto Protocol with the 'meaningful participation' of key developing countries. However, in 2001, the USA unilaterally decided not to ratify the Kyoto Protocol for reasons mostly unrelated to the participation of developing countries. In 2000, at the Hague, developed countries have been unable to achieve an agreement on specific aspects of the treaty – specifically the balance between domestic actions and Kyoto mechanisms, the role of sinks, the implementation of sanctions, and so on. But in 2001, first in Bonn and then in Marrakesh, an agreement on the same issues has been reached between the EU, Japan, a few other developed countries, and the so-called 'transition economies'. However, the environmental effectiveness of this agreement is low and likely to be lower than the one of the agreement that the same countries refused to sign in the Hague in 2000.

Why is it so difficult to achieve an agreement on strategies to control climate change? What is peculiar to this environmental issue that prevents other environmental agreements from being adapted to climate change? At least seven peculiar features characterize the issue of climate change. Even though other environmental problems also possess some of these features, the climate problem is unique both for the intensity of these features and for the strong interdependencies between all of these seven features.

1.1 The Problem is Global

The global nature of the problem implies that climate change control is a public good; that is, the climate that we all enjoy is the product of everyone's behaviour. This provides an important motivation for collective, global action, but also strong incentives to free-riding. Moreover, emissions and consequences are heterogeneous around the world. This exacerbates the basic public-good nature of the problem. Countries are distributed across a spectrum of high emitters–low emitters and high impact–low impacts. Nations with high emissions and low expected impacts have a high potential to control concentrations, but little incentive. On the other hand nations with low emissions and high impacts have great incentive to control emissions, but little capability. While side payments could in principle resolve this dilemma, transaction costs may be significant and present income distributions may lead to unacceptable outcomes.

1.2 The Problem is Long Term

The long-term nature of the problem implies that it is concentrations not emissions that matter. Climate change is related to the concentration of greenhouse gases and not to any individual year's emissions. Carbon dioxide concentrations are closely related to the net accumulation of emissions over long periods of time. Hence strategies to control net emissions must take long periods of time into account in a meaningful way. Moreover, intergenerational transfers are inevitable because the consequences of climate change will be visited primarily on those who are alive in the future. The fact that most of the affected parties are not present to participate in the decision-making process raises complicated ethical questions. Indeed, the implications of their absence are not immediately obvious. Because climate change is anticipated to be greater in the future than it is at present, those who live in the future will reap most of the benefits that accrue to near-term actions to limit emissions. Intergenerational asymmetry can lead to a form of public-good problem in which the willingness to undertake emissions mitigation in the near term may be less than would have been the case if decision makers were infinitely lived.

1.3 Narrowly Defined Technological Solutions are Unavailable

Emissions of greenhouse gases (GHGs) are associated with an extraordinary array of human activities. CO_2 emissions alone are associated with the combustion of fossil fuels and changes in land use, thus affecting activities ranging from agriculture to manufacture. As a consequence, GHG emissions are greatly affected by other, exogenous and non-climate-policy factors. Narrowly defined technological solutions, such as were available to address the problem of stratospheric ozone depletion, are impossible for the climate issue. Hence, once again, a mix of internationally agreed policies is necessary to control climate change, thus raising issues of appropriate negotiation processes, well-designed international regimes, adequate incentives to participation, and so on.

1.4 Policy Interactions are Strong

Future emissions depend to a large degree on the rate and direction of technological developments in a broad array of human endeavours. For example, China's policies to stabilize its population size, taken for reasons unrelated to climate change, will have a profound effect on Chinese emissions of GHGs to the atmosphere. Policies to control non-GHG air pollutants can greatly affect GHG emissions. For example, measures to substitute natural gas and non-carbon-emitting energy forms such as solar and nuclear power for coal in electric power generation, to control local and regional air pollution, can affect GHG emissions as well. On the other hand, some policies that reduce local air pollution, such as scrubbing power plants for sulphur, can reduce power plant efficiency and increase GHG emissions. This raises several forms of interaction between domestic and international policies and makes it difficult to define baselines for negotiations and net costs and benefits from climate policies. It also raises economic domestic conflicts between different economic interests in addition to the international ones.

1.5 Uncertainty is Pervasive

Uncertainty surrounds the climate change problem at every point. Researchers disagree as to the rate and extent of emissions to the atmosphere in the absence of climate change. They disagree as to the change in concentration of GHGs that would result from a given emission rate. They disagree as to the timing, extent and distribution of climate change and sea-level rise for a given concentration of GHGs. And they disagree as to whether some climate change is an aggregate net benefit or net cost to ecosystems and humanity. The potential for an unspecified, low-probability, but catastrophic turn of events haunts the problem. While uncertainties are great, they are not distributed evenly throughout the problem. The implications of emissions mitigation are better known than the more distant-in-time potential benefits from mitigation. Uncertainty and the uneven distribution of uncertainty is another factor that explains the difficulty of agreeing on an effective international regime to control climate change.

1.6 The Consequences are Potentially Irreversible

Many global bio-geochemical processes have long characteristic timescales. Sea-level changes as a consequence of changes in mean global temperature can take more than 1000 years to play out. Similarly, changes in the concentration of GHGs can rise rapidly, but decline slowly. And, even if concentrations can be reduced, the nature of the climate system is such that it might not return to the same climatic state associated with an earlier concentration.

1.7 The Global Institutions Needed to Address the Issue are only Partially Formed

The UNFCCC has been ratified by more than 170 parties and entered into force in 1994. It provides the institutional foundations upon which international climate change negotiations occur. It sets as its ultimate objective the stabilization of the concentration of GHGs in the atmosphere at levels that would prevent dangerous anthropogenic interferences with the climate. But the UNFCCC establishes a process rather than creating the institutions for implementing the objective. The objective has not been quantified. The term ‘dangerous’ has been left open to interpretation by the parties. The Kyoto Protocol of December 1997 represents a further important step in international regime formation under the UNFCCC. The Kyoto Protocol has brought a number of new elements and broadened the context of the decision-making process regarding implementation of climate change policy. The concrete results of Kyoto include the setting of legally binding emission limitations or reduction targets, the expansion of the number of controlled gases from three to six, the adoption of flexible mechanisms, such as the clean development mechanisms (CDM), joint implementation (JI) and emissions trading (ET) to increase the cost-effectiveness of GHGs emission abatement. However, several issues need to be clarified for these mechanisms to be implemented, although some progress was made in Bonn and Marrakesh in 2001. More generally, there is not yet a supranational institution with the power to enforce and monitor the objectives agreed in Kyoto which are therefore going to be implemented on a voluntary basis. Hence, again, a problem of incentives to free-riding and the difficulty of establishing a stable and effective international regime to control climate change. These incentives to free-ride became apparent in 2001, when the USA decided unilaterally to withdraw from the Kyoto Protocol.

Given the above peculiar features of the climate policy problem, a few fundamental questions arise: what is the future of the UNFCCC and of the Kyoto/Bonn Protocol? Does the Kyoto/Bonn Protocol have any chance of being ratified and implemented by a sufficiently large number of countries within a time limit which actually enables these countries to achieve their commitments? Is there any strategy that could enhance the chances of implementation of the Kyoto/Bonn Protocol? Or should countries move to a different type of agreement, as suggested by the USA after their decision not to ratify the Kyoto Protocol?

In order to answer these questions, it is important to assess the costs and benefits of GHG emission reductions in the presence of the above features of the climate change policy problem and under alternative international regimes, namely different structures of the agreements (number and identity of signatories), different policy mixes to achieve the Kyoto commitments, different cooperation policies with developing countries, and so on. This would enable us to identify the incentives, if any, that different countries have in ratifying the Protocol. Given these incentives, it will be possible to draw a scenario on the likely evolutions of climate negotiations.

In particular, we would like to identify which policies can be designed to lower the costs and increase the incentives to ratify a given international agreement on climate. This would add another dimension to the policy discussions of the previous four chapters of this book. Indeed, whereas the previous chapters focused on the environmental and cost-effectiveness of alternative policy mixes, here we focus on the relationship between policy mixes and international regimes, namely on the effectiveness of different policy mixes in determining and increasing the number of countries which ratify and implement the Kyoto/Bonn Protocol, and on the impacts of different international regimes on the costs of climate policy.

Indeed, environmental effectiveness and cost-effectiveness of climate policy are strictly correlated with the number and identity of countries which ratify and implement the Kyoto/Bonn Protocol. The costs and benefits of different policy options (that is, the timing of optimal responses to climate change, the choice between mitigation and adaptation responses, the role of technological innovation and diffusion, the choice between domestic action and the adoption of the Kyoto mechanisms, the effectiveness of negotiated environmental agreements, and so on) crucially depend on the characteristics of the international agreement on climate change which is adopted. In particular, they depend upon two main features of the international regime: the number of ratifying countries, and the size of their quantitative commitment to control GHG emissions.

For example, it is not possible to assess the costs and benefits of the Kyoto/Bonn Protocol or of other potential agreements on climate change independently of the number of signatories to the agreement and of their abatement targets and/or policy commitment. But at the same time the number of signatories is endogenous and depends on the abatement targets and mitigation polices adopted in various countries. This full interdependence between policies, costs/benefits and the structure of the climate agreement is often neglected in the existing climate literature. Let us take an example. There are studies (for example, Carraro, 1999d, 2000; Carraro and Hourcade, 1999; Weyant, 1999) that analyse the costs of implementing the Kyoto Protocol either through a set of domestic policies and measures or through a system of international tradable permits, keeping as given the number of participating countries. But the adoption of either policy crucially affects the number of participating countries, which can be larger or smaller under policies and measures than under tradable permits. And the number (and identity) of the participating countries crucially affects the costs and benefits of different policy strategies. Therefore, a first objective of this chapter is to provide an analysis of the environmental effectiveness and cost-effectiveness of different climate policies by focusing on the link between policy options and structure of the agreement.

The equity issue is also very important to an understanding of which countries will reduce/control their emissions. As a consequence, given what was said above, equity is crucial to adequately assess the costs of emission reductions at the global and country level. It has been argued that some countries are allowed to reduce emissions less than other countries, both within (Kram, 1998) and outside the EU bubble (Metz, 1999; Bosello and Roson, 1999; Rose and Stevens, 1999), and that even when applying flexibility mechanisms some countries will benefit from the agreements more than others (Nordhaus and Boyer, 1999). It has also been argued that some countries can exploit their monopolistic power in a future trading system (Burniaux, 1998). All these remarks address the problem of the optimal burden sharing (the distribution of costs) of climate change control. This problem is strictly related to the features of an international agreement on climate for two main reasons. On the one hand, an increasing number of participating countries reduces the burden for each signatory; on the other hand, an agreement in which the burden is equitably shared is more likely to be signed by a large number of countries (Convery, 1999). Therefore, equity and the structure of the international agreement (number and identity of signatories) are strictly linked. But the number of signatories affects and is affected by costs. Hence equity and efficiency cannot be separated.

All this reinforces our basic remark that an analysis of the costs and benefits of different policy options, and of the distribution of these costs and benefits across countries, cannot be done independently of an analysis of the likely features of the prevailing international regime, that is, of the incentives that lead countries to sign and ratify an international agreement to control GHG emissions and to set quantitative emission targets.

The importance of the structure of the prevailing international regime and its bi-directional links with the costs and benefits of different policy mixes leads to a second set of issues. If a larger number of participating countries reduces the implementation costs, thus opening the way for further extensions of the Protocol, what are the negotiating strategies that could help to increase the number of countries which sign and ratify it? Are there climate agreements different from the Kyoto/Bonn Protocol which provide greater incentives for ratification? Are there negotiating and institutional rules which favour countries' participation? Are there policy strategies that could provide the right incentives for countries to participate in a climate agreement? More specifically, should negotiation focus only on climate change policies or should they be linked to other policy issues? How can transfers help to expand a coalition?

Hence a second objective of this chapter is to analyse which policy strategies and negotiation rules can provide the right incentives for a sufficiently large number of countries to sign and ratify the Kyoto/Bonn Protocol and its possible extensions after 2012 (or other climate agreements). This analysis will enable us to identify a set of policy guidelines that could help the future evolution of climate agreements.

The structure of the chapter is as follows. Section 2 will analyse the interdependencies between international regimes and the costs of different climate policies. Using the many empirical studies recently published, Section 2 will identify both the impacts of different international regimes on the costs of a climate policy and the effects of different climate policies on the structure of the equilibrium climate agreement. Having identified a positive feedback relationship between participants in the agreement and implementation costs, Section 3 will explore which negotiation rules, institutions and norms and policy strategies can be designed to enhance the incentives for countries to sign and ratify a climate agreement. Finally, Section 4 will analyse the existing negotiation controversies using the incentive structure identified in Sections 2 and 3 and will describe possible future evolutions of climate negotiations which go beyond the technicalities debated in The Hague, thus providing a framework to identify long-term scenarios for climate policy.

The analysis of the incentives to sign and ratify a climate agreement – and the Kyoto/Bonn Protocol in particular – will be based on recent results achieved by the theory of international negotiations described in Appendix 1. The relevant literature on climate policy is concisely presented in Appendix 2.

2. INTERNATIONAL REGIMES AND THE COSTS AND BENEFITS OF DIFFERENT POLICY MIXES

In this section, the relationship between the structure of the climate agreement and the costs and benefits of the related climate policy is analysed. Four main types of international regimes are evaluated: no participation, which occurs when no country decides to sign and ratify the agreement; unilateral participation, when a single country decides to achieve its commitment independently of the other countries' decisions; partial agreements, when only a group of countries, but not all countries, sign and ratify the climate agreement; global agreement, when the Protocol is signed and ratified by all world countries. Table 5A.1 in Appendix 2 summarizes the different policy options and international regimes that have been evaluated. Notice that more emphasis has been given to agreements in which a sub-group of countries (for example Annex B countries) act to control their GHG emissions.

2.1 No Participation

This case constitutes the benchmark against which to evaluate the costs and benefits of policies designed to control GHG emissions under alternative structures of the agreement. It is usually named the business-as-usual scenario, because it identifies the values of the main environmental and economic variables when no coalition forms and no action, unilateral or cooperative, is adopted (the IPCC second assessment report (IPCC, 1995) is a good example of this approach). The construction of the business-as-usual scenario is very important in assessing both the profitability and the stability of a climate agreement (that is, whether the coalition which forms when some countries sign the agreement is self-enforcing). A coalition is profitable when welfare after the coalition is formed is larger than in the no-participation case. A coalition is self-enforcing if there are no incentives to leave or enter the coalition (see Appendix 1). The business-as-usual scenario crucially affects these incentives. If the no-participation case is such that emissions are declining and the target can easily be achieved through small emission reductions, then the incentives to join the coalition (sign the agreement) are much higher: that is, a coalition with many countries is more likely to form (Barrett, 1997b). Symmetrically, if large emission reductions are necessary, both abatement becomes more costly and incentives to free-ride increase, thus further increasing costs for cooperating countries (particularly if leakage is high).

A careful definition of the no-participation case is therefore very relevant to assessing the likelihood of large coalitions and thus the efficiency of a climate agreement. But it is also very relevant in terms of equity. When the burden of emission abatement has to be equitably shared, it is important to distribute emissions reductions with reference to the business-as-usual scenario. Each country therefore has an incentive to pretend that its own business-as-usual scenario implies larger emissions than what is actually true (Bohm, 1999; Grubb, 1998). In this way, the actual cost for the country would be lower. An optimistic scenario in which predicted emissions are lower than 'true' emissions (as measured *ex post*) leads countries to agree on low emission reduction targets, but forces countries to reduce more later and to pay abatement costs larger than expected. A pessimistic scenario makes the agreement more difficult because larger emission reductions have to be agreed upon, but countries find themselves in a better situation and pay lower costs *ex post*. Hence, if a country succeeds in convincing the others that its own business-as-usual emissions are larger than the 'true' ones, then this country achieves a relative benefit in terms of less stringent emission targets and lower abatement costs.

The definition of a business-as-usual scenario has therefore a strategic dimension and can hardly be defined as an 'objective' evaluation of future economic and environmental cycles and trends. It is therefore important to collect the largest amount of information from different sources as possible and to identify the scenario more as an average of many scattered pieces of information than as a subjective analysis of likely future events. This may reduce the likelihood of strategic definitions of the business-as-usual scenario and may partly prevent the consequent impacts on the equilibrium coalition and on the assessment of costs and benefits of climate policies.

2.2 Unilateral Participation

There is a wide literature that analyses the costs and sometimes the benefits of introducing policies to control GHG emissions in a single country (Bucholz and Konrad, 1994; Endres and Finus, 1998; Hoel, 1991; Hoel and Schneider, 1997; Porter and van der Linde, 1995). Given the features of the climate problem described in the Introduction, and the results summarized in Appendix 1, this type of exercise may seem unreasonable. Indeed, a single country's action is unlikely to have a relevant impact on global emissions – in particular in the presence of carbon leakage – while having relevant costs for the country. There are, however, two main justifications for undertaking an analysis of unilateral participation. The first is that domestic abatement costs (related to domestic policies and measures) depend very little on the coalition structure. Indeed, only if leakage is large, and if climate policies have a large impact on trade and financial flows, are the costs of domestic abatement policies significantly affected by the size of the coalitions and by the agreed emission targets. Hence it may be useful to compute the costs of unilateral participation as a benchmark case, which identifies costs that can only be reduced when a coalition forms and flexibility mechanisms are implemented among signatory countries. Note the importance of a careful assessment of leakage and of trade and financial repercussions of climate policies (McKibbin et al. 1998). Note also that the above arguments concern costs but not benefits of climate policies. Indeed, the benefits of unilateral participation are likely to be zero or almost zero for all or almost all countries (a possible exception is the USA) given the global nature of the climate problem (Hoel, 1991; Bucholz and Konrad, 1994; Endres and Finus, 1998).

A second reason for undertaking the assessment of the cost of a unilateral participation is that it could help to identify a series of low-cost (or no-cost) options (the so-called low-hanging fruits or no-regret actions) that could be implemented independently of the formation of a climate coalition. It could also help to identify policy mixes that help to restructure the fiscal system and public regulatory and incentive schemes in such a way that emission abatement costs are more than compensated by other economic (non-environmental) benefits (the so-called double dividends).[1]

There are also cases in which unilateral actions have been analysed from a very specific viewpoint. For example, Bucholz and Konrad (1994) analyse the detrimental effect of pre-negotiations actions (more bargaining power can be achieved by unilaterally increasing emissions before negotiating). Endres and Finus (1998) examine the negative effects on negotiations of a higher environmental consciousness in one country, Hoel (1991) analyses the costs of unilateral actions, Hoel and Schneider (1997) the role of social norms, and Porter and van der Linde (1995) focus on the advantage of being a leader by adopting emission reductions before the other countries. All these studies provide some basic information on the incentive structure of climate negotiations and on the heterogeneity of countries across the world.

2.3 Partial Agreements

The case of partial agreements is frequently analysed in the recent empirical literature. The reason is twofold. On the one hand, as shown in Appendix 1, game theory suggests that a partial coalition usually forms at the equilibrium. Hence the climate problem is neither a 'tragedy of the commons' nor a situation in which there are clear incentives to full cooperative emission control. On the other hand, the history of international environmental negotiations, and of climate negotiations in particular, is a history of partial agreements that are slowly broadened as more and more countries decide to join the group of signatories. In the case of climate, moreover, the Kyoto agreement itself can be seen as a first partial climate agreement. Therefore, there have been many papers that have dealt with the costs and benefits of the Kyoto agreement and with the possible strategies to increase the number of countries which commit themselves to emission control targets (see OECD, 1998; Carraro, 1999d, 2000; see also Burniaux, 1998; Capros, 1998; Carraro, 1999b, Ellerman et al., 1998; Grubb and Vrolijk, 1998; Holtzmark, 1998; Manne and Richels, 1998; Mensbrugghe, 1998; Nordhaus and Boyer, 1999; and the surveys by Metz, 1999; Convery, 1999).

Two remarks are important. First, even if most recent analyses deal with the Kyoto agreement, there are studies that try to compute the optimal coalition structures, both in terms of participation and in terms of targets, independently of the decisions taken in Kyoto (a recent attempt is in Nordhaus and Boyer, 1999). Usually the conclusion that can be derived from these papers is that Kyoto is neither economically nor environmentally optimal. However, the notion of optimality is not very useful when analysing coalition formation. Indeed, what matters is the notion of stability of a coalition (see Appendix 1), which identifies which countries have an incentive to join the coalition (sign the agreement) for different membership rules (Carraro and Moriconi, 1998; Botteon and Carraro 1997a, 1997b), business-as-usual scenarios, abatement costs (and therefore climate policies, including the degree of adoption of flexibility mechanisms), environmental benefits (and therefore impacts, adaptation costs, and so on).

Second, the Kyoto agreement can be interpreted as a partial (Carraro, 1998) or as a global agreement (Chander et al., 1999). It is interpreted as a global agreement when all countries are seen as committed to emission targets. Those in Annex B are committed to emission targets with respect to 1990; the others are 'committed' to emissions levels evolving as in the business-as-usual scenario. This second interpretation is useful to show that (i) optimal emissions targets are not necessary because the same outcome can be achieved through a global unrestricted emissions trading scheme among all countries (Chander et al., 1999); (ii) the resulting outcome can be profitable to all countries if an appropriate transfer scheme is adopted (Chander and Tulkens, 1995, 1997; Germain et al., 1997; Markusen, 1975).

As a consequence, even a 'partial', sub-optimal agreement like Kyoto can be transformed into a 'global' optimal agreement (see Section 2.4).

Leaving this ideal world in which perfectly competitive and global market mechanisms are at work, we are left with the analysis of coalitions which, like the coalition formed by Annex 1 countries of the UNFCCC or Annex B countries of the Kyoto Protocol, are partial, that is, formed by a subgroup of the negotiating countries. Are these partial coalitions effective? Are they cost-effective? Are they anyway 'too costly' for the participating countries?

The answer to the first question depends on two main factors: (i) the baseline scenario; and (ii) the degree of leakage. If the baseline scenario is very ambitious and leakage is high, on the one hand countries find it difficult to undertake large emission reductions, and on the other hand their effort is offset by the leakage effect (the increased emissions by free-riding countries).

As for the second and third questions, there are many studies that try to assess the cost for Annex B countries of achieving given emission targets under alternative policy options (see Appendix 2). These policy options include the timing of the mitigation responses (see the special issue of *Energy Economics* edited by Carraro and Hourcade, 1999), the degree of adoption of the Kyoto flexibility mechanisms and their features (see the papers in OECD, 1998; Carraro, 1999d), the role of complementary industrial policies, mainly designed to foster innovation (see Nordhaus, 1997; Goulder and Schneider, 1998; Kopp et al., 1998; Buonanno et al., 1999), the effects of uncertainty about climate impacts or abatement costs (Carraro and Hourcade, 1999), and so on.

The main result can be summarized as follows. Despite their high variability, all studies show that the Kyoto flexibility mechanisms sensibly reduce the costs of compliance, whatever the coalition structure. Shogren (1999) notes that 'it is estimated that any agreement without the cost flexibility provided by trading will at least double the USA costs, … the key is to distribute emissions internationally so as to minimise the costs of climate policy'. Manne and Richels (1999b) state that 'losses in 2010 are two and one-half times higher with the constraint on the purchase of carbon emission rights; international co-operation through trade is essential if we are to reduce mitigation costs'. Hence emissions trading, and more generally the application of the flexibility mechanisms, can reduce overall mitigation costs without reducing the effectiveness of the climate policy (see also Burniaux, 1998; Capros, 1998; Ellerman et al., 1998; Glomsrød et al., 1992; Mensbrugghe, 1998; Nordhaus, 1999; Rose and Stevens, 1999; Hourcade et al., 1999; Tol, 1999a, 1999b; Kverndokk, 1993).

If flexibility mechanisms can fully exploit both their intertemporal (banking or banking and borrowing) and geographical dimensions (global emissions trading), then compliance costs are further lowered. This result is shown in Bosello and Roson (1999) for banking, Westkog (1999) for banking and borrowing, Manne and Richels (1999a, 1999b), McKibbin et al. (1998) and many others for global emissions trading. If in addition the incentives to innovation provided by flexibility mechanisms are taken into account, then compliance costs are even lower (Buonnano et al., 1999).

Hence a well-designed climate policy may have low costs, at least in the short run. Moreover, all the above papers also show that the size of the coalition crucially affects the size of the benefits deriving from the adoption of flexibility mechanisms. The larger the number of participating countries, and the higher the variability of marginal abatement costs across them, the larger the benefits from emissions trading, JI and CDM. Hence, in order to reduce abatement costs and increase environmental benefits, policies, rules and institutions should be designed to achieve the largest possible coalition. This issue will be discussed in detail in Section 3.

2.4 Global Agreement

A global agreement is achieved when all countries commit to control their GHG emissions. The difficulty of achieving a global agreement on climate change has been underlined in the Introduction. Theoretical analyses confirm that the emergence of a global agreement on climate change is quite unlikely (see Appendix 1). This depends upon four main factors:

1. *The heterogeneity of countries with respect to the causes of climate change, the impacts, the mitigation and adaptation costs.* This factor mainly influences the profitability of the decision to sign a climate agreement. Some countries may lose when signing the agreement even when environmental benefits are fully accounted for. As shown by Chander and Tulkens (1995, 1997), there always exists a system of transfers that may make all countries gain. But this opens again the door to the equity problem and the related burden-sharing issue. Equity may have a large impact on the existence and size of a climate coalition. As shown in Botteon and Carraro (1997a, 1997b), and as argued by many policy makers and scientists, the way in which the burden of controlling emissions is shared across countries crucially affects a country's decision to join a coalition. On the one hand, if the burden is not equitably shared, some countries may not find it profitable to sign the agreement. Profitability depends indeed on two main factors: (i) the distribution of costs within the coalition; and (ii) the size of the coalition. It is indeed possible that there is a minimum size of the coalition above which it becomes profitable. And these two factors are strictly interdependent. On the other hand, equity also affects the free-riding incentives. As discussed in the next section, in some cases it may be reasonable for some countries to transfer resources to other countries to induce them to join the coalition on which they would otherwise free-ride. In this case, the final outcome is not equitable – free-riders would gain more than countries in the starting coalition – but it may be environmentally and economically efficient.
2. *The strong incentives to free-ride on the global agreement and the lack of related sanctions.* When all countries agree to control emissions, a defecting country achieves the whole benefit, because its incidence on global emission is marginal (with a few exceptions) and pays no cost. Hence, a defection with respect to a large coalition is the optimal strat-

egy if there are no sanctions. However, credible sanctions are difficult to design (Barrett, 1994). Emissions themselves are hardly a credible sanction, because countries are unlikely to sustain self-damaging policies. Moreover, in this case, asymmetries play a double role. On the one hand, some countries may not gain from signing the environmental agreement; on the other hand, some countries, even when gaining from environmental cooperation, may lose from carrying out the economic sanctions (Barrett, 1997c; Schmidt, 1997).

3. *The absence of environmental leadership.* It has been shown that the process of achieving a global agreement can be a sequential one (Carraro and Siniscalco, 1993). In this case a group of countries takes the leadership, starts reducing/controlling emissions and implements strategies such to induce other countries to follow.[2] The presence of low-cost climate policies and equitable burden sharing (Schmidt and Koschel, 1998) are again important elements for the formation of an initial profitable coalition. As said, our definition of profitability accounts for the environmental benefits of emission control. Hence, on the one hand, benefits should be increased by increasing the number of countries that control emissions. On the other hand, abatement costs should be minimized by exploiting all possible opportunities (including emissions trading). This is a prerequisite to achieve a strong leader coalition that can then exert its leadership through the design of better negotiation rules, the implementation of transfer mechanisms, and the credibility of international issue linkages (see next section). A preliminary modelling of the effects of leadership is in Ellerman et al. (1999), where it is shown how and when developing countries may join a leader coalition formed by Annex B countries.
4. *The focus on a single international climate agreement.* As explained in Appendix 1, if countries may join different coalitions, which means that several agreements can be signed by groups of countries in the same way as countries form trade blocs, then the likelihood that all or almost all countries set emission reduction targets increases (Bloch, 1997; Carraro, 1997, 1998; Yi, 1997, Yi and Shin, 1994). The outcome of negotiations in which more agreements can be signed is usually a situation with several small environmental blocs (Carraro and Moriconi, 1998; Yi, 1997), but this can be considered another step in the right direction. If all or almost all countries set emission reduction targets within their own bloc (for example regional environmental agreements are signed), then, in a subsequent phase, negotiations among blocs may lead to more ambitious emission reductions.

Despite the warning that global agreement may be difficult to reach, many articles analyse the costs of agreements in which all countries participate, in one form or another (see, for example Bosello and Roson, 1999; Capros, 1998; Ellerman et al., 1998; Manne and Richels, 1998; Nordhaus and Boyer, 1999; Schackleton, 1998). The weakest form is the one, already discussed in Section 2.3, in which a few countries commit to emission reductions, but all agree to trade emissions in a single global market. The strongest form is the one in which a central planner is assumed to set optimal emission levels for all world countries. This optimal solution is often proposed as a benchmark for actual negotiations and was often analysed before Kyoto (see the collection of papers in Carraro, 1999d).

More interesting is the attempt made by Peck and Teisberg (1999) to model the negotiations between developed and developing countries to achieve a global agreement. This paper shows the potential for cooperation to be achieved – the Pareto frontier is small, but not empty – but does not analyse the incentives to actually sign the agreement. However, the paper suggests a research direction which at least helps to identify the optimal emission reductions which are profitable for all negotiating countries.

The conclusions that can be derived from this type of empirical analysis are similar to the ones already mentioned when discussing partial agreements. In the scenario in which business-as-usual (BAU) emissions are lower, it is easier to achieve a global agreement because lower emissions reductions are necessary (Barrett, 1997b) and consequently abatement costs are lower. Optimal emissions targets are such as to equalize marginal abatement costs. This optimal, cost-minimizing solution can also be achieved through an unconstrained emissions trading system (Chander et al., 1999).

Hence either emission targets are optimally set, or countries are allowed to trade emissions for any given set of targets through which a global consensus can be achieved. Of course, these two options have different impacts on equity. As shown in Bosello and Roson (1999), starting from the Kyoto targets, global unconstrained emissions trading achieves optimality but reduces equity.

2.5 Policy Lessons

Albeit preliminary and limited by the preliminary stage of development of climate models, some policy guidelines can be derived from the previous analysis.

First, given the strict interdependence between the costs and benefits of different policy options (that is, the timing of optimal responses to climate change, the choice between mitigation and adaptation responses, the role of technological innovation and diffusion, the choice between domestic action and the adoption of 'flexibility mechanisms', the importance of secondary benefits, and so on) and the characteristics of the international agreement on climate change which is adopted (that is, the number of signatories, the size of their quantitative commitment to control GHG emissions), it is impossible to assess costs and benefits of the Kyoto Protocol or of other potential agreements on climate change independently of the number of signatories to the agreement and of their abatement targets and/or policy commitment. But at the same time the number of signatories is endogenous and depends on the abatement targets and mitigation polices adopted in various countries.[3]

Hence it is important to analyse the environmental effectiveness and cost-effectiveness of climate policy jointly with countries' decision to sign and ratify the agreement. Hence the analysis of which strategies can be proposed to reduce the costs of mitigation policies becomes quite different from those usually proposed in recent papers on climate policy (see Table 5A.1 in Appendix 2). The reason is that a country's goal is not only to identify a new climate-friendly technology or an adequate redistribution of costs across sectors, but also to affect other countries' behaviour in order to increase the number of those that share the burden and to make this burden more equitably shared.

In this perspective, two main strategic dimensions have been identified (see Table 5A.1). The policy dimension moves from domestic and rigid (for example norms and measures) policies to global and increasingly flexible policies. The coalition dimension moves from unilateral to global actions. It is not necessary to achieve the optimum along both dimensions. Even a sub-optimal distribution of targets across countries can be consistent with optimal emission reduction schemes if coupled with a global and flexible policy (for example a global system of emissions trading). Alternatively, a global coalition can optimally set emission targets in order to minimize abatement costs even without a global coordinated policy.

However, it is quite difficult to identify as likely a scenario in which either a global coalition forms or a global and perfectly competitive emissions trading system is implemented. Hence we are left with a world in which only a subset of countries may decide to commit themselves to emission control and only a subset of countries, not necessarily the same countries, may decide to participate in a freely competitive trading system (see Section 2.3). In this case, there are three options.

The first is to design policies that induce countries which are not committed to emission control to accept this commitment. This can be done through appropriate policies that, on the one hand reduce the cost of the commitment (for example competitive emissions trading again) and on the other hand increase the benefit of the commitment by means of transfers or by linking the climate agreement to other economic agreements (for example on technological cooperation).

A second option is to recognize that countries' main incentive is to form several coalitions which may adopt different policy measures. In this case, instead of a single global agreement, countries should be left free to negotiate regional climate agreements. In this setting, it may be reasonable to pursue a climate agreement within the EU or between the EU, Eastern Europe and Former Soviet Union countries, because other regional climate agreements are likely to emerge subsequently. Once several regional agreements enter into force, for example in Europe, East Asia and the Americas, the three groups of countries can negotiate to achieve a global agreement.

A third option is to focus on treaty design and check whether there are negotiation and accession rules that could increase the incentives to sign and ratify a climate agreement.

These three options will be analysed in the next section.

3. POLICY DESIGN, THE FUTURE OF THE KYOTO PROTOCOL AND ALTERNATIVE CLIMATE AGREEMENTS

As said above – and as shown by the theoretical results summarized in Appendix 1, which are robust with respect to different specifications of countries' welfare function, and with respect to the burden-sharing rule[4] used in the asymmetric case (Barrett, 1997a; Botteon and Carraro, 1997a) – the attempt to negotiate on emission reductions is unlikely to be completely successful, that is, to achieve the grand coalition, unless more complex policy strategies, in which environmental policy interacts with other policy measures, are adopted. This is why in the environmental economics literature two main sets of economic instruments have been proposed to expand environmental coalitions, that is, to increase the number of signatories of an environmental agreement. These instruments are 'transfers' and 'issue linkage'.

3.1 Transfers

Let us consider transfers first. It is quite natural to propose transfers to compensate those countries that may lose by signing the environmental agreement. In other words, a redistribution mechanism among signatories, from gainers to losers, may provide the basic requirement for a self-enforcing agreement to exist, that is, the profitability of the agreement for all signatories. Therefore, if well designed, transfers can guarantee that no country refuses to sign the agreement because it is not profitable. Moreover, Chander and Tulkens (1995, 1997) and Chander et al. (1999) show that there exist transfers such that not only is each country better off with global cooperation than it is with no cooperation at all (the no-participation case), but it is also better off with global cooperation than it is in any sub-coalition, provided the remaining countries behave non-cooperatively (see also Germain et al., 1997; Markusen, 1975). This result is important because it implies that no country or group of countries has an incentive to exclude other countries from the environmental coalition; that is, the grand coalition is optimal (but it may not be stable).

Transfers play a major role also with respect to the stability issue (Carraro and Siniscalco, 1993; Petrakis and Xepapadeas, 1996; Schmidt, 1997). Indeed, it is not sufficient to guarantee the profitability of the environmental agreement. Incentives to free-ride must also be offset. The possibility of using self-financed transfers to stabilize environmental agreements is analysed in Carraro and Siniscalco (1993) and Hoel (1994), which show that transfers may be successful only if associated with a certain degree of commitment. For example, when countries are fairly symmetric, only if a group of countries is committed to cooperation can another group of uncommitted countries be induced to sign the agreement by a system of transfers (Carraro and Siniscalco, 1993).[5] This gives developed countries the responsibility to lead the expansion of the coalition. However, the amount of resources that would be necessary to induce large developing countries to join the agreement may be such that some developed countries perceive the economic costs of a climate agreement larger than its environmental benefit. In this case, the transfer mechanism would undermine the existence of the leader coalition and would therefore be ineffective. This is why countries in the leader coalition must be strongly committed to cooperation on emission control.

Three types of partial commitment that could serve as possible blue-prints for environmental cooperation can be proposed (of course, other types of institutional mechanisms could be proposed as well): stable coalition commitment when only the *j* countries belonging to the stable coalition commit to cooperation; sequential commitment when the *j* countries are committed to cooperation, and any new signatory, as soon as it enters the expanded coalition, must commit to cooperation as well; external commitment when a subset of non-cooperating countries commits to transfer welfare in order to induce the remaining non-signatories to cooperate, and to guarantee the stability of the resulting coalition. Assuming these alternative commitment schemes, Carraro and Siniscalco (1993) analyse the formal conditions to expand coalitions.[6]

Another general conclusion emerges from the analysis carried out in Carraro and Siniscalco (1993): both the existence of stable coalitions and the possibilities of expanding them depend on the pattern of interdependence among countries. If there is leakage, that is, a non-cooperating country expands its emissions when the coalition restricts them, thus offsetting the effort of the cooperating countries, then environmental benefits from cooperation are low, the incentive to free-ride is high, and conditions for transfers to be effective are unlikely to be met. If, on the contrary, there is no leakage, that is, the free-riders simply enjoy the cleaner environment without paying for it, but do not offset the emission reduction by the cooperating countries, then environmental benefits are larger, free-riding is less profitable and transfers may achieve their goal to expand the coalition.

The stability issue has been often analysed within a theoretical framework in which all countries are identical (symmetric). However, there are a few attempts to analyse the existence of self-enforcing agreements and the role of transfers in the case of asymmetric or heterogeneous countries. This is done both in Barrett (1997a) and in Botteon and Carraro (1997a). These papers show that asymmetries can increase the effectiveness of transfers rather than reducing it. For example, a commitment may not be necessary (in this case an agreement with transfers would also be self-enforcing). Moreover, they address the issue of burden sharing by showing that the way in which gains are redistributed affects both profitability, as previously stressed, and the stability of the agreement, thus modifying the effectiveness of transfers and the role of commitment. In other words, there are two types of transfers: those which make the agreement profitable to all countries and those which make it stable. There are therefore two objectives (profitability and stability) with a single instrument (transfers), a situation that economists immediately recognize as inefficient.

3.2 Issue Linkage

In light of the above, a second approach to address the profitability and stability problems has been proposed. The basic idea is to design a negotiation mechanism in which countries not only negotiate on the environmental issue, but also on another interrelated (economic) issue. For example, Barrett (1995, 1997c) proposes to link climate negotiations to negotiations on trade liberalization, Carraro and Siniscalco (1995, 1997), Katsoulacos (1997) propose to link them to negotiations on R&D cooperation, and Mohr (1995), Mohr et al. (1998) propose to link climate negotiations to international debt.

Again we must distinguish the profitability from the stability problem. The idea of 'issue linkage' was originally proposed by Folmer et al. (1993) and Cesar and De Zeeuw (1996) to solve the problem of asymmetries among countries. The intuition is that some countries gain on a given issue, whereas other countries gain on a second one. By 'linking' the two issues it may be possible that the agreement in which the countries decide to cooperate on both issues is profitable to all of them.

The idea of 'issue linkage' can also be used to achieve the stability goal. Suppose there is no profitability concern (either because countries are symmetric or because a transfer scheme is implemented to make the agreement profitable to all countries). Consider the case in which it is Pareto optimal to link the environment to another economic issue (see Carraro and Siniscalco, 1995 for a formal definition). Then, if stable, the linked agreement is also self-enforcing (no commitment is necessary).

Let us consider the stability of the linked agreement. Suppose the environmental negotiation is linked to the negotiation on R&D cooperation,[7] which involves an excludable positive externality and increases the joint coalition welfare. In this way, the incentive to free-ride on the benefit of a cleaner environment (which is a public good fully appropriable by all countries) is offset by the incentive to appropriate the benefit stemming from the positive R&D externality (which is a club good fully appropriable only by the parties). The latter incentive can stabilize the joint agreement, thus increasing its profitability because countries can reap both the R&D cooperation and the environmental benefit (this second benefit would be lost without the linkage).

This idea is also exploited in Katsoulacos (1997). This paper, which accounts for information asymmetries, provides additional support to the conclusion that 'issue linkage' can be very effective in guaranteeing the stability of an environmental agreement.[8] However, the benefit function which relates gains from cooperation on the issue characterized by excludable benefit with the number of cooperating countries is generally hump-shaped. In other words, the grand coalition is not optimal when the negotiation takes place on this issue. By contrast, negotiations on the environmental issue lead to a monotonically increasing function (this is an implication of Chander and Tulkens, 1997). The behaviour of the benefit function for the joint coalition depends on the relative weight of the two issues. There are cases in which the optimal number of countries in the joint coalition is lower than the number of countries belonging to the stable group of signatories to the joint agreement. Hence three groups of countries may emerge (three roles): those that cooperate, those that would like to cooperate but are excluded from the agreement (and are therefore forced to non-cooperation) and those that prefer not to cooperate. This conclusion holds both when countries are symmetric (Carraro and Siniscalco, 1997) and when they are asymmetric (Botteon and Carraro, 1998). In the asymmetric case a further result arises. A given country *i* may prefer some countries, say *j* and *h*, as partners in the cooperating group, but these countries may want to sign the agreement with country *k*, rather than with *i*. And *k* may prefer *i* and *h* rather than *j*. In this case, an equilibrium may not exist; that is, a stable international environmental agreement may not be signed (Carraro and Siniscalco, 1998).

These latter insights lead to the conclusion that issue linkage may damage environmental protection rather than benefit it. This is the case whenever the incentives to exclude some countries from the linked agreement or the possible political-economy problem that undermines the emergence of an equilibrium dominate the benefits of linking two synergetic (in terms of profitability and stability) issues.

Despite these possible emerging conflicts, issue linkage may be a powerful tool to address the enlargement issue. If developed countries on the one hand increase their financial and technological support to developing countries, and on the other hand make this support conditional on the achievement of given environmental targets, then a number of countries are likely to be induced to join the environmental coalition, that is, to sign a treaty in which they commit themselves to adequate reductions in their emission growth.

Note the importance of technological development and diffusion, which are crucial to create the incentives for issue linkage to be effective. Hence climate-related technical change accomplishes a twofold task. On the one hand, it reduces abatement costs; on the other hand, it supports the emergence of a larger stable coalition; that is, a greater number of countries participate in the cooperative GHG emission reduction.

3.3 Participation Rules

The economic mechanisms described above are not the only possible way through which the EU or other leader coalitions can exert their leadership in climate negotiations in order to achieve an enlargement of the climate coalition. There are other interesting ways, closely related to the institutional design of the negotiation, that is, the rules that govern the negotiation process. Here we would like to address two of them:

1. The possibility of setting a minimum participation rule, that is, a minimum number of signatories below which the treaty is not operational. This provision is already included in the Kyoto Protocol (at least 55 countries whose emissions amount to at least 55 per cent of total emissions from Annex B countries) but we wonder whether this type of rule could be modified to increase its effectiveness.
2. The possibility of giving countries the freedom to negotiate on more than one agreement in the same way in which regional agreements can be established on trade issues. We would therefore have the emergence of climate blocs similar to the trade blocs already at work.

We focus on the minimum participation rule first. The main result, derived in Carraro et al. (1999), is that all players agree on a rule that leads to the formation of a coalition whose size is larger than the one that would form without any minimum participation rule. The basic intuition behind this result is that all players have an incentive to reduce their freedom to free-ride in order to achieve a larger payoff. However, this does not necessarily lead to the grand coalition because all players also have an incentive to undercut on this limitation to their freedom. The balance of these two incentives leads to the formation of a coalition that is larger than the one that forms when no minimum participation rule is introduced, but generally smaller than the grand coalition.

This result is only partly positive. Indeed, the coalition that forms under the minimum participation rule is larger but may not be sufficiently large to be environmentally effective. Indeed, given the results shown above, one may wonder why at Kyoto a more stringent minimum participation rule was not approved. The answer is that, although a minimum participation rule guarantees a larger coalition, free-riders have no incentive to agree on a stringent rule. This is why in the Kyoto Protocol the rule is that the treaty is effective if at least 55 countries producing at least 55 per cent of total emissions commit to emission reductions. A more stringent rule would have forced more countries to commit to emission reductions, but there were not the incentives, because of free-riding, to sign a more stringent rule.

Given the above results, how can a coalition exert its leadership? One possible answer is that economic mechanisms like transfers and issue linkage should be designed to achieve two goals:

- to make the climate agreement profitable for the largest number of countries on the one hand;
- to induce countries to accept a more stringent minimum participation rule on the other.

In this way, the free-riding incentives would be partly, and if possible almost completely, offset, thus leading to the formation of a very large climate coalition.

3.4 Regional Agreements

Another important change of the negotiation rules concerns the possibility that more than one treaty to control GHG emissions is signed by the negotiating countries, that can thus form two or more 'climate blocs'. Indeed, the results proposed above crucially depend on the assumption that countries can negotiate on a single agreement. If instead countries can choose whether to sign a single global agreement or a region-specific agreement, results may drastically change.

This change of assumptions has important implications. Whereas in previous sub-sections only one coalition could emerge, with the free-riding countries behaving as singletons, in this sub-section we allow for the emergence of a complex coalition structure in which several coalitions may emerge at the equilibrium. The problem is to determine which equilibrium structure is most likely – one in which one coalition is formed, or one in which $k > 1$ coalitions, which interact among each others, characterize the equilibrium of the coalition game (see Appendix 1 for a more detailed analysis of this issue).

The implications for environmental negotiations are clear. In the latter case, there would not be one environmental agreement, but k agreements signed by k groups of countries. The multiplicity of coalitions may allow for region-specific agreements in which the characteristics of countries in the region are better reflected by the contents of the agreement.

The free-riding incentive, which naturally arises in a game with positive externalities, can be described as follows in a multi-coalition game. In any coalition structure, members of small coalitions have higher payoffs than members of big coalitions. The limit case is the one of singletons (that is, coalitions formed by one player only), which receive the greatest net benefit from the other coalitions' abatement.[9]

Even if non-singleton coalitions can form, the stability and the profitability condition are defined with respect to an individual viewpoint in keeping with the spirit of a non-cooperative approach. Therefore, a coalition structure π is profitable if any coalition s in π is profitable. A coalition $s \in \pi$ is profitable if each cooperating player belonging to s gets a payoff larger than the one he would get in the singleton structure.

Let us now consider the stability condition, a coalition belonging to any coalition structure: since it is always possible for any cooperating country to deviate to form a singleton, the stability condition described in Section 2 is again a necessary condition. In the single coalition case, this condition was sufficient to define the equilibrium coalition structure, since countries had only two possible strategic choices: joining a coalition (that is, signing the agreement) or behaving as a lone free-rider (singleton). By contrast, in the multi-coalition game, if a country chooses to cooperate, it has also to choose which coalition to join. This is why a further condition on the entire coalition structure has to be added: the intracoalition stability: that is, at the equilibrium no country wants to move from one coalition to another.

The specific results on the size of the coalitions depend on the model structure and in particular on the slope of countries' reaction functions, that is, on the presence of carbon leakage. If there is no leakage and countries are symmetric, then the equilibrium of the multi-coalition game is characterized by many small coalitions, each one satisfying the properties of internal and external stability (this result is shown in Carraro and Moriconi, 1998).

Moreover, in some preliminary analyses of specific examples, it can be shown (see Carraro, 1998) that social welfare when more than one agreement is signed is larger than social welfare when countries are constrained to sign a single agreement. In other words, even if a single agreement is signed by more countries than each of the many agreements, and even if each country's abatement is larger when a single agreement is signed, total abatement is larger when many regional agreements are signed, which are therefore to be preferred from both an economic and an environmental viewpoint.

Is this conclusion general enough to provide sound support for policy recommendations? The answer is no, because there may be other examples in which a single agreement is preferred, at least from an environmental viewpoint, to many small regional agreements. Moreover, the conclusion crucially depends on the choice of the equilibrium concept and on the absence of leakage. However, it provides important insights on the feasible outcomes of environmental negotiation processes. The problem still to be solved can be phrased as follows. If countries are free to choose the number and features of agreements, then the negotiation process is likely to lead to several agreements. As a consequence, if the negotiating agenda focuses on a single agreement, will it reduce the probability of stabilizing climate change? Should countries and international institutions accept that climate change control can be easier to achieve if many regional agreements, which account for the specific characteristics of countries in the region, are proposed? Unfortunately, more research is necessary to provide a reliable answer to these questions.

Note also that these questions and doubts implicitly contain an extension of previous theoretical results, derived for the case of symmetric countries, to the case of asymmetric countries. This makes the situation even worse, because there is no theoretical analysis that can support this type of extension, which can therefore be accepted only as very preliminary. However, results contained in Barrett (1997a), and Botteon and Carraro (1997a, 1997b, 1998) for the case in which a single coalition is assumed at the equilibrium suggest that theoretical results derived for the case of symmetric countries are largely confirmed when countries' asymmetries are introduced into the model. More work on this issue would none the less be very valuable.

The consequence of the results proposed in this section, albeit preliminary and restricted to the case of symmetric countries, is that the structure of the international climate agreement is a crucial dimension of the negotiating process. If all countries negotiate on a single agreement, the incentives to sign are lower than those which characterize a multiple agreement negotiating process. At the equilibrium, the environmental benefit (quality) would also be lower. Should a change of strategy be proposed at the institutional level or will it emerge endogenously? At least this question should receive the same attention that similar questions have received in the case of trade policies.

Note that in practice the most recent evolutions of climate negotiations seem to show an increasing tendency towards the formation of groups (coalitions) which negotiate with each other (for example the EU, the umbrella group, the USA, the Group of 77, AOSIS, and so on). The Kyoto Protocol itself incorporates special provisions for several different groups of countries. The non-Annex B countries have no commitments and can benefit from emission reduction investments through the CDM. The most vulnerable non-Annex B countries can also receive financial assistance for adaptation from the levy imposed on the CDM (and possibly on the other mechanisms). The economies in transition have generally less ambitious commitments and have the flexibility to choose their base year. The EU has the ability under Article 4 to redistribute the emissions reduction burden.

In a framework in which regional agreements can be established, it may become reasonable to pursue a climate agreement within the EU, or between the EU, Eastern Europe and former Soviet Union countries, even without the participation of the USA, because other regional climate agreements are likely to emerge subsequently.[10] Once several regional agreements enter into force, for example in Europe, East Asia and the Americas, then the three groups of countries can negotiate to achieve a global agreement.

4. THE CURRENT POLICY DEBATE AND THE FUTURE OF CLIMATE NEGOTIATIONS: CONCLUSIONS AND SOME EVIDENCE

Current research on climate agreements and their relationships with climate policies has achieved some interesting conclusions that could be used to outline possible evolutions of actual negotiations on climate change control.

The results described in previous sections of the chapter emphasize the difficulty of achieving a climate agreement and the importance of designing policies which are not only cost-effective in reducing emissions, but also effective in providing the right incentives for a large number of countries to participate in the climate agreement.

These policies should be able to offset the negative incentives that characterize climate negotiations and that arise because of the peculiar features of the climate problem described in the Introduction (countries' heterogeneity, free-riding, uncertainties, irreversibilities, technical change, lack of global institutions, and so on).

The sequence of steps that well designed incentive policies should follow can be described as follows.

1. *Policies, mainly those that support climate research, should reduce uncertainty concerning climate phenomena, their origins and their impacts (in particular the distribution of these impacts).* In areas such as climate change, attention has focused on reducing uncertainty via the consolidation of existing scientific knowledge on climate systems and the specification of general areas of thematic interest which merit inclusion in future research agendas. Much less attention has been devoted to the strategies needed to develop and exploit new knowledge, particularly knowledge concerning the socioeconomic dimensions of climate change. This is unfortunate, because there is an urgent need for priorities to be established and difficult choices to be made over a broad span of research activities. Is an enhanced understanding of atmospheric chemistry currently more important, for example, than a finer appreciation of the socioeconomic variables which govern GHG emissions, or does R&D into new and innovative technical advances to combat emissions take precedence? And if, as seems likely, there is a need for all three lines of investigation to be pursued, what should be the balance, who should undertake the work, how should each problem be tackled, and how should the results eventually be utilized?

Most importantly, a better knowledge of the climate problem and of its impacts should enable policy makers to quantify two crucial issues:

- What is the damage induced by expected climate changes? Namely, is there a global benefit from controlling climate changes through GHG reduction?
- How do damages affect different countries in the world? Namely, how should countries distribute the global benefit from climate change control in a way that provides benefits to all countries?

Note that these issues are the consequence of the great heterogeneity with respect to climate impacts and mitigation and adaptation costs that currently characterize the world.

2. *Policies should redistribute benefits from climate change control to make a climate agreement profitable to all participating countries.* In other words, policies should be designed to offset the negative incentives deriving from countries' heterogeneity. Countries with the highest costs and the lowest benefits from climate change control should be given appropriate incentives to participation. This can be done through flexibility in the negotiations or through explicit transfers. For example, the joint presence of hot air and emissions trading should not be seen as a loophole of the agreement but rather a compensation mechanism to induce both the USA and the countries of the Former Soviet Union to participate. Similarly,

provisions that help small island countries in adapting to climate change have the same meaning.

In a recent work, Bosello et al. (2001) have analysed the incentive to sign a climate agreement by using the well-known RICE model developed by Nordhaus and Yang (1996).[11] Given the RICE model and the emission abatement burden sharing implicitly revealed by the Kyoto agreement, Bosello et al. (2001) accomplish two tasks. First, they analyse the features of the Kyoto agreement, both in terms of profitability and in terms of stability. Then they analyse the profitability and stability of any other climate agreements that can be achieved either by maintaining the Kyoto burden sharing or by moving to a more equitable sharing of the costs of abating emissions.

In order to assess the profitability and stability of the Kyoto agreement, Bosello et al. (2001) compare the payoffs which Annex 1 countries achieve when they cooperate to the payoffs when no cooperation takes place and/or when a different agreement is signed (this second comparison is crucial in assessing the free-riding incentives and therefore the stability of the Kyoto agreement). To achieve this goal, they compute the costs and benefits of all possible climate agreements (that is, coalition structures; see Carraro, 2000). These costs and benefits can be obtained by solving the game between the six regions of RICE (USA, Japan, Europe, China, Former Soviet Union, and Rest of the World) using a numerical iterative algorithm. The equilibrium concept used to solve the game is the usual Nash equilibrium in the non-cooperative case. By contrast, when a coalition forms, it is assumed that countries which sign the agreement maximize their joint welfare and play Nash against the free-riding countries. The resulting equilibrium is equivalent to the γ-equilibrium proposed by Chander and Tulkens (1995, 1997).[12]

These optimization experiments, where 203 different coalition structures have been examined, can be analysed from the viewpoint of profitability, namely of the benefits that signatory countries achieve when the agreement is signed. The main result is as follows: the Kyoto agreement[13] is neither weakly nor strongly profitable.

The reason is that the total surplus provided by the agreement from now to 2100 is slightly negative. Moreover, all regions would lose from signing the agreement (see Table 5.1). This result is not surprising as cooperation may not be beneficial in the presence of free-riders (Carraro and Siniscalco, 1993; Chander and Tulkens, 1997; Carraro and Moriconi, 1998).[14] When emissions trading is allowed for, given the cost-effectiveness properties of unconstrained emissions trading, losses are lower (this result confirms the theoretical analysis by Chander et al., 1999), but the Kyoto agreement still remains neither weakly nor strongly profitable (see Table 5.1 again).[15] The reason is that the emission levels attained by non-Annex B countries in the long run are too high. Hence signatory countries pay the cost of emission abatement without

Table 5.1 Winners and losers under the 'Kyoto Forever' agreement

	Relative net gains (benefits from avoided damages minus abatement costs in the case of Kyoto vs the case of non-cooperation) (%)			
	USA	Japan	Europe	Former Soviet Union
Kyoto with domestic abatement only	–0.021	–0.015	–0.009	–0.020
Kyoto with emissions trading	–0.019	+0.008	–0.004	–0.003

getting any benefits. Note that, in the presence of trading, the winner would be Japan, whereas the USA, the EU and the FSU would keep losing from signing the agreement. The reason is that Japan greatly benefits from emissions trading because it is the country with the highest marginal abatement costs.

The above results are quite important because they show that, even in the presence of 'full flexibility' (global emissions trading), there are not the economic incentives to sign the Kyoto Protocol. By comparing economic costs with environmental benefits, only Japan finds it profitable to sign the Kyoto Protocol. All the other countries would lose by adopting it.

Therefore the policy conclusion is that policies should be designed, for example transfer mechanisms, issue linkage and also adaptation policies, in order to make the Kyoto Protocol profitable for all signatories. Most importantly, a revision of the burden sharing, either by redistributing emission abatement efforts or again by means of transfers, may help to introduce the right economic incentives.

3. *Policies should address the free-riding incentives by focusing on equity, but also by implementing a wide range of transfers and linkage strategies.* For example, for a climate agreement to be environmentally effective, there is the need to get developing countries on board: involve them in the whole climate change debate, and make climate change policies part of their national policies. To achieve this goal, equity and efficiency issues need to be carefully integrated. Increased equity would help to increase the number of

participating countries, thus enhancing the efficiency of the agreement. Increased efficiency would reduce each country's cost of controlling emissions, thus facilitating the adoption of transfer and cooperation policies that increase equity as well as the number of participating countries. But this may not be sufficient. Policy integration which takes into account the linkages of environmental policy with energy policies, technology development and diffusion and technological cooperation may be necessary.

As previously shown, the free-riding problem is exacerbated by the presence of carbon leakage. An effective climate policy cannot neglect the fact that its outcome crucially depends on the actions and interactions of several players. The abatement decision of a group of countries, for example Annex B countries, can be costly and ineffective if offset by emission expansions in other countries. This is the well-known carbon leakage issue, which arises because of the free-riding strategic incentives that characterize the climate problem. Carbon leakage is not adequately addressed in the Kyoto Protocol, and future policies should make an effort to develop measures to counteract it.

A quantitative analysis of how the application of different equity principles can be used to offset free-riding is carried out in Bosello et al. (2001). They focus on outcome-based equity principles (see Table 5.2) and in particular on three burden-sharing criteria:

- equal average abatement costs; or
- equal per capita abatement costs; or
- equal abatement costs per unit of GDP,

which are applied to all possible coalition structures. The idea is to verify whether there exist coalition structures which are profitable and stable when countries are certain that the implementation of the climate agreement would meet one of the above burden-sharing criteria.

The main results are summarized in Tables 5.3, 5.4 and 5.5, which show that, even by adopting a burden-sharing rule more equitable than the one implicit in the Kyoto agreement, only a few coalition structures are profitable and only one is stable; namely it emerges as a Nash equilibrium of the game in which countries non-cooperatively decide whether or not to join the coalition. This equilibrium coalition structure is formed by a coalition of three countries and by three free-riders. It can be obtained only if *ex ante* all countries agree that abatement efforts must be such as to equalize abatement costs per capita. This coalition is formed by Japan, the Former Soviet Union and ROW.

Two possible ways of addressing the problem are available. First, policy strategies could be designed to further redistribute the surplus provided by

Table 5.2 Outcome-based equity criteria

Equity principle	Definition	Implied burden-sharing rule
Horizontal	All nations have the right to be treated equally concerning both emission rights and burden-sharing responsibilities	Welfare changes across nations such that welfare costs or net abatement costs as a proportion of GDP or of population are the same in each country. Implementation criterion: *equal welfare costs per unit of GDP or per capita*
Vertical	Welfare gains should vary inversely with national economic well-being; welfare losses should vary directly with GDP. The greater the ability to pay, the greater the economic burden	Emissions reductions such that net abatement costs grow with GDP. Implementation criterion: *equal abatement costs per unit of GDP*
Compensation (Pareto rule)	'Winners' should compensate 'losers' so that both are better off after mitigation	Distribute abatement costs so that no nation suffers a net loss of welfare. Implementation criterion: *strong profitability*

Sources: Adapted from Cazorla and Toman (2000), Tol (2001), Rose and Stevens (1993), Rose et al. (1998) and Schmidt and Koschel (1998).

the cooperative behaviour within a coalition. This would increase the number of strongly profitable coalitions and hence the probability of identifying a stable coalition structure. Transfer schemes designed to make a climate agreement profitable to all countries have been proposed, for example by Chander and Tulkens (1997) and applied to climate models by Yang (2000) and by Eyckmans (2000).

Second, policy strategies could be designed to redistribute the surplus achieved by internally stable coalitions with the goal to induce other countries to enter the coalition. This idea is proposed and analysed in Carraro and

Table 5.3 Profitability: share of profitable coalitions for each burden-sharing criterion (%)

	Single coalitions	Multiple coalitions	Total
Kyoto implicit burden sharing	0	0	0
Equal average abatement costs	0	0	0
Equal per capita abatement costs	6.9	0	2.0
Equal abatement costs per unit of GDP	1.7	0	0.5

Table 5.4 Internal stability (no incentives to leave a coalition): share of internally stable coalitions for each burden-sharing criterion (%)

	Single coalitions	Multiple coalitions	Total
Kyoto implicit burden sharing	0	0	0
Equal average abatement costs	0	0	0
Equal per capita abatement costs	3.4	0	1.0
Equal abatement costs per unit of GDP	1.7	0	0.5

Table 5.5 Stability (no incentives to leave nor to enter a coalition): number of stable coalitions for each burden-sharing criterion

	Single coalitions	Multiple coalitions	Total
Kyoto implicit burden sharing	0	0	0
Equal average abatement costs	0	0	0
Equal per capita abatement costs	1	0	1
Equal abatement costs per unit of GDP	0	0	0

Siniscalco (1993), who show that, with symmetric countries, the transfer mechanisms can be used to broaden the coalition only if all countries in the initial, internally stable coalition are committed to cooperation once the transfer scheme is adopted. The same idea is applied to the RICE model by Bosello et al. (2001) with encouraging results.

In the above-mentioned empirical analysis of the role of transfers, two restrictions are imposed:

- Transfers are self-financed, that is, countries are allowed to transfer only the surplus yielded by their cooperation. Hence, only weakly and strongly profitable coalitions can be broadened through a transfer mechanism.
- The transfer mechanism is Pareto optimal, that is, all countries gain from using transfers to broaden the coalition. This implies the existence of a unanimous consensus towards the transfer mechanism.

Given this latter restriction, the broadened coalition is also weakly or strongly profitable. However, self-financing implies that there may not be enough resources to offset the free-riding incentives of all countries which are not in the initial coalition. Finally, note that even countries in the initial stable coalition may have an incentive to free-ride when other countries join. Hence the transfer mechanism must also be such as to offset these latter incentives to free-ride.

Table 5.6 Internally stable coalitions before and after the use of transfers

	Internally stable coalitions without and with transfers	
	Internally stable coalitions before transfers	Internally stable coalitions after transfers
Equalization of average abatement costs	0	0
Equalization of abatement costs/GDP	USA + CHN	USA + CHN + ROW
Equalization of per capita abatement costs	EU + CHN + ROW JPN + FSU + ROW (stable)	USA + EU + CHN + ROW USA + JPN + EU + FSU + ROW

Table 5.6 presents the results. The first column shows the initial, internally stable coalitions for each of the three burden-sharing criteria analysed in the previous section. The second column shows the largest internally stable coalition that can be achieved through a transfer mechanism starting from the corresponding initial, internally stable coalition.

The results achieved by Bosello et al. (2001) can be summarized as follows:

- No transfer mechanism and no *ex-ante* burden sharing criterion (of the three that we considered) yields an incentive structure or enough resources to achieve the grand coalition, that is, a global agreement on climate change. At least one region free-rides on the agreement.
- The burden-sharing criteria, which are most effective in guaranteeing the achievement of a large coalition without free-riding incentive, are the equalization of per capita abatement costs and the equalization of abatement costs per unit of GDP. For example, with the former, a coalition formed by Japan, FSU and ROW can offset the free-riding incentives of the EU and the USA. With the latter, a coalition formed by China and the USA can induce ROW to sign the climate agreement. Note that the USA needs a compensating transfer to enter a coalition which forms according to the criterion of equal per capita abatement costs, whereas they belong to the initially stable coalitions if the burden-sharing criterion is equal abatement costs per unit of GDP. This is quite intuitive: the USA has high per capita emissions, but relatively small emissions per unit of GDP.

None the less, the global agreement cannot be achieved. Therefore a different strategy, perhaps less focused on equity, could be more helpful in achieving a global climate agreement. For example, does the introduction of emissions trading, that is, the *ex-post* equalization of marginal abatement costs, regardless of the *ex-ante* burden-sharing criterion, provide enough resources which, once transferred to free-riding countries, induce them to sign the climate agreement?

The answer to this question is positive. As shown in Bosello et al. (2001), regardless of the *ex-ante* burden-sharing criterion (equity), and regardless of the initial, profitable coalition, the equalization of marginal abatement costs, coupled with an appropriately designed *ex-post* transfer mechanism, can lead to the grand coalition, that is, a global climate agreement signed by all countries or regions.

In other words, through emissions trading and transfers, *all* profitable coalitions can be broadened to achieve the grand coalition. As a consequence, these results seem to suggest that an excessive focus on equity rules is not fruitful. It is more effective to minimize overall abatement costs via emissions trading and then use the resulting surplus to provide incentives for free-riding countries to join the initial coalition.

4. *In the absence of supranational institutions to regulate emission reductions and in the presence of strong negative incentives to participation in the climate agreement, negotiations should allow for regional and/or domestic cost-effective attempts to control GHG emissions.* For example, domestic

trading schemes could be effectively introduced in addition to, possibly in advance of, international emissions trading. The Resources for the Future (RFF) proposal for the USA could be adopted as an interesting example of a domestic trading permit scheme. The RFF proposal provides for the auctioning of carbon quotas, with carbon fuel energy producers and importers required to buy the carbon quota equivalent of their output or imports respectively, with a ceiling price of $25 per tonne of carbon, the permits to last for two years, and the revenues to be returned to households in compensation for higher fuel prices. Other interesting domestic schemes have recently been introduced in the UK and in Denmark. In addition, in 2001 the EU proposed an internal trading scheme in order to exploit differences in marginal abatement costs across EU countries. Finally, talks have recently started to establish an emissions trading market among the three NAFTA countries. All the above are clear examples of domestic and regional cost-effective attempts to control GHG emissions.

Strategic interactions and preferences for a better and safer environment can also be exploited by firms for business purposes. BP-Amoco indeed represents a very interesting example of 'far-seeing' climate policy from the industry side; BP-Amoco has in fact adopted a very innovative policy, setting its own reduction target: 22 million tonnes reduction by the year 2010, which is equivalent to a 5 per cent reduction with respect to its 1990 baseline. In order to meet this challenge, BP-Amoco intends to develop a strategy that will result in the greatest GHG emission reduction at the lowest cost over a specific timeframe. The implementation of this strategy will require a process of identification of the emissions, by source and type, of measurement and verification of the emissions, and of identification of GHG reduction opportunities across the business. Reduction policies aim at enhancing efficiency, promoting new technology, and preventing flaring and venting. Part of this process is the implementation of a pilot trading system, which was launched in late 1998. Several business units are involved in the scheme; the pilot group has a reduction target of 3 per cent below the 1995 baseline by the year 2003; basic emission rights have been allocated for five annual periods from 1999, with requirements to be met each year; a bank acts as a broker and registers trade across business units; banking, forward sales and purchases are allowed, but no borrowing can take place; trading occurs in CO_2 equivalent units. The BP-Amoco experiment could serve as a pivotal scheme for industry, with the aim to understand how to deliver commitments at the minimum cost. The identification of opportunities across business could also help industry and governments to develop integrated and more effective climate policies.

5. *Policies should support the development of cheaper climate-friendly technologies and their diffusion.* As discussed in the previous chapter, technical

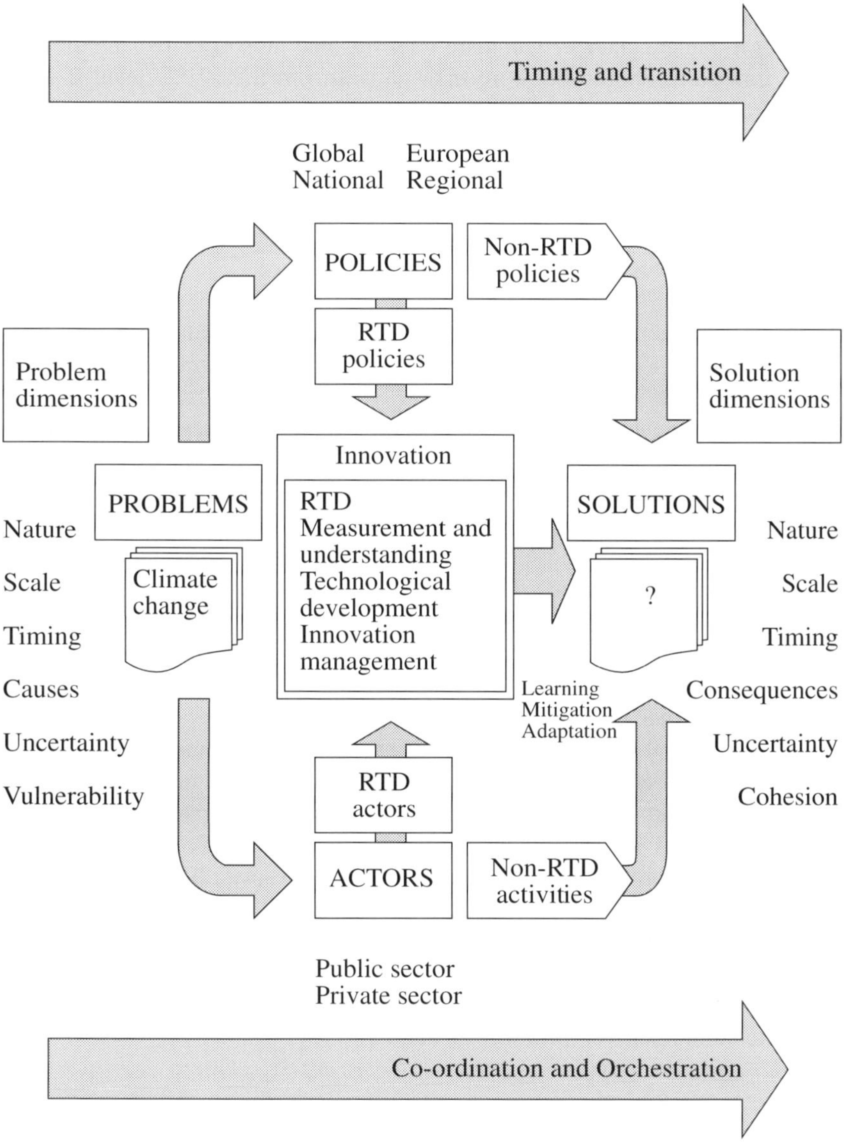

Note: RTD = research and technology development

Source: ETAN (1999).

Figure 5.1 A policy framework

change is likely to be the most important tool to reduce GHG emissions and to control climate change. The main reason is that most scientists predict that GHG emissions should be more than halved from their 1990 level if GHG concentrations have to be stabilized at about 550 ppmv. In addition, it is widely recognized that market failures largely affect the invention and diffusion of new technologies. As a consequence, policies and regulatory measures become necessary. To be effective, climate-change-related innovation policy packages will have to conform to current 'best practice' in terms of innovation policy (see Figure 5.1); that is, they will have to:

- encourage and exploit competition in the development of relevant technologies, using it to drive the innovation process, improve efficiency, reduce costs and increase quality;
- take account of the fact that technological development often manifests itself in terms of the evolution of a number of related technological developments in well-defined clusters or trajectories, often requiring the targeting of support policies on related rather than isolated technologies and on a multitude of parallel developments in the socio-technical infrastructure (see Chapter 4);
- include some measures which redress the problem of inadequate access to risk capital for climate-related innovation, perhaps via mechanisms limiting the liability of the financial institutions providing this capital if a 'climate dividend' is involved;
- pay special attention to the involvement of highly innovative SMEs;
- guide rather than dictate the behaviour of the key actors involved in relevant knowledge generation and innovation systems: stimulating new ideas in areas where none exist; helping to overcome market, information and capability failures where these are blocking the development of technologies thought to be in the collective good; reinforcing and accelerating promising developments on the verge of 'lift-off'; leaving vibrant and dynamic areas to the dictates of market forces; and encouraging the collection and analysis of data on the impact of new technologies to inform future actions.

Finally, technological innovation and diffusion are also driven by relative prices. A low price of fossil fuels and/or low-cost emission reduction possibilities – such as an emissions trading market where a low price prevails – may provide little incentive to technological innovation. This may actually be one of the major problems arising as a consequence of the US decision to withdraw from the Kyoto Protocol (see Buchner et al., 2001).

NOTES

1. See Goulder (1995), Bovenberg (1997), Bosello et al. (2001) for surveys of this literature.
2. See Carraro (1999b) and Grubb (1999) for a more detailed analysis.
3. Hence the weakness of most of the available literature on costs and benefits of climate change policies, which widely neglects the full interdependence between policies, costs/benefits and signatories (more generally the structure of the climate agreement). Let us take an example. There are studies that analyse the costs of implementing the Kyoto Protocol either through a set of domestic policies and measures or through a system of international tradable permits, keeping as given the number of signatories. But the adoption of either policy crucially affects the number of signatories that can be larger or smaller under policies and measures than under tradable permits. And the number (and identity) of signatories crucially affects costs and benefits of different agreements.
4. In the asymmetric case, the rule which is chosen to divide the gains from cooperation among the countries in the coalition (usually called burden-sharing rule) plays a crucial role because it affects the likelihood that each country decides to sign the agreement. The burden-sharing rule is usually taken from cooperative game theory and that of Nash and Shapley is the most used. By contrast, in the symmetric case different rules lead to the same outcome (equal shares).
5. This condition is less stringent when countries are asymmetric. See Botteon and Carraro (1997a).
6. However, recall that the idea of commitment, albeit partial, that is, confined to a group of countries, cannot be entirely consistent with the concept of self-enforcing agreement stressed in the previous sections.
7. See Carraro and Siniscalco (1997) for a full presentation of the model.
8. Information asymmetries are also dealt with in Petrakis and Xepapadeas (1996) and in Chillemi (1997). These papers design implementation mechanisms which can induce countries to sign an environmental agreement even when emission levels are not observable.
9. Yi (1997) shows that the partition function of a multi-coalition game with positive externalities usually satisfies these two properties.
10. This prediction is confirmed by the most recent talks between the USA, Canada and Mexico to establish an emission trading market within the NAFTA.
11. The RICE model is a single-sector optimal growth model that has been extended to incorporate the interactions between economic activities and climate. One such model has been developed for each macro-region into which the world is divided (USA, Japan, Europe, China, Former Soviet Union, and Rest of the World). Within each region a central planner chooses the optimal paths of fixed investment and emission abatement that maximize the present value of per capita consumption. Output (net of climate change) is used for investment and consumption and is produced according to a constant returns Cobb–Douglas technology, which combines the inputs from capital and labour with the level of technology. Population (taken to be equal to full employment) and technology levels grow over time in an exogenous fashion, whereas capital accumulation is governed by the optimal rate of investment. There is a wedge between output gross and net of climate change effects, the size of which is dependent upon the amount of abatement (rate of emission reduction) as well as the change in global temperature. The model is completed by three equations representing emissions (which are related to output and abatement), carbon cycle (which relates concentrations to emissions), and climate module (which relates the change in temperature relative to 1990 levels to carbon concentrations) respectively.
12. The γ-equilibrium of the game between the four Annex B countries and the two non-Annex B countries of RICE is computed as follows. Annex B countries maximize an aggregate utility function which is the weighted sum of their individual utility functions, whereas non-Annex B countries maximize their own utility function, taking as given what the other countries do. In the same way, all possible γ-equilibria of the game have been computed, whatever the size of the coalition and the identity of its members.

13. It is important to stress that the reference to the Kyoto agreement is partly imprecise. Indeed, instead of assuming that the abatement targets are compulsory only by the end of the so-called first commitment period, we assume that these abatement targets are binding until the end of this century. This is the so-called 'Kyoto Forever' hypothesis (Manne and Richels, 1998). All Annex 1 countries are assumed to meet the Kyoto constraints from 2010 onward. The CO_2 concentration levels implicit in this assumption (if RICE is a good description of the world) roughly coincide with those in the A1B scenario (IPCC, 2001) which can be considered the 'median' scenario among those currently proposed.
14. The same result has been found in other fields of economics, for example monetary economics (see Rogoff, 1997)
15. Our conclusion differs from the one in Yang (2000) and Eyckmans (2000), who show that the Kyoto agreement is weakly profitable (but not strongly profitable unless a transfer scheme is introduced).

REFERENCES

Axelrod, A. (1984), *The Evolution of Co-operation*, New York: Basic Books.

Barker, T. (1999), 'Achieving a 10% Cut in Europe's Carbon Dioxide Emissions Using Additional Excise Duties: Coordinated, Uncoordinated and Unilateral Action Using the Econometric Model E3ME', *Economic Systems Research*, **11**(4), 401–21.

Barrett, S. (1992), 'Reaching a CO_2-emission limitation agreement for the Community: implications for equity and cost-effectiveness', *European Economy*, Special Edition, No. 1, pp. 3–23.

Barrett, S. (1994), 'Self-Enforcing International Environmental Agreements', *Oxford Economic Papers*, **46**, 878–94.

Barrett, S. (1995), 'Trade Restrictions in International Environmental Agreements', London Business School.

Barrett, S. (1997a), 'Heterogeneous International Environmental Agreements', in C. Carraro (ed.), *International Environmental Agreements: Strategic Policy Issues*, Cheltenham, UK and Lyme, USA: Edward Elgar.

Barrett, S. (1997b), 'Towards a Theory of International Co-operation', in C. Carraro and D. Siniscalco (eds), *New Directions in the Economic Theory of the Environment*, Cambridge: Cambridge University Press.

Black, J., M.D. Levi and D. de Meza (1992), 'Creating a Good Atmosphere. Minimum Participation for Tackling the Greenhouse Effect', *Economica*, **60**, 281–93.

Bloch, F. (1997), 'Non-co-operative Models of Coalition Formation in Games with Spillovers', in C. Carraro and D. Siniscalco (eds), *New Directions in the Economic Theory of the Environment*, Cambridge: Cambridge University Press.

Bohm, P. (1999), 'International Greenhouse Gas Emission Trading', Nordic Council of Ministers, TemaNord 1999:506.

Bosello, F. and C. Carraro (1999), 'Recycling Energy Taxes. Impacts on Disaggregated Labour Market', *Energy Economics*, **23**, 569–94.

Bosello, F. and R. Roson (1999), 'Carbon Emissions Trading and Equity in International Agreements', in C. Carraro (ed.), *Efficiency and Equity of Climate Change Policy*, Dordrecht: Kluwer.

Bosello, F., C. Carraro and M. Galeotti (2001), 'The Double Dividend Issue: Modelling Strategies and Empirical Findings', *Environment and Development Economics*, 6, 9–45.

Bosello, F., B. Buchner, C. Carraro and D. Raggi (2001), 'Can Equity Enhance Efficiency? Lessons from the Kyoto Protocol', FEEM *Nota di Lavoro* No. 49.2001, Milan.

Botteon, M. and C. Carraro (1997), 'Burden-Sharing and Coalition Stability in Environmental Negotiations with Asymmetric Countries', in C. Carraro (ed.), *International Environmental Agreements: Strategic Policy Issues*, Cheltenham, UK and Lyme, USA: Edward Elgar.

Botteon, M. and C. Carraro (1998), 'Strategies for Environmental Negotiations: Issue Linkage with Heterogeneous Countries', in N. Hanley and H. Folmer (eds), *Game Theory and the Environment*, Cheltenham, UK and Northampton, USA: Edward Elgar.

Bovenberg, A.L. (1997), 'Environmental Policy, Distortionary Labour Taxation and Employment: Pollution Taxes and the Double Dividend', in C. Carraro and D. Siniscalco (eds), *New Directions in the Economic Theory of the Environment*, Cambridge: Cambridge University Press, pp. 69–104.

Brams, S. and M. Kilgour (1988), *Game Theory and National Security*, Oxford: Basil Blackwell.

Buchholz, W. and K.A. Konrad (1994), 'Global Environmental Programs and the Strategy Choices of Technology', *Journal of Economics*, **60**(3), 379–87.

Buchner, B., C. Carraro and I. Cersosimo (2001), 'On the Consequences of the US Withdrawal from the Kyoto Protocol ', FEEM *Nota di Lavoro*, December, Milan.

Buonanno, P., E. Castelnuovo, C. Carraro and M. Galeotti (1999), 'Efficiency and Equity of Emission Trading with Endogenous Environmental Technical Change', in C. Carraro (ed.), *Efficiency and Equity of Climate Change Policy*, Dordrecht: Kluwer.

Burniaux, J.M. (1998), 'How Important is Market Power in Achieving Kyoto?: An Assessment Based on the GREEN Model', paper presented at the OECD Experts Workshop on Climate Change and Economic Modelling: Background Analysis for the Kyoto Protocol, Paris, 17–18 September.

Burniaux, J.-M., J.-P. Martin, G. Nicolletti and J.O. Martins (1992), 'The cost of international agreements to reduce CO_2 emissions', *European Economy*, Special Edition, No. 1, 271–98.

Capros, P. (1998), 'Economic and Energy System Implications of European CO_2 Mitigation Strategy: Synthesis of Results from Model Based Analysis', paper presented at the OECD Experts Workshop on Climate Change and Economic Modelling: Background Analysis for the Kyoto Protocol, Paris, 17–18 September.

Carraro, C. (ed.) (1997), *International Environmental Agreements: Strategic Policy Issues*, Cheltenham, UK and Lyme, USA: Edward Elgar.

Carraro C. (1998), 'Beyond Kyoto : A Game-Theoretic Perspective', in *Proceedings* of the OECD Experts Workshop on Climate Change and Economic Modelling: Background Analysis for the Kyoto Protocol, Paris, 17–18 September.

Carraro, C. (1999a), 'Environmental Conflict, Bargaining and Cooperation' in J. van den Bergh, *Handbook of Natural Resources and the Environment*, Cheltenham, UK and Northampton, USA: Edward Elgar.

Carraro, C. (1999b), 'The Economics of International Coalition Formation and EU Leadership', in M. Grubb and J. Gupta (eds), *European Leadership of Climate Change Regimes*, Dordrecht: Kluwer.

Carraro C. (1999c), 'The Structure of International Environmental Agreements' in C. Carraro (ed.), *International Environmental Agreements on Climate Change*, Dordrecht: Kluwer.

Carraro, C. (ed.) (1999d), *International Environmental Agreements on Climate Change*, Dordrecht: Kluwer.
Carraro, C. (2000), *Efficiency and Equity of Climate Change Policy*, Dordrecht: Kluwer.
Carraro, C. and J.C. Hourcade (eds) (1999), *Optimal Timing of Climate Change Policies*, special issue of *Energy Economics*, Amsterdam: Elsevier.
Carraro, C. and F. Moriconi (1998), 'Endogenous Formation of Environmental Coalitions', presented at the Coalition Theory Network Workshop on 'Coalition Formation: Applications to Economic Issues', Venice, 8–10 January.
Carraro, C. and D. Siniscalco (1992), 'The International Protection of the Environment: Voluntary Agreements among Sovereign Countries', in P. Dasgupta, K.-G. Mäler and A. Vercelli (eds), *The Economics of Transnational Commons*, Oxford: Clarendon Press.
Carraro, C. and D. Siniscalco (1993), 'Strategies for the International Protection of the Environment', *Journal of Public Economics*, **52**, 309–28.
Carraro, C. and D. Siniscalco (1995), 'Policy Coordination for Sustainability: Commitments, Transfers, and Linked Negotiations', in I. Goldin and A. Winters (eds), *The Economics of Sustainable Development*, Cambridge: Cambridge University Press.
Carraro, C. and D. Siniscalco (1997), 'R&D Cooperation and the Stability of International Environmental Agreements', in C. Carraro (ed.), *International Environmental Agreements: Strategic Policy Issues*, Cheltenham, UK and Lyme, USA: Edward Elgar.
Carraro, C. and D. Siniscalco (1998), 'International Environmental Agreements: Incentives and Political Economy', *European Economic Review*, **42**, 561–72.
Carraro, C., F. Moriconi and S. Oreffice (1999), 'α-Rules and Equilibrium Endogenous Coalitions', presented at the fourth CTN Workshop on Coalition Formation, Aix en Provence, 8–10 January.
Cazorla, M. and M. Toman (2000), 'International Equity and Climate Change Policy', *RFF Climate Issue Brief* No. 27, Washington, DC: RFF.
Cesar, H. and A. De Zeeuw (1996), 'Issue Linkage in Global Environmental Problems', in A. Xepapadeas (ed.), *Economic Policy for the Environment and Natural Resources*, Cheltenham, UK and Lyme, USA: Edward Elgar.
Chander, P. and H. Tulkens (1995), 'A Core-Theoretical Solution for the Design of Cooperative Agreements on Trans-Frontier Pollution', *International Tax and Public Finance*, **2**, 279–94.
Chander, P. and H. Tulkens (1997), 'The Core of an Economy with Multilateral Environmental Externalities', *International Journal of Game Theory*, **26**, 379–401.
Chander, P., H. Tulkens, J.P. Van Ypersele and S. Willems (1999), 'The Kyoto Protocol: an Economic and Game Theoretic Interpretation', CLIMNEG working paper No.12, CORE, UCL, Louvain.
Chew, M.S. (1994), 'Farsighted Coalitional Stability', *Journal of Economic Theory*, **54**, 299–325.
Chillemi, O. (1997), 'International Environmental Agreements and Asymmetric Information', in C. Carraro (ed.), *International Environmental Agreements: Strategic Policy Issues*, Cheltenham, UK and Lyme, USA: Edward Elgar.
Convery F.J. (1999), 'Blueprints for a Climate Policy: Based on Selected Scientific Contributions', in C. Carraro (ed.), *Integrating Climate Policies in a European Environment*, special issue of *Integrated Assessment*, Baltzer.
Ellerman, A.D., H.D. Jacoby and A. Decaux (1998), 'Analysis of Post-Kyoto Emis-

sions Trading Using Marginal Abatement Curves', paper released under the Joint Program on the Science and Policy of Global Change, MIT.

Ellerman, A.D., H.D. Jacoby and A. Decaux (1999), 'The Effects on Developing Countries of the Kyoto Protocol, and CO_2 Emissions Trading', paper released under the Joint Program on the Science and Policy of Global Change, MIT.

Endres, E.A. and M. Finus (1998), 'Playing your Better Global Emissions Game: Does it Help to be Green?', *Swiss Journal of Economics and Statistics*, **134**(1), 21–40.

ETAN (1999), 'Climate Change and the Challenge for Research and Technological Development (RTD) Policy', European Commission Directorate General XII, Environment and Climate RTD Programme, Targeted Socio-Economic Research Programme.

Eyckmans, J. (2000), 'On the Farsighted Stability of the Kyoto Protocol, Some Simulation Results', mimeo, University of Leuven.

Folmer, H., P. Van Mouche and S.E. Ragland (1993), 'Interconnected Games and International Environmental Problems', *Environment and Resource Economics*, **3**(4), 313–35.

Germain, M., P. Toint and H. Tulkens (1997), 'Financial transfers to Ensure Cooperation International Optimality in Stock-Pollutant Abatement', CORE Discussion Paper No. 9701, Center for Operations Research and Econometrics, Université de Louvain, Belgium.

Glomsrød, S., H. Vennemo and P. Johnsen (1992), 'Stabilisation of Emissions of CO_2: A Computable General Equilibrium Assessment', *The Scandinavian Journal of Economics*, **94**(1).

Goulder, L.H. (1995), 'Environmental Taxation and the Double Dividend: A Reader's Guide', *International Tax and Public Finance*, **2**, 157–83.

Goulder L.H. and S.H. Schneider (1998), 'Achieving Carbon Dioxide Concentration Targets: What Need To Be Done Now?', *Nature*, **389**, 13–14.

Grubb, M. (1998), 'International Emissions Trading Under the Kyoto Protocol: Core Issues in Implementation', mimeo, RIIA.

Grubb, M. (1999), 'Corrupting the Climate? Economic Theory and the Politics of Kyoto', text of Valedictory Lecture, The Royal Institute of International Affairs, London.

Grubb, M. and J. Gupta (eds) (2000) *European Leadership of Climate Change Regimes*, Dordrecht: Kluwer.

Grubb, M. and C. Vrolijk (1998), 'The Kyoto Protocol: Specific Commitments and Flexibility Mechanisms', *RIIA, EEP Climate Change Briefing*, No. 11.

Hampton, J. (1987), 'Free-Rider Problems in the Production of Collective Goods', *Economics and Philosophy*, **3**, 245–73.

Hardin, G. and J. Baden (1977), *Managing the Commons*, New York: Freeman & Co.

Harrison, G.W. and T.F. Rutherford (1999), 'Burden Sharing, Joint Implementation, and Carbon Coalition' in C. Carraro (ed.), *International Environmental Agreements on Climate Change*, Dordrecht: Kluwer, pp. 77–108.

Hart, S. and M. Kurz (1983), 'Endogenous Formation of Coalitions', *Econometrica*, **51**, 1047–64.

Heal, G. (1994), 'The Formation of Environmental Coalitions', in C. Carraro (ed.), *Trade, Innovation, Environment*, Dordrecht: Kluwer.

Hoel, M. (1991), 'Global Environmental Problems: The Effects of Unilateral Actions Taken by One Country', *Journal of Environmental Economics and Management*, **20**(1), 55–70.

Hoel, M. (1992), 'International Environmental Conventions: the Case of Uniform Reductions of Emissions', *Environmental and Resource Economics*, **2**, 141–59.

Hoel, M. (1994), 'Efficient Climate Policy in the Presence of Free-Riders', *Journal of Environmental Economics and Management*, **27**, 259–74.

Hoel, M. and K. Schneider (1997), 'Incentives to Participate in an International Environmental Agreement', *Environment and Resource Economics*, **9**, 154–70.

Holtzmark, B.J. (1998), 'From the Kyoto Protocol to the Fossil Fuel Markets', in *Proceedings* of the OECD Experts Workshop on Climate Change and Economic Modelling: Background Analysis for the Kyoto Protocol, Paris, 17–18 September.

Hourcade, J.C., M. Minh Ha-Duong and F. Lecoq (1999), 'Dynamic Consistency Problems behind the Kyoto Protocol', forthcoming in the *International Journal of Environment and Pollution*, Special Issue on Methodologies and Issues for Integrated Environmental Assessment.

IPCC (1995), *Second Assessment Report, Vol. 3. Mitigation*, Cambridge: Cambridge University Press.

IPCC (2001), *Third Assessment Report, Vol. 3. Mitigation*, Cambridge: Cambridge University Press.

Jorgenson, D.W. and P.J. Wilcoxen (1993), 'Energy, the environment and economic growth', *Handbook of Natural Resources and Energy Economics*, 3.

Katsoulacos, Y. (1997), 'R&D Spillovers, R&D Co-operation, Innovation and International Environmental Agreements', in C. Carraro (ed.), *International Environmental Agreements: Strategic Policy Issues*, Cheltenham, UK and Lyme, USA: Edward Elgar.

Kopp, R.J., W. Harrington, R.D. Morgenstern, W.A. Pizer and J.S. Shih (1998), 'Diffusion of New Technologies: A Microeconomic Analysis of Firm Decision Making at the Plant Level', Washington, DC: Resources for the Future.

Kram, T. (1998), 'The Energy Technology Sytems Analysis Program: History, the ETSAP Kyoto Statement and Post-Kyoto Analysis', paper presented at the OECD Experts Workshop on Climate Change and Economic Modelling: Background Analysis for the Kyoto Protocol, Paris, 17–18 September.

Kverndokk, S. (1993), 'Global CO_2 Agreements: A Cost-Effective Approach', *The Energy Journal*, **14**(2), 91–112.

Manne, A. and R. Richels (1998), 'Preliminary Analysis of the Costs of the Kyoto Protocol', paper presented at the Conference on Global Carbon Dioxide Abatement, Geneva, 11 June.

Manne, A. and R. Richels (1999a), 'On Stabilizing CO_2 Concentrations – Cost-Effective Emission Reduction Strategies', in C. Carraro (ed.), *International Environmental Agreements on Climate Change*, Dordrecht: Kluwer.

Manne, A. and R. Richels (1999b), 'The Kyoto Protocol: A Cost-effective Strategy for Meeting Environmental Objectives?', in C. Carraro (ed.), *Efficiency and Equity of Climate Change Policy*, Dordrecht: Kluwer.

Mariotti, M. (1997), 'A Model of Agreements in Strategic Form Games', *Journal of Economic Theory*, **73**, 128–39.

Markusen, G.R. (1975), 'Co-operation, Control of International Pollution and Common Property Resources', *Quarterly Journal of Economics*, **89**, 618–32.

McKibbin, J.W., S.E. Schackleton and P.G.J. Wilcoxen (1998), 'What to Expect from an International System of Tradable Permits for Carbon Emissions?', *Brookings Discussion Papers on International Economics*, No. 143.

Mensbrugghe, D. van der (1998), 'A Preliminary Analysis of the Kyoto Protocol Using the OECD GREEN Model', paper presented at the OECD Experts Workshop on Climate Change and Economic Modelling: Background Analysis for the Kyoto Protocol, Paris, 17–18 September.

Metz, B. (1999), 'International Equity in Climate Change Policy', in C. Carraro (ed.), *Integrating Climate Policies in a European Environment*, special issue of *Integrated Assessment*, Baltzer.

Mohr, E. (1995), 'International Environmental Permits Trade and Debt: The Consequences of Country Sovereignty and Cross Default Policies', *Review of International Economics*, (1), 1–19.

Mohr E., J. Heister, W. Plesmann, F. Stähler, T. Stoll and W. Rüdiger (1999), 'Economic and Legal Aspects of International Environmental Agreements', mimeo.

Nordhaus, W.D. (1997), 'Modelling Induced Innovation in Climate-Change Policy: Theory and Estimates in the R&DICE model', paper prepared for the Workshop on Induced Technological Change, IIASA, June 1997.

Nordhaus, W.D. (1999), 'Modelling the Impact of Climate Change', paper presented and the 1999 FEEM–IDEI–INRA Conference, Toulouse, 14–16 June.

Nordhaus, W.D. and J.C. Boyer (1999), 'Requiem for Kyoto: An Economic Analysis of the Protocol', draft, Yale University.

Nordhaus, W.D. and Z. Yang (1996), 'RICE: A Regional Dynamic General Equilibrium Model of Alternative Climate Change Strategies', *The American Economic Review*, **86**(4), 726–41.

OECD (1998), *Proceedings* of the OECD Workshop on Climate Change and Economic Modelling. Background Analysis for the Kyoto Protocol, Paris, 17–18 September.

Oye, K.A. (1986), *Co-operation under Anarchy*, Princeton: Princeton University Press.

Peck, S.C. and T.J. Teisberg (1999), 'CO_2 Concentrations Limits, the Costs and Benefits of Control and the Potential for International Agreements', in C. Carraro (ed.), *International Environmental Agreements on Climate Change*, Dordrecht: Kluwer.

Petrakis, E.A. and A. Xepapadeas (1996), 'Environmental Consciousness and Moral Hazard in International Agreements to Protect the Environment', *Journal of Public Economics*, **60**(1), 95–110.

Porter, M.E. and C. van der Linde (1995), 'Toward a New Conception of Environment–Competitiveness Relationship', *Journal of Economic Perspectives*, **9**(4), 97–118.

Ray, D. and R. Vohra (1997), 'Equilibrium Binding Agreements', *Journal of Economic Theory*, **73**, 30–78.

Rogoff, K. (1997), 'Reputational Constraints on Monetary Policy', *Carnegie-Rochester Conference Series on Public Policy*, **26**, 141–82.

Rose, A. and B. Stevens (1993), 'The Efficiency and Equity of Marketable Permits for CO_2 Emissions', *Resource and Energy Economics*, **15**(1), 117–46.

Rose, A., B. Stevens, J. Edmonds and M. Wise (1998), 'International Equity and Differentiation in Global Warming Policy: An Application to Tradable Emission Permits', *Environmental and Resource Economics*, **12**(1), 25–51.

Rose, A. and B. Stevens (1999), 'A Dynamic Analysis of Fairness in Global Warming Policy: Kyoto, Buenos Aires and Beyond', in C. Carraro (ed.), *Efficiency and Equity of Climate Change Policy*, Dordrecht: Kluwer.

Schackleton, T. (1998), 'HANDOUT: The Potential Effects of International Carbon Emissions Mitigation Under the Kyoto Protocol: What we have learned from the G-Cubed Model', in *Proceedings* of the OECD Workshop on Climate Change and

Economic Modelling. Background Analysis for the Kyoto Protocol, Paris, 17–18 September.

Schmidt, C. (1997), 'Enforcement and Cost-effectiveness of International Agreements: The Role of Side Payments', *Discussionsbeträge des Sonderforschungsbereich* 178, Serie II, No. 350, Universität Konstanz.

Schmidt, T.F.N. and H. Koschel (1998), 'Climate Change Policy and Burden-Sharing Policy in the European Union', *ZEW Discussion Paper Series*, No. 98–12.

Shogren J. (1999), 'Benefits and Costs of Kyoto', in C. Carraro (ed.), *Efficiency and Equity of Climate Change Policy*, Dordrecht: Kluwer.

Taylor, M. (1987), *The Possibility of Cooperation*, Cambridge: Cambridge University Press.

Tol, R. (1999a), 'The Optimal Timing of Greenhouse Gas Emission Abatement, Individual Rationality and Intergenerational Equity', in C. Carraro (ed.), *International Environmental Agreements on Climate Change*, Dordrecht: Kluwer.

Tol, R. (1999b) 'Equitable Cost–Benefit Analysis of Climate Change', in C. Carraro (ed.), *Efficiency and Equity of Climate Change Policy*, Dordrecht: Kluwer.

Tol, R.S.J. (2001), 'Equitable Cost-Benefit Analysis of Climate Change Policies', *Ecological Economics*, 36, 71–85.

Tulkens, H. (1998), 'Cooperation versus Free-riding in International Environmental Affairs: Two Approaches', in H. Folmer and N. Hanley (eds), *Game Theory and the Global Environment*, Cheltenham, UK and Lyme, USA: Edward Elgar.

Wagner, H. (1983), 'The Theory of Games and the Problem of International Cooperation', *American Political Science Review*, **70**, 330–46.

Westkog, H. (1999), 'Intertemporal Emission Trading', paper presented at the EEA Annual Conference, Santiago de Compostela, 26–28 August.

Weyant, J.P. (ed.) (1999), 'The Costs of the Kyoto Protocol: a Multi-Model Evaluation', special issue of *The Energy Journal*.

Yang, Z. (2000), 'Forming Coalitions in International Agreements on Global Climate Change', mimeo, Department of Energy, Environmental and Mineral Economics, The Pennsylvania University.

Yi, S. and H. Shin (1994), 'Endogenous Formation of Coalition in Oligopoly: I. Theory', mimeo, Dartmouth College.

Yi, S. (1997), 'Stable Coalition Structures with Externalities', *Games and Economic Behaviour*, **20**, 201–23.

APPENDIX 1: A SHORT SUMMARY OF THE THEORY OF INTERNATIONAL AGREEMENTS

This appendix reviews some of the most important theoretical results on coalition theory and the theory of international environmental agreements, with a focus on their implications for climate agreements and the related incentives to ratification (see Barrett, 1994, 1997a; Bloch, 1997; Carraro and Moriconi, 1998; Carraro and Siniscalco, 1993; Hart and Kurz, 1983; Hoel, 1992, 1994; Ray and Vohra, 1997; Yi, 1997).

If the goal is to understand which international regime is likely to emerge to control GHG emissions, game theory is certainly the best tool to deal with it. Indeed, game theory has extensively analysed the possibility of coalition formation in the presence of free-riding (that is, when parties have to agree on the provision of a public good). Early contributions (for example Hardin and Baden, 1977) characterized the environmental game among countries as a prisoners' dilemma, inevitably leading to the so-called 'tragedy' of the common property goods. But in the real world, at the same time, a large number of international environmental agreements on the commons was signed, often involving sub-groups of negotiating countries and sometimes involving transfers and other links to other policies (trade, technological cooperation, and so on). It was therefore necessary to develop new models which helped us to understand the logic of coalition formation in the presence of spillovers, and the possibility of increasing welfare by means of appropriate mechanisms and strategies. These new models were developed in the last decade within a non-cooperative game-theoretic framework, and provide interesting indications of the likely outcomes of climate negotiations.

The starting-point is the definition of self-enforcing agreement. Following the seminal papers by Barrett (1994) and Carraro and Siniscalco (1993), we say that an agreement is self-enforcing if it is profitable and stable. An agreement is profitable when all signatory countries increase their welfare with respect to the case in which no agreement is achieved. An agreement is stable if no signatory country would prefer to free-ride on the agreement and no free-rider, if any, would like to join the group of signatories.

Consider first the case in which countries negotiate on a single worldwide agreement. Most papers in the game-theoretic literature on coalition formation applied to environmental agreements (Hoel, 1991, 1992; Carraro and Siniscalco, 1992, 1993; Barrett, 1994, 1997b; Heal, 1994) propose the following conclusions:

- the presence of asymmetries across countries and the incentive to free-ride makes the existence of global self-enforcing agreements, that is,

agreements which are profitable to all countries and stable, quite unlikely (Carraro and Siniscalco, 1993);
- when self-enforcing international environmental agreements exist, they are signed by a limited number of countries (Hoel, 1991, 1994; Carraro and Siniscalco, 1992; Barrett, 1994);
- when the number of signatories is large, the difference between the cooperative behaviour adopted by the coalition and the non-cooperative one is very small (Barrett, 1997b).

These results, which are robust with respect to different specifications of countries' welfare function, and with respect to the burden-sharing rule[1] used in the asymmetric case (Barrett, 1997a, Botteon and Carraro, 1997a), suggest that the attempt to negotiate on effective emission reductions is unlikely to lead to a coalition formed by all or by almost all countries, unless more complex policy strategies, in which environmental policy interacts with other policy measures, are adopted.[2]

For these reasons, in the game-theoretic environmental economics literature, two main sets of instruments have been proposed to expand environmental coalitions, that is, to increase the number of signatories to an environmental agreement. These instruments are 'transfers' and 'issue linkage'. These are analysed in Section 3, which deals with partial agreements and the ways to broaden them.[3]

Consider now the case in which countries are free to sign the agreement proposed by a group of countries or to propose themselves a different one to the same or to other countries (Carraro, 1998). This may lead to the formation of multiple climate agreements, similar to what happens in the case of trade blocs (Bloch, 1997; Carraro and Moriconi, 1998; Yi, 1997). The multiplicity of coalitions may allow for region-specific agreements in which the characteristics of countries in the region are better reflected by the contents of the agreement. Even in this case, game theory provides a clear analysis of the outcome of climate negotiations. Despite the large number of equilibrium concepts,[4] some conclusions seem to be quite robust:

- the equilibrium coalition structure is not formed by a single coalition. In general, many coalitions form at the equilibrium;
- the grand coalition, in which all countries sign the same environmental agreement, is unlikely to be an equilibrium;
- coalitions of different sizes may emerge at the equilibrium (even when countries are symmetric).

The specific results on the size of the coalitions depend on the model structure and in particular on the slope of countries' reaction functions, that is, on the presence of carbon leakage. If there is no leakage and countries are symmetric, then the Nash equilibrium of the multi-coalition game is characterized by many small coalitions, each one satisfying the properties of internal and external stability (this result is shown in Carraro and Moriconi, 1998).

The remaining question is therefore a policy one. Is countries' welfare larger when one or when several coalitions form? And what about environmental effectiveness? The answer is still uncertain, both because theory provides examples in which a single agreement is preferred, at least from an environmental viewpoint, to many small regional agreements (and vice versa), and because empirical studies have not yet convincingly addressed this issue. Moreover, the conclusion crucially depends on the choice of the equilibrium concept and on the size of leakage.

The consequence of the results discussed above is that the structure of the international environmental agreements is a crucial dimension of the negotiating process. If all countries negotiate on a single agreement, the incentives to sign are lower than those that characterize a multiple agreement negotiating process. But at the equilibrium, the environmental benefit (quality) may be higher.

NOTES

1. In the asymmetric case, the rule which is chosen to divide the gains from cooperation among the countries in the coalition (usually called the burden-sharing rule) plays a crucial role because it affects the likelihood that each country decides to sign the agreement. The burden-sharing rule is usually taken from cooperative game theory and that of Nash and Shapley is the most used. By contrast, in the symmetric case different rules lead to the same outcome (equal shares).
2. Surveys of the above literature are proposed in Barrett (1997b), Carraro (1999a), and Tulkens (1998).
3. In the last two decades, political scientists also focused their analyses on the emergence of cooperation in the presence of free-riding (Axelrod, 1984; Brams and Kilgour, 1988; Hampton, 1987; Oye, 1986; Taylor, 1987; Wagner, 1983). Their conclusions are very close to those reached by economists; that is, even in the case of public good provision, a coalition forms at the equilibrium, but some countries are allowed to defect.
4. Unfortunately, game theory is far from having achieved a well-defined non-cooperative theory of coalition formation under the above general assumptions and definitions. There are several stability concepts that can be used and which unfortunately provide different equilibrium coalition structures. Among them, let us recall the concept of equilibrium binding agreements proposed by Ray and Vohra (1997), the concepts of α-stability and β-stability proposed in Hart and Kurz (1983), the sequential stability concept of Bloch (1997), the open-membership stability proposed by Yi (1997) and the far-sighted stability concept used in Chew (1994) and Mariotti (1997).

APPENDIX 2: INTERNATIONAL REGIMES AND POLICY OPTIONS: EXISTING STUDIES ON THE COSTS AND BENEFITS OF DIFFERENT COMBINATIONS

Table 5A.1 reorganizes the existing literature on climate policies in a way that enables us to identify the costs and benefits of different combinations of policy mixes and international regimes. The papers mentioned in Table 5A.1 are only a subset of the available papers (see Carraro, 1998, 1999d, 2000; OECD, 1998; see also Burniaux, 1998; Capros, 1998; Ellerman et al., 1998; Grubb and Vrolijk, 1998; Holtzmark, 1998; Manne and Richels, 1998; Mensbrugghe, 1998; Nordhaus and Boyer, 1999, and the surveys by Metz, 1999 and Convery, 1999, among others). For a more complete survey, see chapter 8 of the IPCC Third Assessment Report.

However, the information contained in Table 5A.1 is sufficient to identify the main dynamics discussed in this chapter, namely:

- An increased number of countries ratifying the Protocol reduces the implementation costs in all participating countries and increases the abated emissions. Hence strategies that increase the size and the number of ratifying countries also increase the environmental and cost-effectiveness of the Protocol.
- A reduced cost of implementing the Protocol increases the incentives for countries to ratify the Protocol. Hence strategies designed to reduce implementation costs also increase the number of participating countries, thus further reducing the cost of achieving the Kyoto commitments.

The models used for the analyses carried out in the papers cited in Table 5A.1 try to capture – albeit in different ways – the peculiar features of the climate change problem as described in the Introduction. Even if they are only partly successful in dealing with uncertainties and irreversibilities, they succeed in taking into account the global and long-term dimensions of the climate problem, the availability of alternative energy technologies, the existence of strong policy interactions and the absence of supranational institutions (in most studies a game-theoretic approach is used to describe countries' interactions).

Table 5A.1 Coalition structures and policy options

	Policy options						
Coalition Structure	Domestic measures only	Coordinated carbon tax	Flexibility mechanisms with ceilings	Free flexibility mechanisms	Flexibility mechanisms with banking	Flexibility mechanisms with R&D	Flexibility mechanisms with market imperfections
No participation	IPCC (1995)						
Unilateral participation	Jorgenson and Wilcoxen (1993) Barrett (1992)						
EU only	Carraro and Siniscalco (1992)	Bosello and Carraro (1999) Barker (1999)	Schmidt and Koschel (1998)				
OECD only	Burniaux et al. (1992)	Capros (1998)		Harrison and Rutherford (1999) Holtzmark (1998) Capros (1998)			
Annex B countries	McKibbin et al. (1998)	Mensbrugghe (1998)	Ellerman et al. (1998, 1999) Buonanno et al. (1999) Manne and Richels (1998)	Ellerman et al. (1998) Grubb and Vrolijk (1998) Holtzmark (1998) McKibbin et al. (1998) Manne and Richels (1999a,1999b) Mensbrugghe (1998) Nordhaus and Boyer (1999) Schackleton (1998)	Bosello-Roson (1999) Westskog (1999)	Nordhaus (1997) Buonanno et al. (1999)	Burniaux (1998) Ellerman et al. (1998)
Double Umbrella				McKibbin et al. (1998) Schackleton (1998)			Ellerman et al. (1998)
All countries	Nordhaus and Yang (1996)		Ellerman et al. (1998) Buonanno et al. (1999)	Bohm (1999) Ellerman et al. (1998) Manne and Richels (1998,1999a,1999b) Nordhaus and Boyer (1999) Schackleton (1998)	Bosello-Roson (1999)	Nordhaus (1997) Westskog (1999) (1999)	Buonanno et al.

Index